Streams and Ground Waters

AQUATIC ECOLOGY Series

Series Editor

James H. Thorp
Department of Biology
University of Louisville
Louisville, Kentucky

Other titles in the series:

Groundwater Ecology
Janine Gilbert, Dan L. Danielopol, Jack A. Stanford

Algal Ecology
R. Jan Stevenson, Max L. Bothwell, Rex L. Lowe

Streams and Ground Waters

Edited by

Jeremy B. Jones
Department of Biological Sciences
University of Nevada
Las Vegas, Nevada

Patrick J. Mulholland
Environmental Sciences Division
Oak Ridge National Laboratory
Oak Ridge, Tennessee

ACADEMIC PRESS

A Harcourt Science and Technology Company

San Diego   San Francisco   New York   Boston   London   Sydney   Tokyo

This book is printed on acid-free paper. ∞

Academic Press
A Harcourt Science and Technology Company
525 B Street, Suite 1900, San Diego, California 92101–4495, USA
http://www.apnet.com

Academic Press
24–28 Oval Road, London NW1 7DX, UK
http://www.hbuk.co.uk/ap/

Library of Congress Catalog Card Number: 99–63444

International Standard Book Number: 0–12–389845–5

PRINTED IN THE UNITED STATES OF AMERICA
99   00   01   02   03   04   BB   9   8   7   6   5   4   3   2   1

Contents

Contributors   xiii
Preface   xvii

SECTION *ONE*

*THE PHYSICAL TEMPLATE: HYDROLOGY,
HYDRAULICS, AND PHYSICAL STRUCTURE*

1   *Quantifying Hydrologic Interactions between
Streams and Their Subsurface Hyporheic Zones*

Judson W. Harvey and Brian J. Wagner

I. Introduction   4

II. Challenge of Investigating Small-Scale Subsurface Processes That May Have Basin-Scale Consequences   6

III. Empirical Approaches to Quantifying Hydrologic Exchange between Streams and Shallow Ground Water   7

IV. Using the Stream-Tracer Approach to Characterize the Hyporheic Zone   17

V. Paradigm Lost? Limitations of the Stream-Tracer Approach as a Means to Quantify Hyporheic Processes   21

VI. Charting New Directions in Hyporheic-Zone Research   38

VII. Conclusion   39
References   41

2 Modeling Surface–Subsurface Hydrologic Interactions

Aaron I. Packman and Kenneth E. Bencala

I. Introduction   46

II. Viewing the Interaction from the Stream   47

III. Viewing the Interaction from the Stream–Bed Interface   51

IV. Viewing the Interaction from the Subsurface   72

V. Challenges   77
References   77

SECTION *TWO*

BIOGEOCHEMISTRY

SUBSYSTEM INTERACTIONS WITH STREAM SURFACE WATER

3 Stream Chemistry and Riparian Zones

Alan R. Hill

I. Introduction   83

II. Riparian Zone Hydrological–Chemical Interactions: An Overview   84

III. Riparian Influences on Stream Chemistry   87

IV. Riparian Zone Influences on Stream Chemistry

in Relation to Watershed Hydrogeology:
A Conceptual Framework   100
V. Future Research Directions   106
References   107

*4   Flood Frequency and Stream–Riparian
Linkages in Arid Lands*

Eugènia Martí, Stuart G. Fisher, John J. Schade,
and Nancy B. Grimm

   I. Introduction   111
  II. Riparian Zones in Arid Catchments   112
 III. Hydrological Linkages in Mesic and Arid
     Catchments   113
 IV. Conceptual Model   115
  V. Case Study: Sycamore Creek   117
 VI. Synthesis   123
VII. Conclusions: Intermediate Disturbance
    and Nutrient Retention   131
    References   133

*5   The Importance of Ground Water to Stream
Ecosystem Function*

Robert M. Holmes

  I. Introduction   137
 II. Influence of Ground Water on Stream
    Functioning   139
III. Summary   144
   References   145

*6   Surface–Subsurface Exchange and Nutrient
Spiraling*

Patrick J. Mulholland and Donald L. DeAngelis

   I. Introduction   149
  II. Empirical Studies   151
 III. A Stream Nutrient Spiraling Model with Subsurface
    Transient Storage   154
 IV. Results of Model Experiments   158
  V. Relevance of Model Experiments   161
 VI. Future Research Needs   163
    References   164

7 *Emergent Biological Patterns and Surface–*
Subsurface Interactions at Landscape Scales

C. M. Pringle and Frank J. Triska

I. Introduction   167
II. The Balance of Physical and Chemical Factors
   on the Geologic Template and Emergent
   Biological Patterns   169
III. Hydrothermal Systems as Models   171
IV. Human Implants on Surface–Subsurface
   Interactions   175
V. Synthesis and Recommendations
   for Future Studies   185
   References   189

SECTION *TWO*
BIOGEOCHEMISTRY

NUTRIENTS AND METABOLISM

8 *Nitrogen Biogeochemistry and Surface–*
Subsurface Exchange in Streams

John H. Duff and Frank J. Triska

I. Introduction   197
II. Nitrogen Forms and Transformation Pathways
   in Fluvial Environments   199
III. Nitrogen Sources in Fluvial Environments   200
IV. Hydrologic Residence in Pristine Streams   201
V. The Redox Environment   203
VI. Ammonium Sorption to Hyporheic Sediments   204
VII. Linking Nitrogen Transformation to Hydrologic
   Exchange in Hyporheic Zone Research   205
VIII. What These Models Tell Us   214
IX. Future Directions for Research   216
   References   217

9 *Stream and Groundwater Influences*
on Phosphorus Biogeochemistry

Susan P. Hendricks and David S. White

I. Introduction   221

II. Sources and Forms of Phosphorus 222
III. Abiotic Phosphorus Retention by Bed
      Sediments 223
IV. Biotic Phosphorus Retention and Release
      within Bed Sediments 224
V. Fluvial Dynamics and Physical Retention
      Mechanisms 225
VI. Phosphorus and Surface–Subsurface Exchange:
      A Conceptual Model 226
VII. Summary and Research
      Needs 232
      References 233

10 *Surface and Subsurface Dissolved Organic Carbon*

Louis A. Kaplan and J. Denis Newbold

I. Introduction 237
II. DOC Concentrations 238
III. Processes within the Hyporheic
      Zone 242
      References 253

11 *Anoxia, Anaerobic Metabolism, and Biogeochemistry of the Stream-water–Groundwater Interface*

Michelle A. Baker, Clifford N. Dahm,
and H. Maurice Valett

I. Introduction 260
II. Methods Used in Studies of Anaerobic
      Metabolism 264
III. Controls on the Establishment
      of Anoxia 266
IV. Biogeochemistry and Evidence for Anaerobic
      Metabolism 269
V. Influence of Subsurface Anaerobic
      Metabolism on Stream Ecosystem
      Processes 276
VI. Conclusions and Future Research
      Directions 278
      References 280

SECTION **THREE**
ORGANISMAL ECOLOGY

12 *Microbial Communities in Hyporheic Sediments*
Stuart Findlay and William V. Sobczak

    I. Introduction 287
    II. Physical and Chemical Environment 288
    III. The Organisms 290
    IV. Respiration 293
    V. Carbon Supply 295
    VI. Alternate Controls on Hyporheic
        Bacteria 300
    VII. Community Composition 301
    VIII. Conclusions and Research Needs 302
        References 304

13 *The Ecology of Hyporheic Meiofauna*
Christine C. Hakenkamp and Margaret A. Palmer

    I. Introduction 307
    II. Meiofaunal Taxa and Their Relative
        Abundance in the Hyporheic Zone 309
    III. Spatial Distribution and Abundance 316
    IV. Meiofauna and the Physical
        Environment 319
    V. Tolerance to Anoxia, Body Size,
        and Biomass 321
    VI. Trophic Roles, Dispersal Dynamics,
        and Response to Spatial–Temporal
        Heterogeneity 322
    VII. Meiofaunal–Microbial Interactions 324
    VIII. Meiofauna and Ecosystem Ecology 325
    IX. Conclusions and Future Research
        Needs 328
        References 329

14 *The Subsurface Macrofauna*
Andrew Boulton

    I. Introduction 337

II. Functional Classification of Subsurface
Macrofauna   338
III. Factors Influencing the Distribution of Subsurface
Macrofauna   345
IV. The Functional Role of Subsurface
Macrofauna   349
V. Potential Role as Biomonitors of Deteriorating
Groundwater Quality   354
VI. Conclusions   355
References   356

15   Lotic Macrophytes and Surface–Subsurface Exchange Processes

David S. White and Susan P. Hendricks

I. Introduction   363
II. Macrophytes and Aquatic Habitats   366
III. Macrophytes as Indicators of Surface and
Subsurface Conditions   367
IV. Macrophytes and Surface- and Subsurface
Water Flow Patterns   369
V. Nutrient Uptake from Sediments   371
VI. Processes at the Rhizosphere   373
VII. Summary and Avenues for Further
Study   375
References   376

16   Subsurface Influences on Surface Biology

C. Lisa Dent, John J. Schade, Nancy B. Grimm,
and Stuart G. Fisher

I. Introduction   381
II. Effects on Primary Producers   382
III. Effects on Microorganisms and Microbial
Processes   386
IV. Effects on Invertebrates   388
V. Effects on Fish   389
VI. Heterogeneity and Scale   391
VII. Human Impacts   394
VIII. Summary   396
References   397

SECTION *FOUR*

SUMMARY AND SYNTHESIS

17 *Surface–Subsurface Interactions: Past, Present, and Future*

Emily H. Stanley and Jeremy B. Jones

    I. Introduction   405
   II. Growth and Development Trajectories   406
  III. Flowpaths and Interfaces   408
  IV. The Spatial Context   409
   V. Synthetic Models and Future Directions   412
     References   414

Index   419

Contributors

Numbers in parentheses indicate the pages on which the authors' contributions begin.

Michelle A. Baker (260), Department of Biology, Utah State University, Utah

Kenneth E. Bencala (46), U.S. Geological Survey, Menlo Park, California 94025

Andrew Boulton (337), Division of Ecosystem Management, University of New England, Armidale 2350, New South Wales, Australia

Clifford N. Dahm (260), Department of Biology, University of New Mexico, Albuquerque, New Mexico 87131

Donald L. DeAngelis (149), National Biological Service, South Florida Field Laboratory, Department of Biology, University of Miami, Coral Gables, Florida 33124

C. Lisa Dent (317), Department of Biology, Arizona State University, Tempe, Arizona 85287

John H. Duff (197), U.S. Geological Survey, Menlo Park, California 94025

Stuart Findlay (287), Institute of Ecosystem Studies, Millbrook, New York 12545

Stuart G. Fisher (111, 381), Department of Biology, Arizona State University, Tempe, Arizona 85287

Nancy B. Grimm (111, 381), Department of Biology, Arizona State University, Tempe, Arizona 85287

Christine Hakenkamp (307), Department of Biology, James Madison University, Harrisonburg, Virginia 22807

Judson W. Harvey (4), U.S. Geological Survey, Reston, Virginia 20192

Susan P. Hendricks (221, 363), Department of Biological Services, Murray State University, Murray, Kentucky 42071

Alan R. Hill (83), Department of Geography, York University, Toronto, Ontario M3J 1P3, Canada

Robert M. Holmes (137), Ecosystems Center, Marine Biological Laboratory, Woods Hole, Massachusetts 02543

Jeremy B. Jones (405), Department of Biological Sciences, University of Nevada Las Vegas, Las Vegas, Nevada 89154

Louis A. Kaplan (237), Stroud Water Research Center, Avondale, Pennsylvania 19311

Eugènia Martí (111), Centre d'Estudis Avancats de Blanes (CSIC), 17 300 Blanes, Girona, Spain

Patrick J. Mulholland (149), Environmental Sciences Division, Oak Ridge National Laboratory, Oak Ridge, Tennessee 37831

J. Denis Newbold (237), Stroud Water Research Center, Avondale, Pennsylvania 19311

Aaron I. Packman (46), Civil and Architectural Engineering Department, Drexel University, Philadelphia, Pennsylvania 19104

Margaret A. Palmer (307), Department of Biology, University of Maryland, College Park, Maryland 20742

C. M. Pringle (167), Institute of Ecology, University of Georgia, Athens, Georgia 30602

John J. Schade (111, 381), Department of Biology, Arizona State University, Tempe, Arizona 85287

William V. Sobczak (287), Institute of Ecosystem Studies, Millbrook, New York 12545

Emily H. Stanley (405), Center for Limnology, University of Wisconsin, Madison, Wisconsin 53706

Frank J. Triska (167, 197), U.S. Geological Survey, Menlo Park, California 94025

H. Maurice Valett (260), Department of Biology, Virgina Polytechnic Institute and State University, Blacksburg, Virginia 24061

Brian J. Wagner (4), U.S. Geological Survey, Menlo Park, California 94025

David S. White (221, 363), Hancock Biological Station and Center for Reservoir Research, Murray State University, Murray, Kentucky 42071

Preface

Historically, streams have been defined by their surface flow and studied as discrete entities within watersheds. Streams, however, are not isolated from their drainage basins but are influenced biogeochemically and ecologically by the flow of water through underlying sediments and discharge of water from adjacent soil and bedrock environments. Beginning in the mid-1960s, and with rapid expansion during the present decade, lotic ecologists have broadened their perspectives on streams. Groundwater environments are now recognized as integral components of streams. They are important habitats for organisms and critical for biogeochemical cycling. To date, however, research on stream and groundwater interactions has largely been confined to studies of individual systems with little synthesis and few generalities produced. Such a synthesis is not trivial, however, given the disparity of research foci ranging from hydrology to biogeochemistry to aquatic organisms. It requires the efforts of multiple investigators with wide-ranging expertise. Nonetheless, this synthesis is imperative given mounting pressures and ever increasing demands on stream and river ecosystems. In *Streams and Ground Waters,* contributing authors have been charged with the task of synthesizing the current state of the field and generating conceptual models or summaries that lead to the generation of testable hypotheses. We have given authors considerable freedom to extrapolate from our current state of understanding and to propose ideas that might not normally be allowed in more conservative journal publications. Through this less restrictive approach, we hope that the book will lead to significant advances in stream ecology by generating new ideas and providing a catalyst for future research.

This book is organized into three sections that examine (1) the hydrology and physical structure of streams, (2) biogeochemistry, and (3) organismal ecology. Foremost in the study of ecological and biogeochemical interactions between streams and gound waters is hydrology. The opening section of the book (Chapters 1 and 2), *the physical template,* focuses on the hydrologic linkages between surface and subsurface waters and provides a framework for interpreting and understanding the resulting biogeochemical and ecological patterns and processes. A central theme of this first section is the use of hydrologic models to examine stream and groundwater interactions. In the second section, *biogeochemistry,* the consequences of hydrologic exchanges between streams and their catchments for organic matter dynamics and nutrient cycling are discussed from two perspectives. First, the interactions between stream, riparian, and groundwater environments are explored. This habitat-specific focus considers not only how the biogeochemistry of streams is influenced by groundwater and riparian linkages (Chapters 3, 5, and 6) but also how these interactions vary spatially across landscapes (Chapter 7) and temporally in response to flooding (Chapter 4). Second, the biogeochemical cycling of specific elements and the role of anoxia and redox potential are examined. This element-specific focus centers on nitrogen (Chapter 8), phosphorus (Chapter 9), and organic carbon (Chapter 10) but includes discussion of other elements such as sulfur and iron, particularly in the context of anaerobiosis and chemoautotrophy (Chapter 11). In the last section, *organismal ecology,* we explore the consequences of surface–subsurface interactions for flora and fauna living in streams. These chapters consider the influence of stream and groundwater linkages on the distribution of microorganisms (Chapter 12), meiofauna (Chapter 13), macroinvertebrates (Chapter 14), and macrophytes (Chapter 15) and the impact of these organisms on stream–groundwater hydrology and stream ecosytem functioning (Chapter 16). Finally, the last chapter provides a synthesis of the preceding contributions (Chapter 17).

We gratefully acknowledge the support we received from the following reviewers: Clifford Dahm, John Duff, Alan Hill, Bryan Harper, Judson Harvey, Robert Holmes, Louis Kaplan, Stanley Smith, Emily Stanley, Maurice Valett, and Philippe Vervier. Support from our home institutions of the University of Nevada, Las Vegas and Oak Ridge National Laboratory (supported by funding from the Department of Energy and managed by Lockheed Martin Energy Research Corporation under Contract DE-AC05-96OR22464 with the United States Department of Energy) is gratefully acknowledged. Finally, special thanks to Rhonda and Cathy for their support throughout this project and our careers.

Jeremy B. Jones
Patrick J. Mulholland

SECTION *ONE*

THE PHYSICAL TEMPLATE: HYDROLOGY, HYDRAULICS, AND PHYSICAL STRUCTURE

1

Quantifying Hydrologic Interactions between Streams and Their Subsurface Hyporheic Zones

Judson W. Harvey* and Brian J. Wagner[†]

*Water Resources Division
United States Geological Survey
Reston, Virginia

[†]Water Resources Division
United States Geological Survey
Menlo Park, California

I. Introduction 4
II. Challenge of Investigating Small-Scale Subsurface Processes That May Have Basin-Scale Consequences 6
III. Empirical Approaches to Quantifying Hydrologic Exchange between Streams and Shallow Ground Water 7
    A. Stream, Hyporheic Zone, and Ground Water: A Reach-Averaged Mass Balance Model 7
    B. Estimating Ground–Water Inflow and Outflow Fluxes from Stream-Flow Discharge Measurements 9
    C. Estimating Hyporheic-Exchange Fluxes from Other Field Data 10
IV. Using the Stream-Tracer Approach to Characterize the Hyporheic Zone 17
    A. Inverse Modeling of Stream-Tracer Data: Advantages of Fitting by Statistical Optimization 19
V. Paradigm Lost? Limitations of the Stream-Tracer Approach as a Means to Quantify Hyporheic Processes 21
    A. Assumptions of the Stream-Tracer Approach to Characterize Hyporheic Zones 21

3

B. Comparing Storage-Zone Modeling Parameters with
Subsurface Measurements in Hyporheic Zones   22
C. Stream Tracers Have a "Window of Detection" for Storage
Processes with Particular Spatial Dimensions and
Timescales   26
D. Designing Better Stream-Tracer Experiments   32
E. Acquiring the Prior Information Needed to Design Tracer
Studies   33
F. Implications for Linking Stream-Tracer Parameters with
Physical Characteristics of Drainage Basins   36
VI. Charting New Directions in Hyporheic-Zone Research   38
VII. Conclusion   39
References   41

I. INTRODUCTION

Water in streams and rivers passes back and forth between the active channel and subsurface (hyporheic) flowpaths. The interaction is rapid enough that, within several kilometers, stream water in the relatively small channels is often completely exchanged with porewater of the hyporheic zone. The importance of hydrologic exchange between streams and hyporheic zones is that it keeps surface water in close contact with chemically reactive mineral coatings and microbial colonies in the subsurface (Fig. 1A), which has the effect of enhancing biogeochemical reactions that influence downstream water quality.

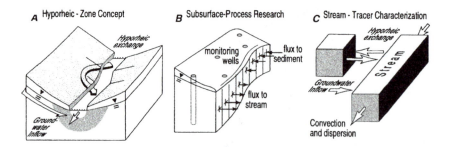

FIGURE 1   Schematic illustration of hyporheic zones (A) and two field approaches to characterize hyporheic zones, using subsurface measurements in wells (B) and modeling the injection and transport of a solute tracer in the stream (C). The two different field techniques average hyporheic conditions over vastly different spatial scales (i.e., centimeters to meters for the subsurface approach and tens to hundreds of meters for the stream-tracer approach). (Modified from Harvey *et al.*, 1996.)

The hyporheic zone was initially identified through observations of stream organisms and unexpectedly high concentrations of dissolved oxygen in shallow ground water beneath streams (an example is the study by Hynes, 1974). White *et al.* (1987) used temperature as a tracer to determine the extent of the hyporheic zone. Triska *et al.* (1989) injected a solute tracer in the stream and delineated the hyporheic zone on the basis of water source; the subsurface zone where wells received at least 10% of their water by input from the stream defined the hyporheic zone. Hydrogeologists approached the problem by defining the hyporheic zone on the basis of flowpath. The hyporheic zone was delineated by subsurface flowpaths that began in the stream and returned to the stream within a stream reach under investigation (Harvey and Bencala, 1993; Wroblicky *et al.*, 1998). Both small and large hyporheic flowpaths are usually present along streams; however, the greatest interaction with the stream usually occurs in relatively short hyporheic flowpaths that return to the stream within centimeters to tens of meters. It is flowpath length and timescale of the interaction that help distinguish hyporheic exchange from the much larger (and longer term) channel and groundwater interactions described by Larkin and Sharp (1992). The hyporheic zone can therefore be viewed as the subset of finer-scale interactions between the channel and ground water that occur within the context of larger-scale patterns of loss and gain of channel water in drainage basins. Size of hyporheic flowpaths (and water fluxes through them) fluctuate in response to seasonal fluctuations in hillslope groundwater levels, increases in stream stage and stream velocity that accompany storms (Wondzell and Swanson, 1996; Angradi and Hood, 1998), and diel fluctuations in groundwater levels caused by evapotranspiration (Harvey *et al.*, 1991).

There are several important reasons to distinguish interactions between streams and hyporheic zones from other types of interactions with ground water. Hyporheic flowpaths, by definition, leave and return to the stream many times within a single study reach, unlike groundwater flowpaths, which enter or leave the channel reach only once. Hyporheic exchange therefore repeatedly brings stream water into close contact with geochemically and microbially active sediment (Findlay, 1995). Relatively rapid inputs of oxygen and organic carbon to hyporheic flowpaths enhance rates of microbially mediated chemical reactions compared to in-stream or groundwater environments (Grimm and Fisher, 1984; Findlay *et al.*, 1993). Examples of the enhancement of microbial activity in the hyporheic zone include nitrification (Triska *et al.*, 1993), microbial uptake of dissolved organic carbon (Findlay *et al.*, 1993), and microbial oxidation of manganese in the hyporheic zone (Harvey and Fuller, 1998).

Experimental injections of solute-tracers into streams provided some of the early insights about hyporheic zones, especially regarding their potential to influence stream ecology and biogeochemistry at larger scales. Researchers such as Bencala and Walters (1983), Newbold *et al.* (1983), Jackman *et al.*

(1984), and Beer and Young (1983) extended and improved the capabilities of riverine transport models (e.g., Valentine and Wood, 1979) to simulate those injections. The updated models emphasized hydrologic retention and chemical reaction in "storage zones" on the sides and the bottom of channels that delay downstream transport and enhance reactions. Storage-zone transport modeling has since been widely applied in studies of solute transport and reaction in streams (e.g., Bencala, 1983; Castro and Hornberger, 1991; Runkel *et al.*, 1996; Valett *et al.*, 1996; Mulholland *et al.*, 1997). Many of the mathematical codes that are now used differ only slightly in their definitions of variables. In this chapter, we discuss the equations as they are presented in the U.S. Geological Survey modeling code OTIS (One-dimensional Transport with Inflow and Storage) by Robert Runkel (1998).

Several trends in investigating the hydrology of hyporheic zones emerged in the 1990s. First, researchers made greater use of hydraulic and hydrogeologic theory to improve intuition about hyporheic processes. Theoretical advancements are the subject of a review by Packman and Bencala in this volume (Chapter 2). Second is the development of innovative empirical approaches, ranging from process investigations at small spatial scales to larger-scale field experiments that seek to quantify cumulative effects. These empirical approaches and the supporting mathematical analyses are reviewed and discussed in this chapter. Our chapter is organized around a central theme—that a clearer recognition of the limitations of current empirical methods will stimulate new ways of measuring and modeling hyporheic processes. The following key issues serve as some of the main discussion points:

1. The challenge of investigating a small-scale hydrologic process that may have consequences for solute transport at much larger spatial scales
2. Fundamental approaches to quantify hydrologic fluxes across streambeds
3. Use of stream tracers to characterize hyporheic processes—is the paradigm lost?
4. Acknowledging that stream tracers have a limited "window of detection" for hyporheic processes
5. Charting new directions—improving the design of stream-tracer experiments and linking stream-tracer modeling parameters with physical characteristics of drainage basins.

II. CHALLENGE OF INVESTIGATING SMALL-SCALE SUBSURFACE PROCESSES THAT MAY HAVE BASIN-SCALE CONSEQUENCES

Some of the most important research questions about hyporheic zones require answers that are "big" in the spatial scale they represent, such as how much solute enters a stream reach with groundwater inflow, how long is solute retained in the stream reach or hyporheic zone, or what is the cumulative effect of chemical reactions in hyporheic zones on downstream chemistry? *In situ* field measurements in hyporheic zones, for the most part, provide results that are "small" in the spatial scale they represent. Examples are measurements of hydraulic head or hydraulic flux near the streambed interface that represent very local conditions. In general, it will be extremely difficult or impossible to make enough point measurements in the subsurface to overcome spatial variability when attempting to characterize cumulative effects.

An important advantage of the stream-tracer approach is the relatively large spatial scale at which the cumulative effects of hydrologic storage and enhanced chemical reactions are characterized. There is a drawback, however—the empirical nature of the parameters that describe the hyporheic zone. The types of storage that are potentially characterized by stream-tracer modeling include hyporheic zones but also include stagnant or recirculating zones in surface water. Therefore, although stream-tracer experiments and modeling provide answers at desired spatial scales, the results are not necessarily process-specific and therefore may not be generalizable.

We illustrate two distinct views of surface–subsurface interactions from the perspective of subsurface measurements and from the perspective of stream tracers (Fig. 1). The disparity in scales and vastly different field techniques used to quantify the processes creates a major challenge for hydrologic investigators. Much of this chapter discusses the difficulty of combining and reconciling measurements of detailed subsurface processes (Fig. 1B) and larger-scale interpretations of cumulative effects based on stream tracers (Fig. 1C).

III. EMPIRICAL APPROACHES TO QUANTIFYING HYDROLOGIC EXCHANGE BETWEEN STREAMS AND SHALLOW GROUND WATER

In this section, we review a broad array of useful field techniques and model calculations to characterize hyporheic zones. Before proceeding with our review and discussion of recent advances, we briefly outline a simple mass balance for stream–aquifer systems that provides a focal point for the wide ranging discussion that follows.

A. Stream, Hyporheic Zone, and Ground Water: A Reach-Averaged Mass Balance Model

A single hyporheic flowpath begins where stream water enters the hyporheic zone and ends where it re-emerges into the stream—a typical length ranges from centimeters to tens of meters. Small-scale hyporheic flowpaths are embedded within the larger-scale groundwater flow system that surrounds a stream or river. A useful tool for the study of stream–subsurface hydrologic interactions is a conceptual model that integrates across spatial scales of surface–subsurface hydrologic interactions, from small-scale interactions of channels with hyporheic flowpaths to larger-scale interactions of channels with the groundwater-flow system. Such a model provides a framework for integrating different data types, such as stream-flow discharge measurements, subsurface flow measurements, and estimates of hyporheic fluxes gained through modeling stream tracer experiments. Figure 2 illustrates the mass balance schematically.

The mass balance approach considers a reach of stream that is long enough (typically 50–500 m) to include many hyporheic flowpaths. For our present purpose, we keep the mass balance simple by assuming that flow is steady and unchanging over time. As a result, precipitation and evapotranspiration are not considered, and neither are any interactions between streams and ground water that operate when stream water levels fluctuate, such as bank storage.

The mass balance equation for those conditions is

$$\frac{dQ}{dx} = q_L^{in} + q_h^{in} - q_L^{out} - q_h^{out}, \tag{1}$$

where x is the downstream channel direction and Q is the volumetric discharge in the channel (m^3s^{-1}). Any change in volumetric discharge in the downstream direction, in a reach without tributaries, is due to fluxes of water across the streambed: q_L^{in} is the reach-averaged groundwater influx per meter of stream ($m^3s^{-1}m^{-1}$), q_L^{out} is the reach-averaged stream water out-

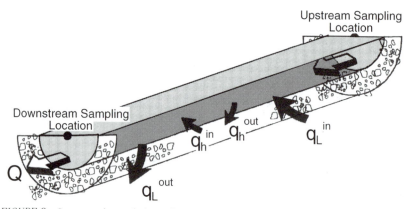

FIGURE 2   Conceptual mass balance for a stream reach that partitions water losses and gains as interactions with ground water (q_L) or as interactions with the hyporheic zone (q_h).

flux to ground water, and q_b^{in} and q_b^{out} are the reach-averaged influxes and outfluxes from hyporheic flowpaths. We recognize that a significant component of subsurface flow, both in ground water and in the hyporheic zone, occurs in a direction parallel to the channel. However, channel-parallel subsurface flow is neglected in this simple mass balance because it is slow relative to flow in the channel, and because the primary objective is to quantify fluxes across the streambed.

B. Estimating Ground–Water Inflow and Outflow Fluxes from Stream-Flow Discharge Measurements

Measurements of stream-flow discharge offer a straightforward and efficient means to constrain the groundwater fluxes of Eq. (1). All that is needed are measurements of stream-flow velocity across sections of the stream according to the stream-flow velocity-gaging method (Carter and Davidian, 1968). Net groundwater fluxes across the streambed are computed as the difference between stream flow at successive cross sections at downstream locations. Such measurements are sometimes referred to as seepage runs. Using velocity-gaging alone, only the *net* exchange of channel water with ground water can be computed (i.e., $q_L^{in} - q_L^{out}$). This approach therefore falls short of identifying both the inflow and outflow components of surface-water exchange with ground water. However, in some situations quantifying both components of groundwater exchange is vital to transport studies because streams sometimes gain water by inflow of ground water in some areas and lose water to ground water in other areas (Winter *et al.*, 1998). Here we explain how, with a few additional measurements, a typical seepage run can be used to estimate both the groundwater inflow rate, q_L^{in}, and the groundwater outflow rate, q_L^{out}.

Our approach combines the velocity-gaging method with a technique that uses a solute tracer injected into the stream, known as the dilution-gaging method (Kilpatrick and Cobb, 1985). To estimate both groundwater inflow and outflow simultaneously, we suggest injecting a solute tracer at the upstream of the reach, measuring stream volumetric discharge at both reach end points by the dilution-gaging method, and then additionally measuring discharge at the downstream end using the velocity-gaging method. Groundwater inflow rate, q_L^{in}, is estimated from the difference between the dilution-gaging measurements at the downstream and upstream ends of the reach (divided by reach length). In contrast, the net groundwater exchange ($q_L^{in} - q_L^{out}$) is estimated by the difference between the velocity-gaging estimate at the downstream end of the reach and the dilution-gaging estimate at the upstream end of the reach (divided by reach length). The final piece of information that is needed, q_L^{out}, is estimated by subtracting the net exchange rate from the groundwater inflow rate. Zellweger *et al.* (1989) interpret the difference between a dilution-gaging method and velocity-gaging method somewhat differently than we do. From similar data, they estimate the flux of sub-surface water flowing parallel to the channel. Although that interpretation may be correct in some instances, we suggest caution. What is actually gaged by the difference in dilution- and velocity-gaging measurements is downward flow across the channel bed in the experimental reach. Stream-flow loss is unlikely to account for all flow beneath the channel, nor is it possible, on the basis of stream-tracer experiments alone, to determine the direction of sub-surface flow once stream water crosses the bed. We believe that our sugges-tion to use dilution- and velocity-gaging data to estimate a groundwater out-flow term, q_L^{out}, for a study reach is consistent with basic assumptions of the techniques and provides an interpretation that is defensible in almost any field setting.

C. Estimating Hyporheic-Exchange Fluxes from Other Field Data

We have explained how groundwater inflow and outflow can be deter-mined from stream-flow measurements, but we have not discussed how to measure the hyporheic exchange fluxes, q_h^{in} and q_h^{out}. Those fluxes cannot be determined from dilution- or velocity-gaging data. Instead, one or more of the following methods is used: (1) Darcy-groundwater-flux calculations, (2) tracer-based approaches, in which hyporheic fluxes are inferred from the movement of an introduced tracer (such as salt) or an environmental tracer (such as water temperature or specific conductivity) that is naturally present at the field site, or (3) direct measurements of water fluxes across the streambed using a device such as a seepage meter.

1. Subsurface Sampling in Hyporheic Investigations

Measurements of hydraulic head and solute-tracer concentrations in shallow ground water are often used as the basis to compute water fluxes across the streambed. Subsurface measurements in hyporheic zones were initially accomplished using standpipes hammered beneath or into the side of streams or in hand-dug, stream-side pits (Hynes, 1974; Grimm and Fish-er, 1984; Bencala *et al.*, 1984). The common occurrence of cobble layers in sediment near streams was an obstacle to emplacing large numbers of sam-pling points—power augers were hardly more effective than hand tools in emplacing wells or piezometers under those conditions. Valuable tech-niques for emplacing small-diameter drivepoints by driving or jack-ham-mering were adapted for hyporheic studies by Wondzell and Swanson (1996) and Geist *et al.* (1998). As emplacement of drivepoints became eas-ier, there was an increasing use of piezometers (screened for only a short length) in place of wells, which provided vertical resolution in hydraulic-head measurements and in chemical-concentration measurements. Ex-panding the number of measurement points allowed investigators to map hyporheic flowpaths in two or three dimensions (Harvey and Bencala,

1993; Wondzell and Swanson, 1996; Wroblicky *et al.*, 1998; Edwards and Priscu, 1997).

2. Darcy Approach to Compute Hyporheic Fluxes

Water fluxes across a streambed are calculated on the basis of two-dimensional contour maps of hydraulic head, estimates of hydraulic conductivity of near-channel sediment, and the basic governing equations for groundwater flow. For example, Wondzell and Swanson (1996) and Wroblicky *et al.* (1998) used the U.S. Geological Survey model code MODFLOW to compute hyporheic fluxes, while Harvey and Bencala (1993) used a finite-difference approximation to the governing equations. Figure 3 shows the typical result of those approaches—a map of hydraulic-head contours and flowpaths, and calculations of water fluxes across the streambed plotted versus distance in the stream. Hydraulic measurements clearly demonstrate that streambed fluxes are directed both into and out of the channel and that the direction of the flux is affected by streambed topography and meandering of the channel. If independent information is available about groundwater fluxes in the reach (e.g., if it can be assumed that only inflow of ground water occurs and no outflow), then it is possible to partition the several components of the total streambed flux into its component parts [i.e., the hyporheic and groundwater terms in Eq. (1)]. The reach-averaged calculation is made by summing up individual calculations along the channel: the sum of fluxes from channel to sediment is then compared with the net flux across the channel bed to compute the reach-averaged hyporheic flux, q_h. The typical result is that the hyporheic component of streambed flux, q_h, is found to be considerably larger than the groundwater component, q_L (Harvey and Bencala, 1993). Figure 3B shows that water fluxes into hyporheic zones decrease during the wet season, due to the opposing force of higher groundwater levels on the lower hill slope. Generally the overall spatial pattern of fluxes into and out of the streambed changes only slightly between wet and dry seasons. Fluxes through meter-scale hyporheic flowpaths typically decline by 30–50% during the wet season, because of higher groundwater heads in the surrounding aquifer (Wondzell and Swanson, 1996; Wroblicky *et al.*, 1998; Harvey *et al.*, 1996). Hyporheic fluxes can decrease by a similar magnitude from day to night, due to increasing groundwater fluxes at night that accompany the cessation of transpiration from shallow ground water (Harvey *et al.*, 1991).

There are many uncertainties in the hydraulic approaches described earlier. In addition to the problem of installing enough instrumentation to describe the complex head distribution, there is considerable uncertainty in the hydraulic conductivity of sediment used in the streambed flux calculations. Hydraulic conductivity of saturated sediments near channels has been estimated using slug tests in piezometers or wells (Morrice *et al.*, 1997), based on assumptions and calculations discussed in references such as Bouwer and

A

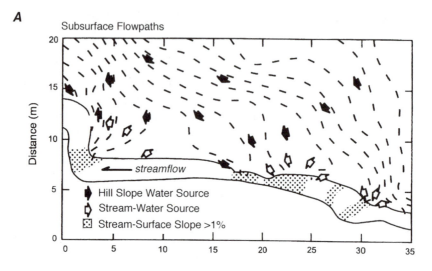

B

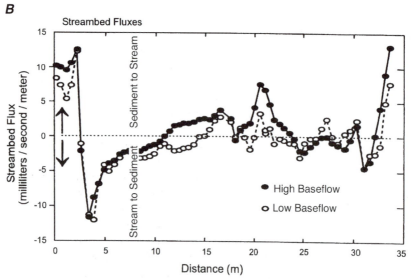

FIGURE 3   Identifying hyporheic flowpaths and fluxes through them by mapping hydraulic-head contours (A) and calculating fluxes across the streambed (B). Plan-view map of stream and hydraulic-head contours (dashed lines) in (A) show that hyporheic flowpaths receive their water from the stream and return the water to the stream a short distance downstream. Meter-scale hyporheic flowpaths are associated with topographic breaks in slope and meanders in the stream—they are embedded within the larger-scale pathways of interaction between the stream and groundwater system. Note the decrease in fluxes into hyporheic flowpaths in (B) during the high baseflow study—when higher groundwater levels create greater opposition to fluxes into the streambed.

Rice (1976). Solute-tracer injections in the subsurface have also been used to estimate hydraulic conductivity, on the basis of travel time between two measurement points, the hydraulic gradient, and the porosity of sediment (Harvey and Bencala, 1993),

$$K = v_s \, n \frac{\Delta l}{\Delta h},\tag{2}$$

where K is the average saturated hydraulic conductivity of the sediment, n is sediment porosity, and v_s is the estimated velocity of the tracer between two wells with a difference in hydraulic head equal to Δh over a path length Δl.

These methods usually are not as effective to estimate K of the coarse sediments that are very close to channels (e.g., 30 cm or less beneath the streambed). In order to determine K in shallow sediments, McMahon *et al.* (1995) installed PVC pipes to a depth of 30 cm in a streambed and operated them as constant-head permeameters. Another means to estimate hydraulic conductivity of shallow hyporheic sediments is to compute K from measurements of the distribution of grain size (Wolf *et al.*, 1991). This method depends on being able to retrieve intact sediment cores from the streambed. Several empirical equations are used to compute K using parameters such as geometric mean diameter of grains, diameter at which 10% of the sample (by weight) is of smaller size, and standard deviation of the grain size. Overall, the grain-size approach to determine K is most useful for well-sorted sands, and estimates are progressively less reliable as grain-size variability increases.

3. Direct Measurement of Water Fluxes across Streambeds

Seepage meters are increasingly being used in streams to determine hydrologic fluxes across the streambed (Jackman *et al.*, 1997; Wroblicky *et al.*, 1998). A seepage meter is an inverted funnel with walls emplaced in the sediment and a plastic bag attached that is prefilled with a measured quantity of water. Over time, a change in volume in the water in the bag represents a vertical flux of water across the streambed. While there is considerable information on performance of seepage meters in test tanks (Belanger and Montgomery, 1992) and in lakes (Lee, 1977; Shaw and Prepas, 1989), there are relatively few detailed evaluations of seepage meter performance in streams. A laboratory flume study showed that increasing surface-water velocities stretched seepage bags and created a slight suction inside the seepage bags relative to surrounding water. The effect on seepage meter measurements was to bias results toward unnaturally high values (Libelo and MacIntyre, 1994). Another innovative technique to estimate fluxes through hyporheic flowpaths is based on calibrated dissolution of plaster of Paris standards buried at shallow depths in the streambed (Angradi and Hood, 1998).

4. Hydrologic Fluxes Inferred from the Appearance of Tracers in the Subsurface

Evidence for stream-water movement into shallow ground water was provided by early field studies that injected solute tracers into streams (Bencala *et al.*, 1984). Flow velocities and residence times in hyporheic zones are usually estimated from subsurface data by monitoring the arrival time at wells of solute tracers injected in the stream (Triska *et al.*, 1993). After tracer concentrations reach a plateau in the subsurface, hyporheic zones are delineated as the zone where greater than 10% of subsurface water was contributed from the stream, which is easily computed from standard mixing equations (Triska *et al.*, 1989). Figure 4 shows a typical result from a tracer

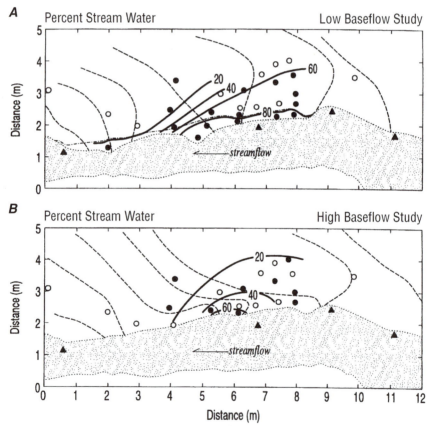

FIGURE 4 Identification of hyporheic flowpaths using chemical tracers. In this plan-view map of the stream, solid contours are percent stream water, dashed contours are hydraulic head, closed circles are wells sampled for chemistry, and triangles are staff gages. Note the less extensive penetration of stream water into hyporheic flowpaths during the higher baseflow study, when higher groundwater levels oppose stream-water inputs to the subsurface. (Modified from Harvey *et al.*, 1996.)

study in the hyporheic zone. Note that the hyporheic zone in Fig. 4 is largest in dimension during the dry season when groundwater levels and stream flow are relatively low and decreases in size when conditions are wetter, such as when groundwater inflows that oppose hyporheic fluxes are higher.

As more studies are undertaken in lower-gradient streams with finer streambed sediments, the need for even finer spatial resolution in detecting hyporheic flows is required. In those situations, the installation of wells becomes impractical. Even narrow diameter piezometers begin to be ineffective samplers below a scale of approximately 20 cm. Diffusion equilibrators (Hesslein, 1976; Carignan, 1984) are an important method in lakes and wetlands, but they are not often used in stream sediments due to concerns about their performance, including (1) inability to resample frequently to establish arrival times of solute at different depths in the hyporheic zone because of diffusion limitations and (2) the possibility that equilibrator body itself might be so large that it interferes with natural pathways of hyporheic flow through the sediment. Recently, Duff *et al.* (1998) developed a miniature drivepoint sampler (MINIPOINT) for acquiring small-volume, close-interval, porewater samples without disturbing the natural flow conditions. By pumping at very low rates (<6 ml min^{-1}), porewater samples are obtained at a vertical resolution of 2.5 cm through the hyporheic zone.

Even sampling with centimeter-scale resolution may not be sufficient for all hyporheic investigations. One means to increase resolution of sampling might be diffusion sampling using gel probes, a technique that has been extensively tested in lake-bottom sediments (Davison *et al.*, 1994). Further development of the gel probe technique for use in hyporheic zones has the potential to push the resolution of porewater chemical measurements toward the millimeter-scale level.

Recently there has been renewed interest in using temperature as a tracer to characterize stream interactions with subsurface water. The usual application of the technique requires a natural fluctuation in temperature within the stream that will propagate into the streambed at a rate that is sensitive to the direction and magnitude of groundwater flow. The vertical flux of ground water is determined by adjusting the groundwater velocity parameter to fit the streambed temperature data (Lapham, 1989; Silliman *et al.*, 1995). By far the easiest approach is to model heat transport in one dimension below the streambed, although model codes for two- or three-dimensional modeling are available. Constantz *et al.* (1994) considered the nonlinear relation between temperature, hydraulic conductivity, and vertical flux of ground water. Their results indicate that daily temperature swings of 7° in surface water could affect hydraulic conductivity (because hydraulic conductivity is a property of both fluid viscosity and sediment characteristics) by as much as 25%. Streambed temperature fluctuations have not, to our knowledge, been used as a tracer to quantify hyporheic exchange fluxes. This potential application of temperature as a tracer is especially promising where rapid temperature fluctuations in shallow porewater beneath stream-beds could serve as a signal of fast-timescale exchange between the stream and hyporheic zone.

5. Using Tracers to Quantify Chemical Reactions in Hyporheic Zones

Several investigators have used solute tracers to quantify *in situ* chemical reactions in hyporheic zones. Triska *et al.* (1993) injected nitrate and bromide into a hyporheic flowpath at Little Lost Man Creek in California and determined the net uptake of nitrate. Hill and Lymburner (1998) used background contributions of chloride as an environmental tracer to determine the fate of nitrate in the hyporheic zone of a headwater stream in Ontario. Findlay *et al.* (1993) used a similar approach to determine uptake of dissolved organic carbon in a hyporheic flowpath through a gravel bar in a headwater stream in the northeastern United States.

Some early studies illustrate the importance of good hydrologic constraints on field-tracer tests. For example, not accounting for mixing between surface water and ground water when calculating solute uptake in a subsurface tracer test is likely to result in an overestimate of chemical uptake rates. Related work on oxygen uptake in subsurface flowpaths was conducted in Europe in subsurface flowpaths that were recharged by river water flowing toward water supply wells. Bourg and Bertin (1993) and Bertin and Bourg (1994) used radon concentrations along a flowpath to quantify subsurface flow rate and specific conductivity to quantify mixing between recharged surface water and ground water. A simple model was then used to calculate the extent of dissolved oxygen uptake and release of dissolved manganese along the flowpath. Recently, Harvey and Fuller (1998) determined the rate of microbially mediated uptake of dissolved manganese in the hyporheic zone of Pinal Creek, Arizona, a southwestern U.S. stream contaminated by mine drainage. That study obtained *in situ* measurements of reactive uptake in hyporheic flowpaths simultaneously with reach-scale stream-tracer injections, allowing cumulative effects of hyporheic reactions in the drainage basin to be quantified.

6. Stream-Tracer Injections as a Means to Quantify Hyporheic Zone Fluxes

The stream-tracer approach offers an alternative to making point measurements at numerous locations to characterize the hyporheic zone. Using that method assumes that information about hyporheic zones and other storage processes are evident as an "imprint" in the record of changing solute-tracer concentrations in the stream at a point located tens to hundreds of meters downstream of the solute-tracer injection. The average characteristics of hydrologic storage in hyporheic zones (and other storage zones) are extracted from the experimental data by simulating results with a transport model. The result from modeling tracer experiments is a set of parameters that quantifies the average hydrologic flux between streams and storage zones, average residence time of water in storage zones, and average dimension of storage zones. An averaging approach to characterize reach-scale processes is, in our opinion, so important to progress that we devote a considerable portion of the remainder of this chapter to reviewing and discussing the advantages and limitations of the stream-tracer approach.

IV. USING THE STREAM-TRACER APPROACH TO CHARACTERIZE THE HYPORHEIC ZONE

In a typical stream-tracer experiment, a nonreactive solute tracer such as chloride or bromide is injected into a stream at a constant rate, and concentrations of the tracer are monitored over time at points downstream. A simple one-dimensional model for transport in the stream is used to simulate the field data. The purpose is to link the field measurements to four general processes: advection and longitudinal dispersion in the channel, groundwater inflow, and hydrologic retention in surface or subsurface zones (referred to as storage zones). We use the term *storage* to describe any physical, nonequilibrium process that retains water and solute and then releases it back to the active central channel. Storage processes include (but are not limited to) water and solute exchange with hyporheic zones and exchange with stagnant water on channel sides, on the bottom of pools, or in recirculating eddies. By "nonequilibrium process" we mean any physical process in the stream channel or subsurface that causes solute retention, but that cannot be simulated by adjusting the longitudinal dispersion coefficient in a transport model. Later in the chapter, we will discuss the nature of nonequilibrium transport processes and their importance with regard to designing successful tracer studies.

The commonly used equations to model one-dimensional transport in streams with groundwater exchange and storage are

$$\frac{\partial C}{\partial t} = -\frac{Q}{A}\frac{\partial C}{\partial x} + \frac{1}{A}\frac{\partial}{\partial x}\left(AD\frac{\partial C}{\partial x}\right) + \frac{q_L^{in}}{A}(C_L - C) + \alpha(C_s - C) - \lambda C, \tag{3}$$

$$\frac{\partial C_s}{\partial t} = \alpha\frac{A}{A_s}(C - C_s) - \lambda_s C_s, \tag{4}$$

where t and x are time and direction along the stream; C, C_s, and C_L are concentrations in the stream, storage zones, and ground water, respectively (mg L^{-1}); Q is the in-stream volumetric flow rate (m^{3}s^{-1}); D is the longitudinal dispersion coefficient in the stream (m^{2}s^{-1}); A and A_s are the stream and storage-zone cross-sectional areas, respectively (m^2); α is the storage-exchange coefficient (s^{-1}), and λ and λ_s are first-order rate constants (s^{-1}) describing net uptake of a reactive solute by a biological or geochemical process occurring in streamflow or in the storage zone, respectively.

The process of hydrologic storage in streams is often represented in transport models as a simple mass transfer of solute between the channel and storage compartments. Note that storage zones are characterized in Eqs. (3) and (4) by a rate constant, α, which defines the rate at which stream water is exchanged for water in storage zones, and by a cross-sectional area, A_s, which is a measure of a stream's physical capacity for storage. Harvey *et al.* (1996) showed how the parameters of the storage zone model in Eqs. (3) and (4) can be related to the hyporheic fluxes defined by Eq. (1),

$$q_s = \alpha A, \tag{5}$$

$$t_s = \frac{A_s}{\alpha A}, \tag{6}$$

where q_s is the storage-exchange flux (i.e., the average flux of water through storage zones per unit length of stream; $m^3 s^{-1} m^{-1}$), and t_s is the average residence time of water in storage zones. Mathematically, the storage-exchange flux, q_s, in Eq. (5) is identical with the hyporheic-exchange flux q_h presented in Eq. (1). A different subscript is used in Eq. (6) to acknowledge that a flux computed from stream-tracer modeling parameters might characterize storage in stagnant surface-water zones in addition to, or instead of, in hyporheic zones.

Another important hydrologic parameter that is fundamental to stream-transport studies is the turnover length for stream water through hyporheic flowpaths,

$$L_s = \frac{u}{\alpha}. \tag{7}$$

The turnover length is computed from the parameters of Eq. (4) by dividing the stream volumetric flow rate by stream area (to compute average stream velocity, u) and then dividing the velocity by the storage-exchange coefficient, α. The turnover length characterizes how much distance is required to route stream flow through storage zones. It can also be thought of as the average distance traveled downstream by a parcel of water before entering a storage zone.

The effect of chemical reactions in storage zones can also be characterized by stream-tracer injections and modeling. Harvey and Fuller (1998) proposed a significance factor for chemical reactions in the hyporheic zone

$$RSF = \frac{\lambda_s t_s L}{L_s}, \tag{8}$$

where λ_s is the reaction rate constant in the hyporheic zone, t_s is the hydrologic residence time in the hyporheic zone, L_s is the turnover length for stream water through hyporheic flowpaths, and L is the length of the stream reach under consideration. Values of the dimensionless index greater than 0.2 appear to characterize systems where chemical reactions in the hyporheic zone are fast enough and flow through the hyporheic zone significant enough to exert a cumulative influence on downstream chemistry.

A. Inverse Modeling of Stream-Tracer Data: Advantages of Fitting by Statistical Optimization

Before discussing stream-tracer modeling specifically, we briefly distinguish between two general categories of modeling studies—forward and inverse modeling. In forward modeling, the parameters of the model are specified beforehand on the basis of independent studies, and the model is run to determine the spatial and temporal patterns of solute transport. Forward modeling simulations are often compared with field data, although precisely matching simulations and field data are not usually emphasized. Rather, the purpose of forward modeling is usually to increase intuition about processes, often through the use of sensitivity analysis to identify the relative importance of different controlling factors (e.g., Choi *et al.,* 1998). Inverse modeling, as the name implies, is the inverse of forward modeling. Inverse modeling relies on precisely fitting simulations to data in order to select the best-fit values of the parameters (i.e., the values that, when inserted into the model, best match measured spatial and temporal patterns of solute transport). Most of the solute-transport investigations cited in this chapter use the inverse modeling approach. Using the inverse method to determine transport parameters in stream–groundwater systems assumes that each transport process is uniquely imprinted on the tracer data. Wagner and Harvey (1997) showed how those process "imprints" can be described in terms of regions of sensitivity of the modeled tracer concentration to changes in parameter values. Figure 5 shows qualitatively those regions of sensitivity on a typical simulation of tracer rise in concentration to a plateau and fall (after the tracer has been turned off) to background concentrations.

The effect of storage zones is imprinted on the "shoulder" and "tail" of the tracer history (Fig. 5). The rate of exchange, α, is reflected in the early curvature of the shoulder and tail, while the storage-zone area, A_s, is reflected in the slope at which the shoulder and tail approach the plateau or background concentration. Figure 5 highlights some important principles about parameter reliability. First, the imprint for each process is associated with a particular segment of the concentration history. Second, information about a parameter is most efficiently gained by experiments that sample at points where the imprint of that parameter—the parameter sensitivity—is greatest. Third, whereas sampling the appropriate regions of sensitivity is necessary to acquire reliable parameters, it does not guarantee success. Details of the tracer experiment, such as length of the experimental reach and the sampling frequency, also play

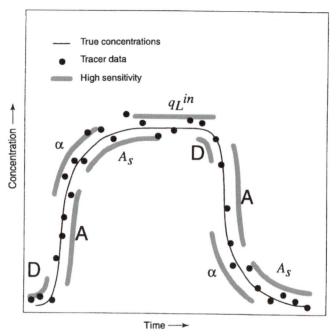

FIGURE 5 Regions of sensitivity in a model simulation of stream-tracer measurements. Regions of high sensitivity are shown as shaded bars for each parameter of the transport model. A, stream cross-sectional area; D, longitudinal dispersion coefficient in the stream; q_L, groundwater inflow; α, stream/storage-zone exchange coefficient; and A_s, storage-zone cross-sectional area.

an important role in determining the reliability of parameters. Later in this chapter, we will discuss how an improperly designed experiment can lead to erroneous parameters that appear to simulate the tracer data quite well.

For several reasons, using the trial-and-error method (see Stream Solute Workshop, 1990) to select best-fit parameter values is not the best approach to simulate field data. First, although the trial-and-error approach can always lead to parameters that "fit" the tracer data, it cannot formally address the problem of parameter nonidentifiability. Simply stated, non-identifiability occurs when different sets of parameter values result in the same model output. We are increasingly encountering situations where nonidentifiability is a problem for the parameter group D, α, and A_s—in those situations many different combinations of the parameters seemingly provide good fits to the tracer data but bear little or no relation to the transport processes that were acting to produce those data. Second, trial-and-error calibration is never fully free of investigator biases, even when parameter sets that are identified appear to be reasonable. Finally, trial-and-error calibration does not provide a mechanism for assessing the uncertainty of final parameter estimates.

A growing number of investigators are using statistical optimization as the technique for modeling stream-tracer experiments in the inverse mode (Wagner and Gorelick, 1986; Hart, 1995; Harvey *et al.*, 1996). In fact, a popular code that solves Eqs. (3) and (4), the OTIS model by Runkel (1998), was updated from a previous version to include an optimization package. Optimization uses mathematical search algorithms to sort through an essentially unlimited number of parameter combinations to identify those parameter values that best reproduce the tracer data; uncertainty analysis uses tools of statistics to quantitatively assess the reliability of model parameter estimates. The advantages of this approach are threefold. First, it provides parameters more quickly and efficiently than trial and error while minimizing investigator bias. Second, it provides an assessment of parameter uncertainty, which greatly increases the value of the parameters. Third, it can be used to evaluate and compare the effectiveness of alternative tracer test designs and to develop guidelines for planning tracer studies (Wagner and Harvey, 1997). We will return to the idea of tracer test design later in the chapter.

V. PARADIGM LOST? LIMITATIONS OF THE STREAM-TRACER APPROACH AS A MEANS TO QUANTIFY HYPORHEIC PROCESSES

Streams interact with many types of storage zones, some in the stream and some in the subsurface, with various sizes and with timescales of exchange that span from minutes to tens of days or weeks. Several significant problems arise in evaluating stream-tracer characterizations of storage—such as distinguishing hyporheic zones from in-stream storage zones, or determining whether a particular tracer experiment is only sensitive to a subset of the storage zones that are actually present. How can we be assured that our interpretations based on stream-tracer experiments are valid? We begin our answer by reviewing basic assumptions of stream-tracer modeling. Next, we discuss how stream-tracer modeling results can be compared with *in situ* measurements in hyporheic zones. Next we evaluate limitations and biases of stream-tracer methods from a theoretical perspective, and we support our conclusions using tracer data collected from a large number of experiments in different streams across the country. Finally, we argue how all the preceding information can be combined toward a goal of designing better tracer experiments that will increase the reliability of results.

A. Assumptions of the Stream-Tracer Approach to Characterize Hyporheic Zones

The first assumption is that hyporheic exchange can be represented as a first-order mass transfer process. There is, in fact, ample evidence that a first-order mass transfer approximation is a perfectly adequate model, because it implies an exponential distribution of residence times of the type often observed in groundwater systems. Some field studies indicate that more complexity is possible for hyporheic zones. For example, Castro and Hornberger (1991) and Harvey *et al.* (1996) found two different exponentially distributed populations of hyporheic flowpaths (one with relatively short and one with relatively long subsurface travel times)—fast flow occurred through gravel bars (average travel time of hours) and slower flow occurred through alluvial flowpaths (average travel time ranging from tens of hours to days).

The stream-tracer model discussed in this chapter [e.g., Eqs. (3) and (4)] is not flexible in allowing more than one "class" of hyporheic flowpaths to be considered. There are, however, alternative models that are more flexible in considering multiple classes of subsurface storage zones (see Beer and Young, 1983). For example, Castro and Hornberger (1991) used the alternative type of transport model that had no restrictions on the potential number of storage zones to investigate hyporheic processes in a mountain catchment in Virginia. The use of statistical optimization with that model identified two dominant classes of storage zones through gravel bars and alluvium in that system. Recent work by Worman (1998) and Choi (1998) extend further the idea of modeling multiple storage zones.

A second implicit assumption of the stream-tracer approach is that storage zones uniquely characterize hyporheic exchange (as opposed to in-stream storage zones). In mountain streams where the dimension of the storage zone is often much larger than the stream, that argument may be easy to defend (Mulholland *et al.*, 1997). However, at higher stream flow in the same stream, or in larger, faster-flowing streams with little slope variation, storage areas are often considerably smaller than stream areas. Procedures to distinguish the contribution of hyporheic zones from surface-water storage zones are less clear-cut in those cases.

B. Comparing Storage-Zone Modeling Parameters with Subsurface Measurements in Hyporheic Zones

In this section, we discuss how reach-averaged storage-zone parameters can be appropriately scaled so that they can be compared with *in situ* subsurface measurements in hyporheic zones. The main purpose of comparing model-derived parameters with *in situ* measurements is to identify order-of-magnitude types of disagreements between data types—usually this will mean identifying those situations in which the stream-tracer method is definitely not characterizing certain classes of hyporheic flowpaths, or not identifying any hyporheic flowpaths at all. For example, Harvey *et al.* (1996) compared storage timescales and subsurface tracer travel times in a mountain pool-and-riffle stream and demonstrated that stream-tracer modeling could not detect meter-scale hyporheic flowpaths through the alluvium—the residence time of hyporheic water in alluvium was simply too long relative to the transport time in the experimental channel reach. However, the much faster hyporheic flow through gravel bars was easily detected from stream-tracer results in the same experiment.

The hyporheic cross-sectional area determined by modeling is an empirical parameter that is a fundamentally different quantity than what can be measured in the field. How can we compare it to something measurable? Because each field site is different, and because the quantities to be compared are inherently empirical, there will never be a single comparison that is appropriate for all streams. Here we recommend a few simple ways in which stream-tracer modeling results can be compared with *in situ* subsurface measurements. Figure 6 illustrates two very simple characterizations of storage area beneath and to the side of streams. Figure 6A illustrates a stream channel in which the storage-zone cross-sectional area is considerably smaller than the cross-sectional area of the stream ($A_s < A$), and where stream width,

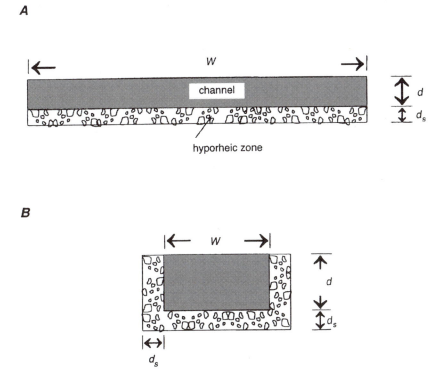

FIGURE 6 Two simple characterizations of the spatial dimension of the hyporheic zone, d_s, are shown in (A) and (B). The hyporheic dimension is computed using Eq. (9) or (10), respectively, on the basis of measurements of stream width, depth, and porosity of the sediment.

w, is much greater than stream depth, d (e.g., $w/d > 20$). A simple approximation of the hyporheic-zone depth in that setting uses the storage cross-sectional area, stream width, and the porosity of the channel

$$d_s = \frac{A_s}{w\,n},$$

(9)

where n is channel-sediment porosity.

Equation (9) assumes that storage in hyporheic zones primarily occurs vertically beneath the channel rather in zones beneath the channel sides. That assumption certainly is not valid where stream depth is a significant fraction of stream width or in broad stream valleys filled with sediments with very high hydraulic conductivity. For example, in some systems, alluvial deposits of well-sorted cobbles, pebbles, and coarse gravel to the sides of the channel might be the principal hyporheic zone, especially if the channel sediments are not vertically extensive.

An approximation for the dimension of the hyporheic zone that is more general than Eq. (9) assumes that the hyporheic zone is as laterally extensive as vertically extensive. For that situation, the hyporheic dimension is

$$d_s = \frac{2A_s}{n\left[(w + 2d) + \sqrt{(w + 2d)^2 + 8A_s}\right]}.$$

(10)

Equation (10) was developed by first writing an equation for the cross-sectional area of storage that assumes the hyporheic zone extends an equal distance in all directions from a rectangular channel. When rearranged to solve for d_s, the equation took the form of a quadratic, which was solved using an accurate method (Press *et al.*, 1986) to produce Eq. (10). Equation (10) is more broadly applicable than Eq. (9) because it is more likely to be a good approximation in channels where $w/d \ll 20$.

It should be noted that both Eqs. (9) and (10) assume that the source of all water in the hyporheic zone is the stream. Although that assumption is consistent with the concept expressed in Eqs. (3) and (4) it is not really an accurate approximation of the field situation where mixing between stream water and ground water occurs in hyporheic flowpaths (Triska *et al.*, 1989). Thus, Eqs. (9) and (10) are highly simplified approximations that should not be overly relied upon.

In situ measurements of the dimension of the hyporheic zone can be made in conjunction with stream-tracer injections (after many hours, when tracer concentrations have everywhere achieved a constant-plateau concentration). Figure 7 shows an example from Pinal Creek, Arizona, where the hyporheic zone dimension, d_s, was calculated from Eq. (9) and compared with measured depths of bromide penetration into the streambed at a number of locations along the stream. No attempt was made to adjust field or

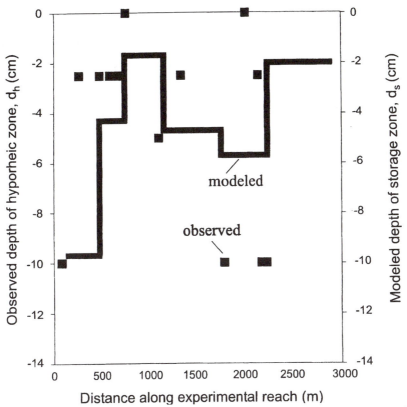

FIGURE 7   The depth of penetration of a stream-solute tracer (bromide) into porewater in the streambed of Pinal Creek, Arizona, is compared with the depth of the storage zone determined independently at a larger spatial scale by modeling bromide transport in the stream. (Modified from Harvey and Fuller, 1998.)

modeling results to account for the effect of mixing between stream water and ground water in hyporheic flowpaths. Note that the reach-by-reach estimates of storage depth show some of the same trends observed *in situ* but not the same extent of variability.

Possibly a better way to test whether a stream-tracer experiment is sensitive to the hyporheic zone is to compare the timescale of storage exchange determined by modeling (Eq. 6) with residence-time estimates determined *in situ* in the hyporheic zone. Figure 8 compares measurements of the concentration of a stream tracer in two types of hyporheic zones (gravel bars and deeper alluvium). The comparison indicates that although the stream-tracer experiment characterizes hyporheic exchange with gravel bars reasonably well, it does not adequately characterize exchange with longer flowpaths through less conductive alluvium. Graphical results can be further summa-

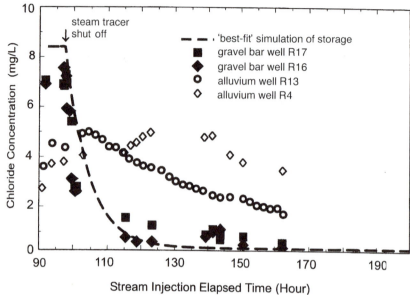

FIGURE 8   Flushing of a chloride tracer from streamside wells following shutoff of the injection of the tracer in the stream. The timescale of flushing of the tracer from the wells characterizes the residence time of stream water in hyporheic flowpaths. The best-fit simulation of storage was obtained independently by simulating tracer measurements in the stream. Although the stream-tracer simulation characterized hyporheic exchange with gravel bars reasonably well, it did not characterize hyporheic exchange with deeper alluvium.

rized by calculating the characteristic time for a tracer to enter hyporheic zones or be flushed from hyporheic zones. A simple means to estimate the fluid residence time in the hyporheic zone involves calculating the elapsed time, following inception or cutoff of a stream injection, for the tracer concentration to increase (or decrease) to a concentration equal to half the plateau concentration (Triska *et al.*, 1993; Harvey and Fuller, 1998).

C. Stream Tracers Have a "Window of Detection" for Storage Processes with Particular Spatial Dimensions and Timescales

Hydrologic characteristics of the hyporheic zone can vary substantially from reach to reach or between drainage basins, depending on many factors. The causes of variability can be lumped into two broad classes, the factors that drive the exchange of water across the bed (e.g., overpressures caused by flow over a rough streambed) and the factors that determine the resistance to exchange (e.g., sediment hydraulic conductivity) in the streambed. Figure 9 illustrates the broad spectrum of possible interactions between channels and hyporheic zones. Different scales of surface–subsurface interaction could operate simultaneously in a given stream; yet, it is becoming increasingly evident that stream-tracer experiments are not sensitive to all those interactions. The stream-tracer approach is inherently limited in the dimensions and timescales of hyporheic exchange that can be detected. Figure 9 illustrates the broad spectrum of possible surface–subsurface interactions and indicates subsets that are potentially accessible through stream-tracer experimentation. Hyporheic exchange that occurs over very large spatial scales and long timescales, such as flow through channel-point bars or abandoned channels, usually is not going to be accessible to stream-tracer experimentation (Fig. 9A). The window of detection is generally weighted toward the faster components of exchange (minutes to tens of hours) that occur within relatively short (centimeters to meters) hyporheic flowpaths, through

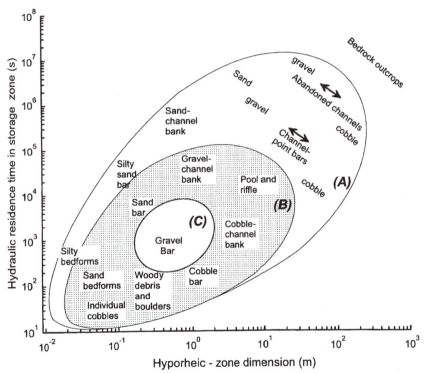

FIGURE 9 Spatial and temporal domain of hyporheic interactions and relation to roughness features in channels. The part of the domain with longer timescales and larger spatial scales of hyporheic interactions (A) is generally inaccessible to detection by stream tracers. The region that is typically accessible by stream tracers is shown by B. For a given experiment the window of detection is relatively small, depending on local transport characteristics and characteristics of the tracer experiment such as choice of experimental reach length. Region C shows a typical window of detection for a single tracer experiment.

coarse gravel bars in the channel center or on the channel margin (Fig. 9B). Due to reasons to be explained, a particular experiment generally has a window of detection that is quite small in comparison with the possible ranges of detectable scales (Fig. 9C).

The window of detection of storage processes depends on two general categories of factors, (1) characteristics of the stream including storage-zone dimension and exchange rate as well as velocity of stream flow and volumetric flow rate and (2) the design aspects of the experiment, such as the length of the experimental reach, the amount of tracer injected, and the time period over which the injection occurs. Of those, stream velocity is a critical aspect that can interact with experimental reach length to determine which timescales and length scales of storage exchange can potentially be characterized. To better understand this limitation, we must examine exactly how seemingly arbitrary choices, such as the choice of experimental reach length, can affect the outcome of an experiment.

The design of the experiment (e.g., reach length selected by the investigator) interacts with stream velocity by controlling the opportunity for stream water (and tracer) to interact with hyporheic zones. Too little interaction and there will be too little sensitivity to detect storage; too much interaction and there will arise a condition known as equilibrium mixing after which storage processes acquire an imprint that is identical to that of longitudinal dispersion in the channel. Only an intermediate amount of interaction between tracer and storage zones in the selected reach achieves the goal of identifying reliable parameters through modeling. Later, we explain how the investigator can plan experiments that promote an optimal amount of interaction between tracer and storage zones (i.e., the amount that will minimize uncertainty of parameter estimates).

The extent of interaction between tracer and storage zones in a particular experimental reach is determined by the balance between the following factors: stream velocity, the length of the study reach, and the true average properties of the hyporheic zones as represented by a cross-sectional area and by an exchange coefficient. All those factors are part of an experimental Damkohler number, DaI, which is a dimensionless grouping that expresses the degree of balance (or dominance) between downstream transport processes and storage processes in the tracer mass balance. DaI is written

$$DaI = \frac{\alpha(1 + A/A_s)L}{u},\tag{11}$$

where u is stream velocity, L is length of the experimental stream reach, and α, A_s, and A are the storage-exchange coefficient and the storage-zone and stream cross-sectional areas, respectively.

Sensitivity to storage processes and reliability of parameter estimates is highest when DaI is approximately equal to 1 (Wagner and Harvey, 1997).

When *DaI* is much smaller than 1 (i.e., when the transport timescale in the reach is short relative to the timescale of storage), very little tracer has interacted with the storage zones and the imprint of storage exchange on the tracer data is simply not identifiable. When *DaI* is much larger than 1 (e.g., when the transport timescale in the reach is long relative to the timescale of storage), parameter uncertainty rises because tracer spreading caused by storage exchange has reached the equilibrium stage that follows after all tracer has interacted with storage zones. When equilibrium mixing has been fully achieved, the imprint of storage-zone exchange is no longer separable from the imprint of many other processes, including mixing processes that are caused by velocity variations in the stream. The problem is that equilibrium mixing with storage zones can be modeled equally well by adjusting storage parameters α and A_s or by adjusting the longitudinal dispersion coefficient *D*. In other words, many combinations of storage parameters and dispersion coefficients will fit the data equally well when *DaI* values are much greater than 1.

The investigator has some ability to influence the outcome of an experiment, in terms of the reliability of the results. The investigator's control is exercised mainly through choice of the length and choice of the time period of the injection. The influence of choosing a reach length is easily illustrated by varying reach length in a numerical experiment and computing uncertainty of storage parameters (Fig. 10), which shows that if the reach is too short (*DaI* << 1) or too long (*DaI* >> 1), storage zone parameter uncertainty increases to unacceptable levels. The measure of parameter uncertainty that is used in Fig. 10 is the coefficient of variation (CV). The CV is a dimensionless measure of parameter uncertainty determined by dividing the standard deviation of the parameter estimate by the best-fit parameter estimate for that simulation. A CV of 1 (plotted as 10^0 in Fig. 10) indicates that uncertainty is as large as the parameter estimate itself! A highly reliable parameter estimate should have a CV equal to or less than 0.1 (i.e., 10% of the parameter's value). Typical CV values are in the range of $0.1-0.3$ for storage parameters, which in our opinion still represent reasonable levels of uncertainty.

We demonstrate the generality of our point about a limited window of detection using data from a number of tracer experiments conducted over the past 30 years all over the United States. In Fig. 11, the hydrologic residence times in storage zones were plotted as a function of the hydrologic residence time in the experimental reach. Note that most of the experimental data fall within a window of detection bounded by intermediate values of *DaI* ($0.5-5$). Within that range, the timescale of retention in storage zones is similar to the timescale of transport in the channel. In short reaches (or in reaches with very high stream velocity), there is only time available to detect rapid interactions with small storage zones; in long reaches (or in reaches with very low stream velocity), there is more time available for the tracer

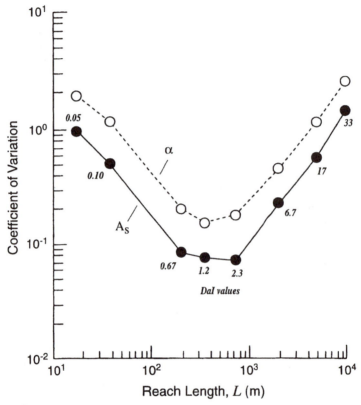

FIGURE 10   Effect of variations in experimental reach length on uncertainty of the storage parameters determined by modeling stream-tracer experiments. Uncertainty is expressed as a coefficient of variation (standard deviation of parameter estimate divided by best-fit parameter estimate). Results were created by varying reach length seven times in a representative simulation, each time re-estimating the parameters of the simulation by inverse modeling. Results show that parameters are reasonably well estimated (±30%) for only a relatively narrow range of the Damkohler number, a dimensionless grouping of physical characteristics, such as stream velocity, and experimental conditions, such as length of the experimental reach. (Modified from Wagner and Harvey, 1997.)

to interact with larger storage zones that retain tracer longer. Thus, the timescales and spatial dimensions of storage that are observed through tracer experiments are determined, in part, by the investigator's choice of reach length.

The main conclusion to draw from Figs. 9 and 11 is that only certain timescales and dimensions of hyporheic processes are accessible through stream-tracer experimentation. For practical reasons, this will usually mean that only the relatively fast exchange pathways through hyporheic flowpaths near channels will be detected, whereas larger and longer timescale interactions with hyporheic flowpaths will remain "invisible." Mostly, it is the hyporheic exchange associated with small roughness features such as bedforms, bars, and pools and riffles that is potentially quantifiable through stream-tracer experimentation. Hyporheic exchange associated with abandoned channel meanders or very large channel-point bars generally will only be accessible through very long-term injections in low-flow channels.

To summarize this section, the stream-tracer approach remains without question the best method to characterize the cumulative effects of storage processes in drainage basins. The main weaknesses are that specific hydro-

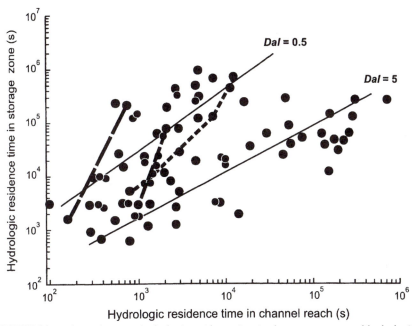

FIGURE 11   Relation between hydrologic residence time in the storage zone and hydrologic residence time in the experimental channel. Data are summarized from published data describing over 50 tracer experiments conducted in streams and rivers throughout the United States. The area bounded by intermediate values of the Damkohler number represents the window of reliable detection for storage-zone processes. Dashed lines connect results from repeat experiments for different stream-flow conditions in three stream reaches (short dashes, Morrice *et al.,* 1997; intermediate dashes, D'Angelo *et al.,* 1993; long dashes, Harvey *et al.,* 1996). Order of magnitude decreases in storage-zone residence timescales at higher stream flows, when hydrologic residence times in the channel reach decreased, are explained in part just by the shifting sensitivity toward shorter timescale storage processes. Caution is therefore advisable as part of any physical interpretation on storage-zone parameters—interpretations need to consider the bias in detection of storage processes that depends in large part on the hydrologic residence time in the experimental-channel reach.

logic or chemical processes cannot be isolated, results are not transferable to longer reaches or differnet flow regimes, and results only represent a limited spectrum of the true spectrum of hyporheic flowpaths present. As a means to increase physical interpretations, we suggested that statistical optimization should more routinely be used to choose parameter values and estimate their uncertainty. Uncertainties as large as 30% are typical for storage-zone parameters—parameters with larger uncertainties should be evaluated with skepticism. We also recommended that hyporheic zone dimensions and residence timescales be independently measured and compared for order of magnitude agreement with stream-tracer modeling results. Parameterizations of stream-tracer models should be tested on the basis of the preceding criteria before acceptance or rejection as measures of the hyporheic zone. We also discussed the fact that the results of stream-tracer experiments are biased to some extent, reflecting an interaction between experimental setup and actual field conditions. Although the experimental Damkohler number provides some guidance to assess the bias and reliability of results, we will have to accept that biases are, to some extent, inherent to the method and are difficult to separate from the physical interpretations that we wish to draw. Later in this chapter, we discuss implications of these biases for developing reliable physical interpretations of storage parameters.

D. Designing Better Stream-Tracer Experiments

We introduced the concept of experimental design in the previous section and suggested how, through the design process, an investigator might be able to increase the probability of obtaining reliable storage parameters in an upcoming experiment (e.g., by adjusting reach length). We now continue our discussion of design principles as they apply to our general objective of learning more about the hyporheic zone.

The goal of design is more than maximizing sensitivity of tracer data to storage zone processes. We would also like, to the extent possible, to design experiments that are able to distinguish and isolate the effect of hyporheic storage processes from in-stream storage processes. As a first step, it will be helpful to locate the stream-tracer injection point just far enough upstream of the experimental reach so that surface-water storage processes (i.e., along channel sites, along the bottom, or behind boulders) undergo complete mixing and achieve the state of equilibrium mixing before the tracer enters the experimental reach. If equilibrium mixing conditions in surface water are established somewhere between the point of injection and the upstream sampling point in the study reach, then further retention of tracer in surface-water storage zones in the experimental reach can easily be simulated by adjusting the longitudinal dispersion parameter, D, in Eq. (3)—this ensures that the storage parameters in Eq. (3) and (4) are available to uniquely characterize hyporheic exchange. How long an upstream mixing reach is needed? Rutherford (1994) provided some general guidance for selecting a mixing length for tracer in the surface channel,

$$L_e = \beta \ (ub^2/k_z),\tag{12}$$

where u is stream velocity, b is channel width, k_z is the transverse dispersion coefficient (often estimated as $0.23du^*$ where d is channel depth and u^* is the shear velocity), and β is a constant in the range of 1–10 for rough channels. Clearly Eq. (12) can only serve as a rough guideline because of the considerable uncertainty in the constant β.

The other major design concern is the length of the experimental reach. Equation (11) for DaI can sometimes be used to help design an experiment with a good chance of substantially improving the estimates of storage processes. To compute the experimental reach length, we suggest rearranging Eq. (11) to solve for reach length after having set DaI equal to 1. The difficulty is that the investigator must have prior estimates of the stream velocity, the timescale of storage, and the dimension of the storage zone in order to solve for reach length. Consider what we know before an experiment is performed. Preliminary estimates of storage-zone properties might be available from previous tracer experiments in the same stream or in similar streams, and stream velocity can commonly be estimated using simple dye experiments. Or the case might be that the investigator has a special interest in a particular timescale or spatial scale of storage that will determine the prior estimates to be used in designing the experiment.

E. Acquiring the Prior Information Needed to Design Tracer Studies

Our stream-tracer design problem uses the Damkohler number to estimate a reach length that will increase the reliability of parameters estimates. Optimizing reach length is the goal because reach length, in addition to the amount of tracer and the time of the tracer injection, is one of the few parameters that are under the control of the investigator. As a design tool, the Damkohler number shares the same problem with experimental design studies throughout environmental science in that the information needed to improve designs is a function of the parameters that the tracer test is being designed to estimate. This common problem is sometimes referred to as the paradox of experimental design. For us, beating the paradox means obtaining some preliminary, low-cost information about the stream of interest, including preliminary estimates of storage dimensions, exchange rate, groundwater inflow, and stream velocity. These could come from a variety of sources, such as estimates from earlier tracer experiments or from preliminary field measurements at the site of interest. A second approach is to define DaI based on data from "similar" sites. A third approach is to use readily obtainable data about the field site (slope, etc.) that are known to be strongly correlated to one or more of the parameters that define DaI.

In the following example, we show how prior information can be obtained for several sites where prior fieldwork is very limited. To construct the design, we need prior estimates of the components of DaI such as the stream-water velocity, u, the stream cross-sectional area A, the storage-exchange rate α, and the storage zone cross-sectional area A_s. An estimate of the stream-water velocity can be obtained using a dye tracer (i.e., rhodamine-WT) to gage the time to breakthrough at the tracer sampling sites. It is a simple matter to convert dye observations to an estimate of stream-water velocity. Stream cross-sectional area can then be estimated from velocity if an estimate of stream discharge can be obtained. Now consider the storage parameters A_s and α. Obtaining prior estimates of these parameters is a more difficult matter because reliable measurements are difficult to make in a single field visit. One approach that probably could be utilized more often is estimating the extent of the hyporheic zone by measuring streambed temperatures (White *et al.*, 1987). Another approach is to estimate A_s using correlations based on simple characteristics of streams. For example, an analysis of results from many experiments conducted throughout the United States (Fig. 12), showed that the ratio A_s/A covaries with a dimensionless

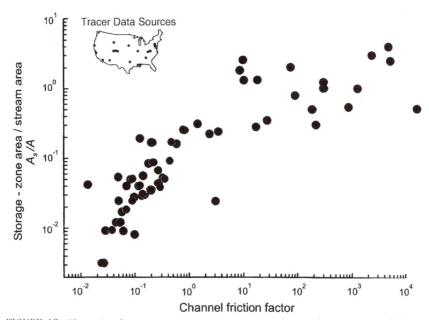

FIGURE 12   The ratio of storage-zone area to stream cross-sectional area, A_s/A, exhibits a strong positive relation with channel friction factor. Data are summarized from published data on more than 50 tracer experiments conducted in streams and rivers throughout the United States.

measure of roughness in channels known as the channel friction factor. The channel friction factor, f, is a function the depth of flow, the streambed slope, stream velocity, and gravitational acceleration:

$$\text{Channel friction factor} = \frac{8gds}{u^2}, \qquad (13)$$

where g is gravitational acceleration, d is stream depth, s is streambed slope, and u is stream velocity.

Once the stream friction factor has been estimated for a new study site, we can estimate (or at least place bounds on) the ratio A_s/A using the trend shown in Fig. 12. Although the storage-exchange rate α is more difficult to estimate from correlations, there is less of a problem in selecting an estimate because the range of α values from previous experiments is relatively small compared with variation of the ratio A_s/A. Future work will undoubtedly provide a better source of information for placing bounds on the exchange rate.

We conclude this section with a simple demonstration of the design process. For this example, we determine optimal reach lengths given some prior information about two streams that are representative of certain classes of streams. The first representative stream type can be described as a high-gradient mountain stream. In that stream type, there is relatively slow flow in a steep channel (~5% slope) through pools and riffles with coarse sediment (cobbles and gravel with fine gravel and well-sorted sands intermixed). The second representative stream type can be described as a mountain valley stream. In that stream type, there is relatively fast flow in a channel of intermediate slope (~0.5% slope) through a gravel or sand-bed stream without major pools and riffles. There are some significant meanders and some braiding of subchannels in the second stream type, and sand bars and cobbles are the primary roughness units in that stream type.

To acquire prior estimates of channel storage characteristics in the two representative streams, we first define friction factors for the channels using representative estimates of slope, depth of flow, and velocity. We then use the values of channel friction factors and the information in Fig. 12 to estimate the representative values of A_s/A for each setting. Finally, we select characteristic values of α from the results of existing tracer studies. With those data, we were able to define Damkohler numbers for the high- and intermediate-gradient settings as a function of the experimental reach length, L. We then calculate for each setting the reach length that resulted in a Damkohler number of 1.0. The results of this comparison are summarized in Table I.

Our analysis indicates that the optimal reach length—the reach length with a Damkohler number of 1.0—for the average high-gradient stream is 220 m, which compares reasonably well with the study scales that are typically reported in the literature for these settings (usually 50–500 m). For

TABLE I   Damkohler Analysis for Two Representative Stream Types

Parameter	Stream type	
	Mountain stream	*Mountain valley stream*
Gradient (slope, %)	5	0.5
Velocity (u; m s^{-1})	0.1	0.5
Storage area ratio (A_s/A)	0.8	0.1
Exchange coefficient (α; s^{-1})	0.0002	0.0008
Damkohler number (DaI)	0.0045·L	0.018·L
Optimal reach length (m)	220	57
Typical reach length (m)	100	1000

mountain valley streams, the optimal reach length is 57 m. This reach length is shorter than the kilometer-scale studies that are often performed in mountain valley streams. The purpose of this example is not to suggest that a reach length of 57 m can always be used in mountain valley streams—actual designs must be done on a case-by-case basis using the best available information. However, our example does suggest that a kilometer-scale reach of stream will commonly be too long to reliably determine storage characteristics in fast-flowing streams with relatively small storage zones.

A final design concern to keep in mind is that an optimal reach length for estimating storage parameters might not also be the best reach for estimating other parameters, such as groundwater inflow or chemical reaction. Recall that as experimental reach length increases, uncertainty of storage parameters initially declines and then later increases (Fig. 10). Uncertainty of groundwater inflow or chemical reaction parameters, on the other hand, is more likely to decline as reach length is increased. Therefore, even with excellent preliminary estimates, it is unlikely that a design for a single experiment will ensure reliable estimates of all the relevant parameters.

F. Implications for Linking Stream-Tracer Parameters with Physical Characteristics of Drainage Basins

Stream-tracer experimentation and modeling of hydrologic retention has now been an active area of research for three decades. More than 50 sets of model-fitted parameters exist from experiments in approximately 40 stream and rivers distributed nationwide. This data set allows a characterization of transport and storage timescales across basins with different physical characteristics or within the same drainage basin but between different time periods. Examples from the data set are shown in Figs. 11 and 12. One goal is to use the data to regionalize concepts about hydrologic retention in streams and expected on transport of chemical constituents. For example,

Fig. 11 shows a fairly tight positive relation between the ratio of storage to stream cross-sectional areas and the dimensionless channel friction factor.

Given that much of this chapter focuses on methodological limitations, it will not be a surprise that we consider the empirical nature of the stream-tracer parameters and the limited timescales of detection to be key barriers to improving our physical interpretations. If physical interpretations are sought by relating storage-zone parameters and channel physical characteristics, then the possible biases of the stream-tracer approach toward detecting shorter timescales of storage will need to be closely considered. For example, a number of investigators have repeated stream-tracer experiments at two or more times during the year (Legrand-Marq and Laudelout, 1985; D'Angelo *et al.*, 1993; Harvey *et al.*, 1996; Morrice *et al.*, 1997; Hart *et al.*, 1999). Tracer data from most of those studies indicated substantial decreases in storage cross-sectional areas at higher streamflow, both in absolute terms (A_s) and in relative terms (in ratio with the stream cross-sectional area, A_s/A). The stream water exchange coefficients α tended to increase substantially at higher streamflow, and thus the hydrologic residence times in the storage zone $A_s/(\alpha A)$ tended to decrease substantially at higher streamflow. That shift can be seen in Fig. 11—the dashed lines connect repeat experiments at different levels of streamflow at three different sites in the United States (D'Angelo *et al.*, 1993; Harvey *et al.*, 1996; Morrice *et al.*, 1997).

It is tempting to draw conclusions, on the basis of changes in storage parameters measured in repeat tracer experiments, about the factors that control hyporheic zones. In fact, some of the authors cited previously did interpret seasonal changes in storage parameters as measures of actual seasonal changes in hyporheic-zone characteristics. In contrast, our work in a mountain stream in Colorado showed that changes in storage parameters between seasons mainly reflected the shifting sensitivity of the stream-tracer method (Harvey *et al.*, 1996). Our interpretation was supported by subsurface measurements that showed that although hyporheic flows did adjust to changing groundwater conditions, those fluctuations (30–50%) were much smaller than indicated by stream-tracer experiments. Other hydrogeologic studies corroborated our subsurface observations, by indicating only relatively small declines in hyporheic-zone dimensions from dry to wet seasons (Wroblicky *et al.*, 1998; Wondzell and Swanson, 1996).

We believe that caution is warranted whenever attempting to ascribe physical significance to seasonal changes in storage-zone parameters—due to the shift in observable characteristics of storage that accompany changes in flow conditions. Order of magnitude declines in the storage-zone residence timescales shown in Fig. 11 appear to be too large to account for actual changes in hyporheic zones that are observed by subsurface measurements. Instead, as streamflow increases, the sensitivity of the tracer approach shifts toward the smaller, faster-turnover component of the distribution of hyporheic flowpaths. The larger hyporheic flowpaths are still present but are not detected by the stream-tracer experiment conducted at higher flow. Therefore, any physical interpretations of storage parameters, either between seasons or between drainage basins, must consider the different flow characteristics and resulting biases of the stream-tracer method.

VI. CHARTING NEW DIRECTIONS IN HYPORHEIC-ZONE RESEARCH

The questions in hyporheic-zone research continue to be refined, as introductory queries such as are chemical reactions enhanced in hyporheic zones give way to advanced questions such as what is the principal physical or chemical factor that limits the extent of a particular chemical reaction in the hyporheic zone? At the same time, the geographical range of hyporheic investigations is expanding beyond forested sites in uninhabited areas to include streams and rivers in more populated areas that are affected by agricultural, industrial, and residential pollution. Future investigations of the hyporheic zone are likely to focus more often on natural remediation such as the enhancement of chemical reactions in hyporheic zones that transform dissolved contaminants such as nutrients, metals, pesticides, and volatile organic compounds (e.g., McMahon *et al.*, 1995; Kim *et al.*, 1995).

New hydrological questions are also emerging as research is undertaken in more diverse settings. For example, at study sites away from humid and high-elevation areas, streams commonly lose water to the groundwater system. Yet, because of the limited scope of stream types investigated in the past, questions such as do hyporheic zones exist in losing streams have not yet been widely addressed. There is also a need to integrate hyporheic and riparian studies over longer timescales than the period of an experimental injection. For example, Squillace *et al.* (1993) investigated the role of bank storage in riparian areas on storage and transformation of atrazine but did not undertake any investigations of the hyporheic zone in the river itself. It would be of interest to gain an understanding of the relative roles of hyporheic and bank-storage processes in contributing to retention or transformation of contaminants.

Refinement of the research questions calls for parallel refinements models and field techniques. Runkel *et al.* (1998) recently took the important step of modifying a commonly used numerical solution for Eqs. (3) and (4) for unsteady flow conditions, and Worman (1998) and Choi (1998) made progress in modeling storage-zone processes in multiple storage zones with differing physical characteristics and chemical reactions.

The importance of relatively small-scale hyporheic flowpaths is increasingly being recognized. Small-scale hyporheic flowpaths typically have relatively fast hydrological exchange rates and, consequently, over a given reach, could be responsible for more throughflow and chemical retention than relatively large hyporheic flowpaths (Harvey and Fuller, 1998). As interest in the role of finer-scale pathways of hyporheic exchange grows, the scales of field measurements are reduced far below the scales that are accessible using traditional hydrogeological instrumentation. The centimeter-scale sampling approach developed by Duff *et al.* (1998) is likely to see increased use: further refinements using microelectrodes or gel probes may eventually push observations toward the millimeter scale.

Still there remains a significant gap between the relatively detailed physical understanding of driving forces that cause surface–subsurface exchange in laboratory flumes and the more empirical measurements at field sites (see Packman and Bencala, Chapter 2 of this book). In particular, it is a detailed understanding of the controlling processes at the scale of the smallest roughness features in streams that drives exchange—the sand waves, small sand bars, cobbles, and other small roughness features—that still eludes field investigators.

Continued development is expected in field experiments and modeling transport behavior at the reach scale. Important work remains in the area of improving the tracer-test design to increase the reliability of tracer tests (Wagner and Harvey, 1997), in particular toward improving estimation of groundwater inflows and their concentrations and toward chemical reaction-rate constants in the storage zones. Progress is also being made by careful comparisons between tracer studies done at different times of year or in different drainage basins (Hart *et al.*, 1999). We also expect increased emphasis in extending models in the direction of more physically and chemically based formulations. For example, the use of detailed chemical submodels with equilibrium speciation and kinetic terms has become more common (Runkel *et al.*, 1996; Choi *et al.*, 1998). Reach-scale transport models are no longer only being used as a tool to estimate parameters (i.e., inverse modeling). Instead, forward models of transport processes are also being undertaken, using parameters that are independently estimated. The purpose of forward modeling is to obtain insight into how various physical and chemical processes interact. Using sensitivity analysis as a tool, it is usually possible to rank by importance the many factors that influence transport and reaction at the scale of the stream reach or drainage basin (Choi *et al.*, 1998).

VII. CONCLUSION

Characteristics of hyporheic zones vary widely along streams and rivers of the United States. The length of hyporheic flowpaths, from the point where stream flow enters the subsurface to the point where it reenters the stream, ranges from centimeters to hundreds of meters, reflecting site-to-site variability in streambed slope, slope variation, hydraulic conductivity of the sediment, and curvature of the stream. Several classes of hyporheic flowpaths are often active in a single stream system, each with substantially different exchange timescales. *One of the major points of this chapter is that any particular study method is likely to provide insight into only a small spectrum of the exchange pathways that may be active.*

Some investigators approach hyporheic studies with an interest in the larger-scale, longer-term interactions between stream water and ground water. Those studies collect distributed measurements of hydraulic head and hydraulic conductivity and apply the tools of hydrogeology to define the flow systems. Others approach the topic with a specific interest in subsurface pathways of interaction that are very near to the stream, relying less on tools of hydrogeology and more on tools and techniques that are specially developed for use near interfaces (e.g., seepage meters or mini-drivepoint samplers). Finally there are those investigators with questions about the overall effects of subsurface processes on downstream volumetric flow and chemistry, without specific concerns about the fundamental subsurface processes.

The choice of an approach to characterize hyporheic processes depends foremost on the specific goal of an investigation. If the purpose is to characterize processes and time scales that affect solute mass balance at the scale of the stream reach, then stream tracers are the preferred tool. If the goal is to understand specific subsurface processes or reactions in hyporheic zones, disregarding for the moment the cumulative influence in the reach, then subsurface measurements are the tool. We argue that the objectives of many studies fall somewhere between these goals and are therefore best performed using a combination of approaches. The multiple scales of inquiry provide a potential means to corroborate or refute hypotheses using independent data. Multiscale studies also link the small-scale causes of hyporheic processes with the larger-scale consequences in the drainage basin.

A second major point of this chapter is the need for wider recognition of the limitations of the stream-tracer approach, especially as a means to quantify hyporheic processes. As the popularity of using stream-tracer methods grows, so does the risk of overinterpreting results. Stream-tracer experiments can inform us about timescales and length scales of storage, but not about the processes themselves. Understanding hyporheic processes comes from complementary *in situ* studies in the subsurface. A well-designed stream-tracer study can sometimes isolate hydrologic retention in hyporheic zones (as opposed to surface-storage zones), but even then stream tracers are still only likely to be sensitive to a subset of the actual timescales of hyporheic exchange. Due to those methodological constraints, slowly exchanging hyporheic flowpaths (and biogeochemical processes within them) usually will not be detectable by stream tracers. On the positive side, the timescales of hyporheic retention that are accessible through stream tracers are the ones that are most likely to influence downstream water chemistry. *Stream tracers are best at telling us about the timescales of storage that affect solute mass balance at the reach scale.* Investigators that keep the limitations of stream tracers in mind are using a tool that is ideally suited for characterizing the subset of storage processes and chemical reactions that are most likely to affect biogeochemical mass balances at the scale of drainage basins. Because tracer experimentation is an efficient and flexible approach that can be implemented almost everywhere that surface flow is channelized, the stream-tracer approach eventually promises to help identify, through common databases, the simple relations that help explain basin to basin differences in storage processes across very large regions.

REFERENCES

Angradi, T., and R. Hood. 1998. An application of the plaster dissolution method for quantifying water velocity in the shallow hyporheic zone of an Appalachian stream system. *Freshwater Biology* **39**:301–315.

Beer, T., and P. C. Young. 1983. Longitudinal dispersion in natural streams. *Journal of Environmental Engineering* **109**:1049–1067.

Belanger, T. V., and M. T. Montgomery. 1992. Seepage meter errors. *Limnology and Oceanography* **37**:1787–1795.

Bencala, K. E. 1983. Simulation of solute transport in a mountain pool and riffle stream with a kinetic mass transfer model for sorption. *Water Resources Research* **19**:732–738.

Bencala, K. E., and R. A. Walters. 1983. Simulation of solute transport in a mountain pool-and-riffle stream: A transient storage model. *Water Resources Research* **19**:718–724.

Bencala, K. E., V. C. Kennedy, G. W. Zellweger, A. P. Jackman, and R. J. Avanzino. 1984. Interactions of solutes and streambed sediments. 1. An experimental analysis of cation and anion transport in a mountain stream. *Water Resources Research* **20**:1797–1803.

Bertin, C., and A. C. M. Bourg. 1994. Radon-222 and chloride as natural tracers of the infiltration of river water into an alluvial aquifer in which there is significant river/groundwater mixing. *Environmental Science and Technology* **28**:794–798.

Bourg, A. C. M., and C. Bertin. 1993. Biogeochemical processes during the infiltration of river water into an alluvial aquifer. *Environmental Science and Technology* **27**:661–666.

Bouwer, H., and R. C. Rice. 1976. A slug test for determining hydraulic conductivity of unconfined aquifers with completely or partially penetrating wells. *Water Resources Research* **12**:423–428.

Carignan, R. 1984. Interstitial water sampling by dialysis: Methodological notes. *Limnology and Oceanography* **29**:671–673.

Carter, R. W., and J. Davidian. 1968. General procedures for gaging streams. *U.S. Geological Survey, Techniques of Water Resources Investigations*, Book 3, Chapter A-6.

Castro, N. M., and G. M. Hornberger. 1991. Surface-subsurface water interactions in an alluviated mountain stream channel. *Water Resources Research* **27**:1613–1621.

Choi, J. 1998. Transport modeling of metal contaminants in a stream-aquifer system. Ph.D. Dissertation, University of Arizona, Tucson.

Choi, J., M. H. Conklin, and J. W. Harvey. 1998. Modeling CO_2 degassing and pH in a stream-aquifer system. *Journal of Hydrology* **209**:297–310.

Constantz, J., C. L. Thomas, and G. Zellweger. 1994. Influence of diurnal variations in stream temperature on streamflow loss and groundwater recharge. *Water Resources Research* **30**:3253–3264.

D'Angelo, D. J., J. R. Webster, S. V. Gregory, and J. L. Meyer. 1993. Transient storage in Appalachian and Cascade mountain streams as related to hydraulic characteristics. *Journal of the North American Benthological Society* **12**:223–235.

Davison, W., Z. Zhang, and G. W. Grime. 1994. Performance characteristics of gel probes used for measuring chemistry of pore waters. *Environmental Science and Technology* **28**:1623–1631.

Duff, J. H., F. Murphy, C. C. Fuller, F. J. Triska, J. W. Harvey, and A. P. Jackman. 1998. A mini drivepoint sampler for measuring porewater solute concentrations in the hyporheic zone of sand-bottom streams. *Limnology and Oceanography* **43**:1378–1383.

Edwards, R. L., and J. C. Priscu. 1997. Three-dimensional structure of hyporheic flowpaths within a floodplain backchannel in a Pacific northwest coastal river. *Bulletin of the American Society of Limnology and Oceanography, Annual Aquatic Sciences Meeting*, February 10–14.

Findlay, S. 1995. Importance of surface-subsurface exchange in stream ecosystems: The hyporheic zone. *Limnology and Oceanography* **40**:159–164.

Findlay, S., D. Strayer, C. Goumbala, and K. Gould. 1993. Metabolism of stream water dissolved organic carbon in the shallow hyporheic zone. *Limnology and Oceanography* **38**:1493–1499.

Geist, D. R., M. C. Joy, D. R. Lee, and T. Gonser. 1998. A method for installing piezometers in large cobble bed rivers. *Groundwater Monitoring and Remediation* **18**:78–82.

Grimm, N. B., and S. G. Fisher. 1984. Exchange between interstitial and surface water: Implications for stream metabolism and nutrient cycling. *Hydrobiologia* **111**:219–228.

Hart, D. R. 1995. Parameter estimation and stochastic interpretation of the transient storage model for solute transport in streams. *Water Resources Research* **31**:323–328.

Hart, D. R., P. J. Mulholland, E. R. Marzolf, D. L. DeAngelis, and S. P. Hendricks. 1999. Relationships between hydraulic parameters in a small stream under varying flow and seasonal conditions. *Hydrological Processes* (in press).

Harvey, J. W., and K. E. Bencala. 1993. The effect of streambed topography on surface-subsurface water interactions in mountain catchments. *Water Resources Research* **29**:89–98.

Harvey, J. W., and C. C. Fuller. 1998. Effect of enhanced manganese oxidation in the hyporheic zone on basin-scale geochemical mass balance. *Water Resources Research* **34**:623–636.

Harvey, J. W., K. E. Bencala, and G. W. Zellweger. 1991. Preliminary investigation of the effect of hillslope hydrology on the mechanics of solute exchange between streams and subsurface gravbel zones. *Water-Resources Investigation, (U.S. Geological Survey), Report* **91-4034**, 413–418.

Harvey, J. W., B. J. Wagner, and K. E. Bencala. 1996. Evaluating the reliability of the stream tracer approach to characterize stream-subsurface water exchange. *Water Resources Research* **32**:2441–2451.

Hesslein, R. H. 1976. An in-situ sampler for close interval pore water studies. *Limnology and Oceanography* **21**:912–914.

Hill, A. R., and D. J. Lymburner. 1998. Hyporheic zone chemistry and stream-subsurface exchange in two groundwater-fed streams. *Canadian Journal of Fisheries and Aquatic Sciences* **55**:495–506.

Hynes, H. B. N. 1974. Further studies on the distribution of stream animals within the substratum. *Limnology and Oceanography* **19**:92–99.

Jackman, A. P., R. A. Walters, and V. C. Kennedy, 1984. Transport and concentration controls for chloride, strontium, potassium and lead in Uvas Creek, a small cobble-bed stream in Santa Clara County, California, U.S.A. 2. Mathematical modeling. *Journal of Hydrology* **75**:111–141.

Jackman, A. P., F. J. Triska, and J. H. Duff. 1997. Hydrologic examination of ground-water discharge into the upper Shingobee River. *Water-Resources Investigations (U.S. Geological Survey), Report* **96-4215**, 137–142.

Kilpatrick, F. A., and E. D. Cobb. 1985. Measurement of discharge using tracers. *U.S. Geological Survey, Techniques in Water Resources Investigations*, Book 3, Chapter A-16.

Kim, H., H. F. Hemond, L. R. Krumholz, and B. A. Cohen. 1995. *In-situ* biodegradation of toluene in a contaminated stream. 1. Field studies. *Environmental Science and Technology* 29:108–116.

Lapham, W. W. 1989. Use of temperature profiles beneath streams to determine rates of vertical ground-water flow and vertical hydraulic conductivity. *U.S. Geological Survey Water-Supply Paper* **2337**.

Larkin, R. G., and J. M. Sharp. 1992. On the relationship between river basin geomorphology, aquifer hydraulics, and ground-water flow direction in alluvial aquifers. *Geological Society of America Bulletin* **104**:1608–1620.

Lee, D. R. 1977. A device for measuring seepage flux in lakes and estuaries. *Limnology and Oceanography* **22**:140–147.

Legrand-Marq, C., and H. Laudelout. 1985. Longitudinal dispersion in a forest stream. *Journal of Hydrology* **78**:317–324.

Libelo, E. L., and W. G. MacIntyre. 1994. Effects of surface-water movement on seepage-meter measurements of flow through the sediment-water interface. *Applied Hydrogeology* **2**:49–54.

McMahon, P. B., J. A. Tindall, J. A. Collins, K. J. Lull, and J. R. Nuttle. 1995. Hydrologic and geochemical effects on oxygen uptake in bottom sediments of an effluent-dominated river. *Water Resources Research* **31**:2561–2570.

Morrice, J. A., H. M. Valett, C. N. Dahm, and M. E. Campana. 1997. Alluvial characteristics, groundwater-surface water exchange, and hydrologic retention in headwater streams. *Hydrological Processes* **11**:253–267.

Mulholland, P. J., E. R. Marzolf, J. R. Webster, D. R. Hart, and S. P. Hendricks. 1997. Evidence that hyporheic zones increase heterotrophic metabolism and phosphorus uptake in forest streams. *Limnology and Oceanography* **42**:443–451.

Newbold, J. D., J. W. Elwood, R. V. O'Neill, and A. L. Sheldon. 1983. Phosphorus dynamics in a woodland stream ecosystem: A study of nutrient spiralling. *Ecology* **64**:1249–1265.

Press, W. H., B. P. Flannery, S. A. Teukolsky, and W. T. Wetterling. 1986. "Numerical Recipes: The Art of Scientific Computing," 1st ed. Cambridge University Press, Cambridge, UK.

Runkel, R. L. 1998. One-dimensional transport with inflow and storage (OTIS): A solute transport model for streams and rivers. *Water-Resources Investigations (U.S. Geological Survey), Report* **98-4018**.

Runkel, R. L., D. M. McKnight, K. E. Bencala, and S. C. Chapra. 1996. Reactive solute transport in streams. 2. Simulation of a pH modification experiment. *Water Resources Research* **32**:419–430.

Runkel, R. L., D. M. McKnight, and E. D. Andrews. 1998. Analysis of transient storage subject to unsteady flow: Diel flow variation in an Antarctic stream. *Journal of the North American Benthological Society* **17**:143–154.

Rutherford, J. C. 1994. "River Mixing." Wiley, Chichester.

Shaw, R. D., and E. E. Prepas. 1989. Anomalous, short term influx of water into seepage meters. *Limnology and Oceanography* **34**:1343–1351.

Silliman, S. E., J. Ramirez, and R. L. McCabe. 1995. Quantifying downflow through creek sediments using temperature time series: One-dimensional solution incorporating measured surface temperature. *Journal of Hydrology* **167**:99–119.

Squillace, P. J., E. M. Thurman, and E. T. Furlong. 1993. Groundwater as a nonpoint source of atrazine and deethylatrazine in a river during base flow conditions. *Water Resources Research* **29**:1719–1729.

Stream Solute Workshop. 1990. Concepts and methods for assessing solute dynamics in stream ecosystems. *Journal of the North American Benthological Society* **9**:95–119.

Triska, F. J., V. C. Kennedy, R. J. Avanzino, G. W. Zellweger, and K. E. Bencala. 1989. Retention and transport of nutrients in a third-order stream in Northwestern California: Hyporheic processes. *Ecology* **70**:1893–1905.

Triska, F. J., J. H. Duff, and R. J. Avanzino. 1993. The role of water exchange between a stream channel and its hyporheic zone in nitrogen cycling at the terrestrial-aquatic interface. *Hydrobiologia* **251**:167–184.

Valentine, E. M., and I. P. Wood. 1979. Experiments in longitudinal dispersion with dead zones. *Journal of Hydraulic Engineering* **105**:999–1016.

Valett, H. M., J. A. Morrice, C. N. Dahm, and M. E. Campana. 1996. Parent lithology, surface-groundwater exchange and nitrate retention in headwater streams. *Limnology and Oceanography* **41**:333–345.

Wagner, B. J., and S. M. Gorelick. 1986. A statistical methodology for estimating transport parameters: theory and applications to one-dimensional advective dispersion systems. *Water Resources Research* **22**:1303–1315.

Wagner, B. J., and J. W. Harvey. 1997. Experimental design for estimating parameters of rate-limited mass transfer: Analysis of stream tracer studies. *Water Resources Research* **33**: 1731–1741.

White, D. S., C. H. Elzinga, and S. P. Hendricks. 1987. Temperature patterns within the hyporheic zone of a northern Michigan river. *Journal of the North American Benthological Society* **6**:85–91.

Winter, T. C., J. W. Harvey, O. L. Franke, and W. A. Alley. 1998. Groundwater and surface water: A single resource. *U.S. Geological Survey Circular* **1139**.

Wolf, S. H., M. A. Celia, and K. M. Hess. 1991. Evaluation of hydraulic conductivities calculated from multiport-permeameter measurements. *Ground Water* **29**:516–525.

Wondzell, S. M., and F. J. Swanson. 1996. Seasonal and storm dynamics of the hyporheic zone of a 4th-order mountain stream. I: Hydrologic processes. *Journal of the North American Benthological Society* **15**:3–19.

Worman, A. 1998. Analytical solution and timescale for transport of reacting solutes in rivers and streams. *Water Resources Research* **34**:2703–2716.

Wroblicky, G. J., M. E. Campana, H. M. Valett, and C. N. Dahm. 1998. Seasonal variation in surface-subsurface water exchange and lateral hyporheic area of two stream-aquifer systems. *Water Resources Research* **34**:317–328.

Zellweger, G. W., R. J. Avanzino, and K. E. Bencala. 1989. Comparison of tracer-dilution and current-meter discharge measurements in a small gravel-bed stream, Little Lost Man Creek, California. *Water-Resources Investigations (U.S. Geological Survey), Report* **89-4150**.

2

Modeling Surface– Subsurface Hydrological Interactions

Aaron I. Packman* and Kenneth E. Bencala[†]

*Civil and Architectural Engineering
Drexel University
Philadelphia, Pennsylvania

[†]Water Resources Division
United States Geological Survey
Menlo Park, California

I. Introduction   46
II. Viewing the Interaction from the Stream   47
    A. The Transient Storage Model   48
    B. Case Studies   49
    C. Connection to Stream Ecology   50
III. Viewing the Interaction from the Stream–Bed Interface   51
    A. Use of the Recirculating Flume to Study Hyporheic
        Exchange   51
    B. Case Studies   53
    C. Reactive Solute and Particulate Exchange   66
    D. Summary   72
IV. Viewing the Interaction from the Subsurface   72
    A. Case Studies   73
    B. Summary   75
V. Challenges   77
    References   77

I. INTRODUCTION

For stream ecosystems, the significance of interactions of surface and sub-surface waters results from the inherent biogeochemical transitions extant between the stream, hyporheic zone, and groundwater aquifer. Similarly, there is also a sharp physical transition at the stream water–streambed interface from a regime of open channel flow to one of flow in porous media. Quantitative assessment of surface–subsurface hydrologic interactions is an essential step in understanding and interpreting the exchange of oxygen, nutrients, and all other constituents between surface water and subsurface zones.

Previous comprehensive papers have viewed the interactions of surface and subsurface waters primarily as river–aquifer interconnections (e.g., Sharp, 1988; Larkin and Sharp; 1992, Winter, 1995). Water flow from the stream to subsurface may, or may not, return to the stream, and return flows can occur over a wide range of spatial and temporal scales. Thus, surface–subsurface interactions can be further classified according to the scale of interest as (1) hyporheic flow, (2) bank storage (Pinder and Sauer, 1971; Squillace *et al.*, 1993), or (3) one-way movement between the surface and the subsurface (Constantz *et al.*, 1994, Zellweger 1994). In this review, we focus on surface–subsurface interaction as hyporheic flow—the localized movement of stream water into shallow groundwater flowpaths that return to the stream (Harvey *et al.*, 1996). This terminology emphasizes the relationship between water flow and the hyporheic zone identified by stream ecologists. Even within this context, however, it is important to note that hyporheic exchange occurs due to different physical processes over a range of spatial scales, and that analysis of hyporheic flow must be consistent with the scale of the problem being considered. Our goal in this review is to explain, in the context of stream ecology, the significant distinctions in conceptual basis and modeling method of several views of the study of hyporheic exchange.

The challenge in modeling surface–subsurface hydrologic interactions arises from the differing nature of water flow in the stream and subsurface. Hydrologic modeling methods can be classified according to "where-you-stand" (Bencala, 1993); that is, are you viewing the interaction:

1. from the stream,
2. from the stream–bed interface, or
3. from the subsurface?

Upon accepting this classification, the conceptual bases and the modes of investigation from which the models are formulated are quite distinct (Bencala *et al.*, 1993):

1. idealized exchange, with operational parameterization inferred from field experiments in streams,

2. environmental fluid mechanics, relying on laboratory flume studies for process-level detail, and
3. continuum hydrology, represented by large-scale numerical simulations of flow.

Each of these approaches is appropriate for the examination of hyporheic exchange at a different physical and temporal scale.

The first method of modeling surface–subsurface hydrologic interactions, which we characterize as "viewing the interaction from the stream," is based on idealizing all hyporheic exchange occurring in a stream system in order to model net tracer exchange between the stream and subsurface. The purpose of this methodology is to understand how hyporheic exchange impacts tracer concentrations in the stream, and the resulting models have been used to interpret the results of field tracer experiments. The second methodology, which we term "viewing the interaction from the stream/bed interface," is based on modeling the flowpaths that penetrate the surface of the streambed. The goal of this methodology is to idealize the stream and near-stream subsurface in order to calculate the local porewater flows in the streambed and the resulting interfacial transport. Models of this type have been used to predict tracer exchange in flume experiments but have not yet been applied to many streams. The third and final methodology, which we characterize as "viewing the interaction from the subsurface," is based on idealizing groundwater flows both in the hyporheic zone and larger aquifer system. The prime purpose here is to understand how stream features and aquifer processes affect hyporheic flow patterns. This methodology takes advantage of the existence of well-developed numerical groundwater flow modeling packages and has been used to describe the interaction of hyporheic flows with reach-scale groundwater flow patterns at several field sites. Each of these conceptual approaches will be discussed in turn, with the associated models presented in sufficient detail so that their utility can be ascertained.

Additional studies have been directed at modeling the exchange of reactive solutes and colloidal particles. This work has primarily been done in recirculating flumes with chemically well-defined and homogeneous media. These experiments and the resulting models will be described in the section on exchange through the stream–bed interface.

II. VIEWING THE INTERACTION FROM THE STREAM

Much of the work in the ecological study of surface–subsurface interactions focuses on solute transport and reaction processes along a stream channel. Such work includes the study of nutrients, carbon, and dissolved oxygen. The effects of hydrological interactions between the channel and the subsurface are easiest to envision by considering the steady injection into the channel of a dissolved tracer. The tracer pulse will, in effect, "spread-out" or show "tailing," and the overall residence time of the tracer in the stream will be extended due to water movement into (and then out of) the subsurface. Solutes are carried along concurrently. The details of the hydrologic interactions are complex. However, the spreading of the solute and the increase in solute residence can be modeled as idealized exchange processes.

A. The Transient Storage Model

The simplest idealization of such an exchange process is that of "transient storage" as presented in Bencala and Walters (1983). In the context of surface–subsurface interactions, the transient storage process describes water moving from the flowing stream channel into subsurface zones in which waters are well-mixed, but not transported downstream. There is no net gain of water in the subsurface; a "packet" of water, of equal volume, exchanges and moves into the flowing stream channel. The exchanging packets of water carry solutes back and forth between the stream channel and the subsurface. Thus, solute from the stream channel water will be stored in the subsurface zone; but only on a transient basis. The model equations are

$$\frac{\partial C}{\partial t} + \frac{Q}{A}\frac{\partial C}{\partial x} = \frac{1}{A}\frac{\partial}{\partial x}\left(AD\frac{\partial C}{\partial x}\right) + \frac{q_L}{A}(C_L - C) + \alpha(C_S - C), \qquad (1)$$

$$\frac{\partial}{\partial t}C_S = -\alpha\frac{A}{A_S}(C_S - C), \qquad (2)$$

where C is the solute concentration in the stream (mg L^{-1}), Q is the volumetric flow rate of the stream (m^{3}s^{-1}), A is the cross-sectional area of the stream (m^2), D is the dispersion coefficient (m^{2}s^{-1}), q_L is the lateral volumetric inflow rate (per length; m^{3}s^{-1}m^{-1}), C_L is the solute concentration in lateral inflow (mg L^{-1}), A_S is the cross-sectional area of the storage zone (m^2), C_S is the solute concentration in the storage zone (mg L^{-1}), α is the stream storage exchange coefficient (s^{-1}), t is the time (s), and x is the distance along the stream channel (m). Equation (1) represents a mass balance for solute in the stream, including advection and dispersion in the stream itself, lateral inflow, and transient subsurface storage. Equation (2) represents a mass balance for solute in the storage zones.

The physical mechanics of transient storage are strictly idealized. In the natural environment, there is no stirring paddle moving water between the stream channel and a well-mixed subsurface zone. In the natural system, exchange occurs along well-defined hyporheic flowpaths, as documented by Harvey and Bencala (1993). The transient storage concept yields concentration-time histories similar to the net effect of solute transport along these hyporheic flowpaths. Thus, when viewing the transport of solutes from within the stream, the transient storage concepts can provide an effective characterization of those exchange processes that influence downstream solute transport. In essence, the effects of all hyporheic exchange flows that are important over the timescale of tracer passage are included in the model. Harvey *et al.* (1996) present field measurements showing concentration-time histories along hyporheic flowpaths which are, and which are not, effectively characterized. Wagner and Harvey (1997) develop guidelines for designing instream experiments sensitive to the effects of hyporheic flows. New results by Harvey and Fuller (1998) demonstrate that the transient storage model parameters can be related to physical stream characteristics in at least some cases.

The water exchange processes and solute storage effects represented in the transient storage concept do not result exclusively from hyporheic flow. Storage zones of water, not moving downstream but mixing relatively slowly with the main flow, could also be located along the edges of slowing-moving pools or behind protruding logs, boulders, and vegetation. For stream-catchment systems in which the transient storage does result from hyporheic flow, Harvey *et al.* (1996) have derived three quantitative estimates representing the implicit "modeling" of the hydrologic interactions using the transient storage concept. These are

q_S, the hyporheic water exchange flux (per length) equal to αA, presented as q_S/q_L, the ratio of hyporheic flux to the net groundwater flux (per length);

t_S, the characteristic time of water residence in the hyporheic zone equal to $A_S/\alpha A$; and

L_S, the characteristic channel length for exchange (surface water turnover length) equal to $(Q/A)/\alpha$.

B. Case Studies

Values of these hydrologic characterizations are presented in Table I for several reaches of four mountain streams. For each of the streams, the spatial variation in these characteristic values is typically an order of magnitude. Such variation, in part, reflects the natural heterogeneities within stream-catchment systems. Additionally, however, this variation reminds us that these values are merely average, reach-scale characterizations of complex physical processes. If we look in aggregate at all the reaches of a given stream, then we can conclude that all three hydrologic characterizations indicate a greater degree of surface–subsurface hydrologic interaction in Little Lost Man Creek than in the Snake River, with Uvas Creek and St. Kevin Gulch falling in between. Making broad generalizations, we can conclude that surface–subsurface hydrologic interaction is substantial in Little Lost Man Creek, with (1) hyporheic flux exceeding net groundwater flux, (2) residence times in storage zones of several hours, and (3) the length of channel needed for exchange being < 1 km. At the other extreme, even though surface–

TABLE I   Characterization of Surface–Subsurface Hydrologic Interactions Based on Parameters from Transient Storage Analysis

Stream	Reach (m)	Discharge (L s^{-1})	Subsurface flux ratio, q_s/q_L	Storage time, t_s (h)	Exchange length, L_s (km)
LLM	5–62	7.4	3	2	0.5
	62–126	7.6	13	14	0.2
	126–184	7.8	41	3	0.1
	184–327	8.2	4	14	0.7
Uvas	105–281	13.6	6	9	1
	281–433	14.0	1	28	3
StK	26–484	8	0.5	7	4
	526–948	17	0.8	9	6
	948–1557	20	2	7	3
Snake	0–628	260	0.2	2	17
	628–1365	320	0.3	0.3	12
	1365–1487	340	7	0.6	0.4
	1487–1605	360	1	0.8	2
	1605–2845	390	5	0.3	3
	2913–3192	650	0.3	0.3	12
	3192–3889	680	1	0.3	11
	3889–5231	800	2	0.8	5

Stream: LLM, Little Lost Man Creek, Redwood National Park, California (Bencala 1984); Uvas, Uvas Creek, Santa Clara County, California (Bencala and Walters 1983); StK, St. Kevin Gulch, Leadville, Colorado (Broshears *et al.*, 1993); Snake, Snake River, Summit County, Colorado (Bencala *et al.*, 1990).

Reach: Summary presented for net gaining reaches, not dominated by surface tributaries. Individual reaches are identified by the locations given in the primary references.

Subsurface flux ratio = q_s/q_L (Harvey *et al.*, 1996: hyporheic flux/net groundwater flux). $q_s = \alpha A$, the hyporheic water exchange flux (per length).

Storage time (Harvey *et al.*, 1996: storage zone residence time).

Exchange length = u/α (Harvey *et al.*, 1996: characteristic channel length for exchange). $u = Q/A$.

subsurface hydrologic interactions are readily observed in the Snake River, they are not a dominant feature of stream-catchment hydrology, with (1) hyporheic flux and net groundwater flux being roughly equivalent, (2) storage time on the order of 0.5 hours, and (3) channel exchange length on the order of 10 km. Detailed studies in which hydrologic characterizations based on the transient storage concept were applied to several streams can be found in D'Angelo *et al.* (1993) and Morrice *et al.* (1997).

C. Connection to Stream Ecology

For interpreting stream ecological processes, the transient storage concept of idealized exchange has been used to provide quantitative characterization of surface–subsurface hydrology, no doubt because it is highly idealized (e.g., Rutherford, 1994; Allan, 1995). With physical transport processes represented as transient storage, Kim *et al.* (1992) and Tate *et al.* (1995) present reactive simulation analyses of nutrient-cycling experiments. Other approaches have also been taken in the analyses of nutrient-cycling experiments. In the work of Valett *et al.* (1996) and Mulholland *et al.* (1997), nutrient concentrations are not explicitly included in reactive simulations; rather measures of nutrient uptake are compared to hydrologic characterizations derived from the transient storage analysis.

The in-stream view is also effectively characterized in simple models of diffusion into the streambed or advective underflow pathways (Jackman *et al.*, 1984) and by time series analysis (Castro and Hornberger, 1991). All these methods for the study of surface–subsurface hydrologic interaction rely upon operational parameterization inferred from field experiments.

III. VIEWING THE INTERACTION FROM THE STREAM–BED INTERFACE

The approach of viewing the interaction from the stream–bed interface involves the use of fluid mechanics to calculate the flux of water through the bed surface. In this manner, hyporheic exchange is examined at the interface where surface water is initially mixed with subsurface water, and the processes that control this exchange locally are investigated in order to determine their impact on the overall exchange. Fundamental descriptions of the relevant small-scale hydrodynamic processes have been developed, and this information has been used to calculate net stream–subsurface exchange. In addition, the transport of both reactive solutes and suspended particles has been modeled by considering these substances to be primarily carried along with the exchanging water but also subject to some additional processes.

To date, the scope of this type of model has been limited to exchange between an idealized stream with a regular cross section and the sediment bed immediately underlying the stream. Larger elements of natural streams such as meanders and changes in slope are generally ignored when considering the local exchange with the streambed. Some additional complexities typically found in the natural environment, such as heterogeneity in the bed sediment, have also been omitted from the current models. Thus, even though these models are useful because they include process-level understanding, their application has been limited.

A. Use of the Recirculating Flume to Study Hyporheic Exchange

The recirculating flume is the primary experimental apparatus used to study near-surface exchange. A recirculating flume is composed of a long rectangular channel with a pump at the downstream end and a return pipe to provide recirculation. To mimic conditions in a real stream, appropriate streambed material is placed in the channel, and water is continuously recirculated over the bed. This results in sediment transport and the development of bedforms, such as dunes and ripples. After running for a sufficient time, the flow will become uniform along the flume and all hydraulic parameters will approach a steady state. That is, the sediment transport rate and flow depth will become essentially constant over the entire flume, and the average bedform shape will no longer change. The resulting steady flow over the sediment bed closely resembles a reach of a real stream. Processes such as sediment transport, bedform development, and exchange with the streambed are determined by the local velocity profile, which will be similar to that in a natural stream with the same average depth and velocity. Thus, the flume is a good experimental system to study processes that occur at the streambed. However, the flume walls obviously restrict the flow to a straight channel so that bank or meander processes are not included in this system.

Nonrecirculating flumes have also been used as a laboratory apparatus to study stream processes. These flumes differ from recirculating flumes only in that they do not have a closed-circuit return system for water and sediment. Nonrecirculating flumes are more difficult to use because sediment and water must be continually supplied at the upstream end of the flume. However, this allows variation of the input conditions (e.g., water chemistry, flow rate, sediment supply) over the course of the experiment.

Flume experiments offer many advantages for the study of hyporheic exchange. Primarily, they allow examination of exchange processes under controlled, easily observable conditions. This provides a considerable advantage when a new problem is being approached because the highly controlled conditions allow the researcher to eliminate complexities that tend to obscure individual processes. For the study of hyporheic exchange, flumes allow close examination of porewater flow through the bed sediment. It would be extremely difficult, if not impossible, to obtain adequate visualization of these processes in the natural environment. In addition, full control of the stream conditions allows the researcher to determine the importance of individual parameters or processes, such as the manner in which stream parameters affect the overall exchange.

Data sets from flume experiments can be used to test models for exchange between the stream and streambed. In the laboratory, tracers are easily added to the stream in order to investigate hyporheic exchange. Exchange between the stream and bed can be observed by monitoring the tracer concentration in both the stream and the bed. Most experiments done to date have involved conservative tracers, either dye or nonreactive solutes. However, some experiments have been done with reactive tracers, including metal ions and colloidal particles. Models for the exchange of both conservative and reactive tracers have been developed in conjunction with the experimental programs.

B. Case Studies

Experimental studies can be classified according to the type of bed sediment material (typically gravel or sand). This classification is useful because there is a distinct difference in the exchange with beds of each of these compositions. With gravel beds, stream turbulence can penetrate a significant distance into the bed and influence porewater motion. However, sand beds will generally limit turbulent penetration to a very thin layer at the bed surface and have laminar porewater flow. In other words, flows through gravel beds will be more intimately connected with the stream, in the sense that the upper layers of gravel beds have induced flows with velocities that approach the stream velocity, whereas induced flows in sand beds are always much less than the stream velocity. The experiments conducted on each of these types of stream systems will be discussed, along with the associated models for exchange with each type of streambed.

1. Study of Exchange with Gravel Beds

Work done to date indicates that a stream flowing over a gravel bed can induce a flow in the bed in two ways. First, stream flow over bedforms results in a pressure distribution at the bed surface which drives flow in the bed. Second, interaction (coupling) of the stream and porewater flow results in a slip velocity at the bed surface and an exponentially decreasing flow in the bed. However, no model has been developed for the velocity field in the bed or the exchange between the stream and the bed in the general case where both of these flows occur simultaneously. Flows in gravel beds are complicated because there is generally a region deep in the bed where Darcy flow occurs and another region near the bed surface where the flow is turbulent and Darcy's law does not apply.

Tracer movement in gravel streambeds was first studied by Thibodeaux and Boyle (1987). Their study involved injecting dye into a gravel bed in a flume in order to observe flow patterns in the bed. The bed material had an average diameter of 8 mm, and the bed surface was covered with artificially formed bedforms. Two types of dune-shaped bedforms were used—larger ones of length 55 cm and height 5 cm, and smaller ones with length 25 cm and height 2.5 cm. Flow over the bedforms induced a considerable flow in the bed (Fig. 1). This induced flow was attributed to the pressure difference between the upstream and downstream faces of the bedform.

Shimizu *et al.* (1990) investigated the interaction between turbulent stream flow over a flat gravel bed and the seepage flow in the bed. They conducted experiments in a recirculating flume with a homogeneous flat bed composed of gravel-sized glass beads, which had a diameter of either 2.97 or 1.70 cm. Injected salt water was used as a tracer to obtain the velocity profile in the bed and the vertical dispersion coefficient. The porewater velocity profiles were similar in all experiments, with low flow rates

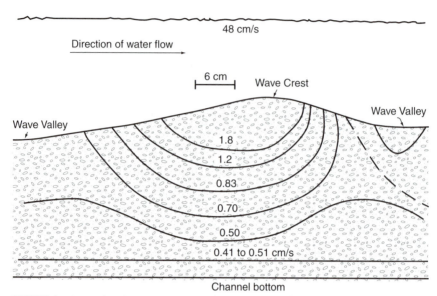

FIGURE 1 Flow induced under a gravel bedform. Scale marker = 6 cm. Reprinted with permission from Thibodeaux and Boyle (1987) *Nature* **325**(6102): 341–343. Copyright 1987 Macmillan Magazines Limited.

deep in the bed and increasing velocities in the upper several centimeters (Fig. 2).

The flow in the bed was assumed to include velocity components both independent of and induced by the stream flow. The independent component of velocity is driven by the channel slope and can be expected to follow Darcy's law. The component of velocity induced by the stream flow results from acceleration (drag) of the porewater flow at the bed surface by the stream flow. The induced flow is turbulent in nature, with a turbulent momentum flux related to the mass exchange across the stream/bed interface. The measured mass dispersion coefficient provided a reasonable estimate for the turbulent eddy kinematic viscosity in the bed, but the velocity at the bed surface (slip velocity) could only be determined by fitting measured velocity profiles.

Mendoza and Zhou performed a rigorous theoretical analysis of the coupling between the flow in the stream and the flow through the gravel bed (Mendoza and Zhou, 1992; Zhou, 1992; Zhou and Mendoza, 1993, 1995). The effect of the bed on the stream flow was considered as a perturbation from the solution for flow over an impenetrable bed (Mendoza and Zhou, 1992). The presence of the porous bed allows a nonzero flow at the bed surface (slip velocity) and exchange of momentum with the pore water in the bed. The momentum exchange can be expressed as an induced Reynolds stress at the interface. This also results in an induced pressure, which can drive exchange with the bed.

The effect on the subsurface flow was analyzed by developing the Navier-Stokes equations of motion for porewater flow at both the microscopic and macroscopic scales (Zhou and Mendoza, 1993). The stream induces flow near the bed surface, and the porewater velocity decays to the slope-driven Darcy velocity deep in the bed. It is shown that the decay of velocity is exponential so that

$$U = U_1 + U_2 = U_1 + (U_s - U_1) e^{-\alpha y} \tag{3}$$

where U is the bulk pore water velocity, U_1 is the portion of the porewater velocity driven by the slope of the bed, U_2 is the portion of the porewater velocity driven by the stream flow, U_s is the velocity at the surface of the bed (slip velocity), α is a constant related to the microscopic flow through the pores (the value of α is apparently a property of the sediment and independent of the porewater velocity), and y is the depth in the bed (a positive number). The induced flow in the bed is controlled by the conditions at the

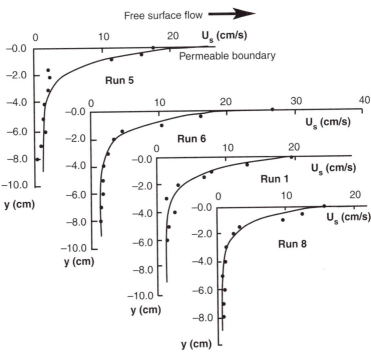

FIGURE 2 Porewater velocity profiles induced by stream flow over a flat gravel bed. Reprinted with permission from Shimizu *et al.* (1990) *Journal of Hydrosciences and Hydrologic Engineering* 8:69–78. Copyright by Japanese Society of Civil Engineering.

stream–bed interface. Matching the momentum transfer at the interface results in

$$\mu \, dU/dy - \rho <u_x u_y>> = \mu \, dW/dy - \rho <w_x w_y>, \qquad (4)$$

where μ is the dynamic viscosity, ρ is the fluid density, u_x and u_y represent instantaneous pore water velocities, W is the bulk stream velocity, w_x and w_y are instantaneous stream velocities, and $< >$ indicates an ensemble average. The first term on both sides of Eq. (4) is the viscous stress, and the second term is a Reynolds stress term for turbulent flow. In the bed, the Reynolds stress results from both turbulence and flow around sediment grains in the bed. This can be seen by writing the instantaneous porewater velocity as

$$u = U + u_s + \hat{u}, \qquad (5)$$

where U is the steady bulk velocity, u_s represents the steady deviation from the bulk velocity at the microscale caused by flow around grains in the bed, and $\hat{u}$ is the turbulent fluctuation. Substituting

$$<u_x u_y> = <u_{xs} u_{ys}> + <\hat{u}_x \hat{u}_y> \qquad (6)$$

into Eq. (4) and assuming that the turbulence will be the same on both sides of the interface yields

$$\mu \, dU/dy - \rho <u_{xs} u_{ys}> = \mu \, dW/dy. \qquad (7)$$

This indicates that flow around the bed grains causes the slope of the velocity profile to be different in the stream and in the bed at the interface.

Despite the understanding of the problem developed with the preceding analysis, no fundamental estimate has been developed for the slip velocity or the momentum transfer across the interface. Instead, measured slip velocities were used as a boundary condition, and porewater velocity profiles were fit using the constant α in Eq. (3). Good fits were obtained to the data of Shimizu *et al.* (1990), indicating that the model captures the physics of the induced flow in the bed, even though the interfacial transfer has not been described explicitly. The velocity and Reynolds stress profiles in the stream flow were similarly fit to experimental data (Mendoza and Zhou, 1992). Our current understanding of both the stream and porewater flow in gravel-bed streams is sufficient to predict these flows given the correct boundary condition at the stream–bed interface but insufficient to predict the interfacial transfer itself. Furthermore, no study has been conducted to determine the flow patterns in the bed in the general case where there is both an induced slip velocity at the bed surface and bedform-driven flow.

2. Study of Exchange with Sand Beds

Flows in sand streambeds differ from those in gravel beds in that the smaller pore spaces restrict stream-driven turbulence to a very small layer near the surface of the bed. As a result, there is no significant coupling of the stream and porewater flows (although this coupling will still occur very near the bed surface, as noted by Mendoza, 1988.) Similarly, the permeability of sands is sufficiently low to typically limit porewater velocities to the Darcy realm.

A variety of types of bedforms will develop in response to the stream flow over a sand bed. These are generally classified according to their shape into ripples, dunes, transition flat bed, and antidunes (Vanoni, 1975). Under a given flow, one stable set of bedforms will develop as the roughness of the movable bed adjusts to the flow conditions. Sediment transport results in these bedforms continually moving and changing shape, but the average bedform size and geometry will be maintained as long as the stream flow is steady and uniform. Sand streambeds will typically be covered with dunes, which have the shape shown previously in Fig. 1.

It is apparent that near-surface exchange between a stream and a sand bed is generally controlled by bedform-driven processes. Bedforms are responsible for two distinct exchange processes. Following the terminology of Elliott, pumping is the exchange due to advective porewater flow, and turnover is the exchange resulting from trapping and release of porewater (Elliott, 1990; Elliott and Brooks, 1997a,b).

Pumping exchange occurs because stream flow over bedforms produces a pressure disturbance at the bed surface. Bedforms protrude into the stream flow, resulting in drag due to high pressure at the upstream side of each bedform and low pressure in the recirculation zone behind each bedform. The periodic high- and low-pressure regions at the bed surface produce subsurface pressure gradients that drive advective porewater flow into, through, and out of the bed. These advective flow patters occur in the bed below each bedform, producing an overall pumping exchange.

Turnover exchange occurs because bed sediment transport causes bedforms to move downstream. As bed sediment is scoured from the upstream side of a dune, porewater that was previously trapped in the bedform becomes mixed with stream water. The scoured sediment is then deposited at the downstream face of the bedform in order to maintain the stable bedform shape. In the deposition region, stream water becomes trapped in the bedform. The overall trapping and release of water results in a net turnover exchange between the stream and bed.

In order to develop analytical models for bedform-driven exchange, it is necessary to consider an idealized system that captures the essence of the natural system but simplifies computation. For the bed-exchange problem, the model system is a wide stream in steady, uniform flow over a set of regular, two-dimensional, fully developed, triangular bedforms. This, in essence, reduces the problem to steady two-dimensional flow over an idealized bedform shape. The validity and range of applicability of these assumptions has been discussed elsewhere (most completely in Elliott, 1990). Thus, the stream can be characterized by some constant depth, velocity, and slope, and the bed-forms can be characterized by a constant wavelength, height, and velocity. In addition, the bed sediment is assumed to be homogeneous and isotropic. These idealized conditions are easily obtained in the laboratory but will not generally be present in nature. However, the idealized models for exchange processes can be applied to more general conditions (e.g., irregular bed-forms) through various numerical modeling methods. Most of the applications to date have involved a straight stream reach with uniform flow, regular bedforms, and steady-state conditions, which merely requires numerical integration to relate stream–subsurface flux to the change in in-stream concentration over time.

a. Models for Pumping Exchange Modeling of pumping exchange has focused on applying potential flow theory to the porewater system. Essentially, this involves calculating the porewater velocity from Darcy's law using the piezometric head distribution at the surface of the bed as a boundary condition. It should be noted that this method is only valid if the porewater velocities are very low, with a Reynolds number based on the grain diameter of <10. Thus, this method can clearly not be applied to flows in gravel beds, where the turbulent interaction with the stream flow produces a higher velocity region near the bed surface.

Early pumping flow models were purely numerical in nature. The calculations were done by establishing an element grid with the correct two-dimensional bedform geometry, applying the known pressure distribution at the surface, and calculating the resulting Darcy porewater flow in the underlying bed (Savant *et al.*, 1987; Elliott, 1990). The pressure distributions used in these calculations were taken from experimental studies of flow over bedforms (Vittal, *et al.*, 1977; Fehlman, 1985; Shen *et al.*, 1990). The velocity profiles obtained from these calculations qualitatively agree with the results from dye injections under bedforms (Savant *et al.*, 1987; Thibodeaux and Boyle, 1987; Elliott, 1990; Mendoza, 1988). Elliott developed an analytical model for pumping flows, calculated the net pumping exchange between the stream and bed, and compared these model predictions with experimental results from dye studies in a laboratory flume (Elliott, 1990; Elliott and Brooks, 1997a,b). The piezometric head distribution at the bed surface is approximated by a sinusoidal profile

$$h = h_{\mathrm{m}} \sin kx, \tag{8}$$

where h is the dynamic head, h_{m} is the half-amplitude of the head variation, k is the wavenumber of the head disturbance equal to $2\pi/\lambda$ where λ is the bedform wavelength, and x is a downstream coordinate parallel to the bed surface. The relative positions of the sinusoidal head profile and the bedform shape are shown in Fig. 3. The variable h_m is taken from a correlation based on the data of Fehlman (1985):

$$h_m = 0.28\frac{U^2}{2g}\begin{cases}\left(\dfrac{H/d}{0.34}\right)^{3/8} & H/d \leq 0.34, \\[2ex] \left(\dfrac{H/d}{0.34}\right)^{3/2} & H/d \geq 0.34,\end{cases} \tag{9}$$

where U is the stream velocity, g is the acceleration due to gravity, H is the bedform height, and d is the stream depth. Given this pressure distribution, the porewater field can be determined by solving Laplace's equation

$$\nabla^2 h = 0 \tag{10}$$

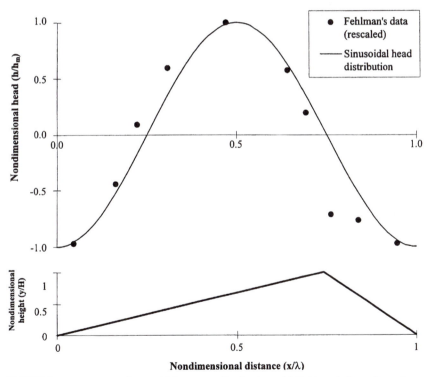

FIGURE 3 Comparison of sinusoidal pressure distribution with Fehlman's data, shown in relation to a triangular bedform. Elliott's numerical pumping calculation used the actual pressure distribution and geometry, whereas the analytical model for pumping exchange applies the sinusoidal distribution over a flat bed. Reprinted with permission from Packman (1997). Copyright by A.I. Packman.

in the bed with appropriate boundary conditions and then applying Darcy's law

$$\bar{u} = -K \nabla h, \tag{11}$$

where K is the hydraulic conductivity. Elliott solved this problem by assuming that the bed was infinitely deep and that the solution was periodic in the bed under each bedform. In addition, the small effect of the bedform geometry on the solution is ignored, so the sinusoidal pressure distribution is applied over a bed that is assumed to be flat. This is a reasonably good assumption because dunes typically have a length:height ratio on the order of 10:1. The resulting components of the porewater velocity are

$$u = -u_m \cos(kx) \, e^{ky}, \tag{12}$$

$$v = -u_m \sin(kx) \, e^{ky}, \tag{13}$$

where u and v are Darcy velocities in the x and y directions, respectively (that is, parallel and perpendicular to the bed surface), and u_m is the maximum velocity

$$u_m = Kkh_m. \tag{14}$$

For a finite bed, the solution is

$$u = u_m \cos(kx) \, [\tanh(kd_b)\sinh(ky) + \cosh(ky)], \tag{15}$$

$$v = u_m \sin(kx) \, [\tanh(kd_b)\cosh(ky) + \sinh(ky)], \tag{16}$$

where d_b is the depth of the bed (Packman, 1997; Packman *et al.*, 1999a). The velocity profiles can be nondimensionalized by using scales related to the bedform geometry

$$x^* = kx, \tag{17}$$

$$y^* = ky, \tag{18}$$

$$d_b{}^* = kd_b, \tag{19}$$

$$u^* = u/kKh_m = u/u_m, \tag{20}$$

$$v^* = u/kKh_m = v/u_m, \tag{21}$$

$$t^*/\theta = k^2 Kh_m t/\theta = ku_m t/\theta. \tag{22}$$

Note that the dimensionless time, t^*, will often appear with θ, the porosity of the bed, because velocities are calculated as Darcy velocities and a factor of $1/\theta$ is required to convert to porewater velocities. The normalized velocity profiles are

$$u^* = -\cos x^* \, e^{y^*}, \tag{23}$$

$$v^* = -\sin x^* \, e^{y^*}, \tag{24}$$

and

$$u^* = -\cos x^* \,(\tanh d_{\mathrm{b}}^* \,\sinh y^* + \cosh y^*), \tag{25}$$

$$v^* = -\sin x^* \,(\tanh d_{\mathrm{b}}^* \,\cosh y^* + \sinh y^*). \tag{26}$$

The dimensionless streamlines for pumping flows in infinite and finite beds are given in Fig. 4. This nondimensionalization is quite useful because it al-

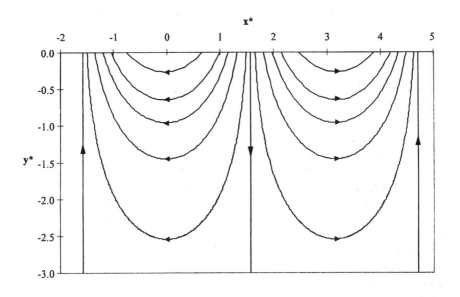

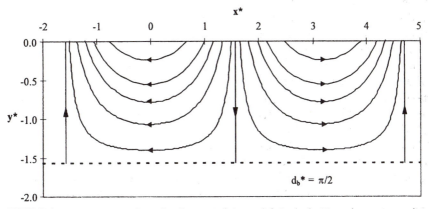

FIGURE 4 Pumping streamlines for flow in infinite and finite beds. Note that axes are distorted. (After Elliott, 1990; Packman, 1997.)

lows a single, dimensionless solution to be applied to problems of different scales. In particular, all the dimensionless parameters in the velocity profile scale with some aspect of the bedform size, indicating that the flow patterns are similar under any bedform (with the assumed geometry). The magnitude of the subsurface flow, on the other hand, depends on both the bedform size and stream discharge [the porewater velocity field scales with h_m, given in Eq. 9].

An additional velocity component occurs due to the slope of the stream. This underflow velocity may simply be superimposed on the preceding velocity distribution. The underflow velocity occurs in the x-direction and is given by

$$u_u = KS, \tag{27}$$

where S is the slope of the stream and streambed. Underflow is also nondimensionalized by the maximum pumping velocity

$$u_u^* = u_u/u_m. \tag{28}$$

In flume experiments, u_u^* was typically observed to be between 0.03 and 0.10, indicating that the subsurface flow is dominated by the bedform-induced component (Elliott and Brooks, 1997b; Packman et al., 1999b).

The pumping velocity field can be used to determine the net exchange between the stream and porewater. Elliott modeled exchange with the residence time function, $R(t)$, the fraction of a tracer pulse which entered the bed at time t_0 that remains at a later time t. The shape of the residence time function reflects the motion of a tracer pulse introduced at the bed surface through and then out of the bed. $R(t)$ can be obtained from the pumping velocity field by use of a numerical particle-tracking model. The curve of R versus time for the finite bed case is shown in Fig. 5.

The net exchange of water or a conservative solute can be calculated by (1) determining the residence time function for the given bedform geometry and stream conditions, and then (2) integrating the residence time function over time. Exchange is conveniently expressed as an equivalent penetration into the bed, $M(t)$, which can be converted to a change of stream concentration, $C(t)$. Predictions of this model have been compared with the results of a large number of experiments on exchange of conservative tracers (experiments with dye: Elliott, 1990; Elliott and Brooks, 1997b; experiments with dissolved lithium chloride: Eylers, 1994; Eylers et al., 1995; Packman, 1997; Packman et al., 1999b). The model does a good job of predicting exchange with shallow beds (Packman et al., 1999b) and initial exchange with deep beds (Elliott and Brooks, 1997b). However, exchange with deep beds is underpredicted at long times, probably due to poor simulation of flows deep in the bed. A typical result for exchange with a shallow bed is shown in Fig. 6.

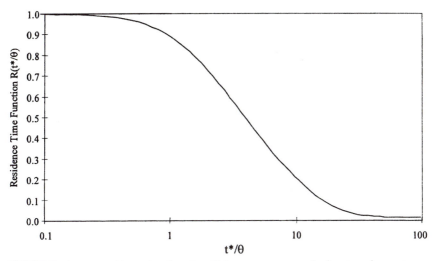

FIGURE 5 Average residence time function. This curve represents the fraction of tracer introduced at the bed surface at time $t^* = 0$ that is still present at time t^*/θ. (After Elliott, 1990.)

b. Models for Turnover Exchange Calculating the exchange due to turnover is straightforward. Because the extent of exchange is directly proportional to the cross-sectional area of the bed that is scoured or newly deposited, the rate of volumetric exchange can be determined geometrically. The transport of a conservative tracer from the stream to an uncontaminated bed is conveniently expressed as an equivalent penetration depth (M), which is just the average depth of penetration of tracer into the bed. The motion of a regular, triangular bedform with velocity U_b results in an equivalent penetration of:

$$M = \begin{cases} \dfrac{H}{2}\left(1-\left(1-\dfrac{U_b t}{\lambda}\right)^2\right) & \text{for } t < \dfrac{\lambda}{U_b}, \\[2ex] \dfrac{H}{2} & \text{for } t \geq \dfrac{\lambda}{U_b}. \end{cases} \tag{29}$$

U_b can be nondimensionalized with the porewater velocity

$$U_b^* = \frac{U_b}{u_m/\theta} = \frac{\theta U_b}{u_m}. \tag{30}$$

Note that turnover can only cause exchange down to the maximum scour depth, $H/2$. Natural bedforms actually have a distribution of heights, and

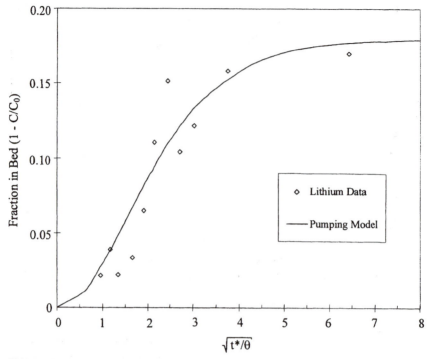

FIGURE 6 Application of pumping model to exchange of lithium (a conservative tracer) between a stream and a shallow sand bed ($d_b{}^* = 2.0$) in a laboratory flume. The model prediction is not fitted to the data but is an entirely independent estimate based on the average bedform geometry and the stream depth and velocity. (After Packman, 1997.)

the maximum scour will be caused by the largest bedform rather than the average bedform as assumed in the model. The bedform size distribution can be included in the model, but this is only necessary when the bedform size distribution is wide (Elliott, 1990; Elliott and Brooks, 1997a; Packman, 1997).

c. Models for the General Case: Combined Pumping and Turnover The models discussed in the last two sections describe the exchange due to pumping and turnover independently. Each of these models is applicable in certain limited cases. The pumping model is applicable when bed sediment transport is zero but there are bedforms left on the bed from a higher flow. The turnover model is applicable either when the bed is shallow and the bedforms move very rapidly, or when the bed sediment has very low permeability and subsurface flows are very small. However, in the general case, both pumping and turnover occur simultaneously. It is difficult to model the general case because combined pumping and turnover produce very complex porewater flow patterns near the bed surface. For example, some of the water pumped into the upstream side of the bedform will immediately be released by turnover. As a result of this complexity, modeling has focused on solutions for asymptotic regions of either slow or fast bedform motion, with slow and fast defined by the relative magnitudes of the exchange processes. The relative importance of pumping and turnover is characterized by the dimensionless bedform velocity, U_b^* defined above in Eq. (30), so that bedforms are considered slow-moving if $U_b^* \ll 1$, fast-moving if $U_b^* \gg 1$, and intermediate if $U_b^* \sim 1$. Exchange has a different character in each of these cases, as evidenced by the dye penetration fronts presented in Fig. 7.

Elliott developed a model for combined exchange that uses a Lagrangian framework that travels with bedforms as they move downstream (Elliott and Brooks, 1997a). The streamlines shown in Fig. 4 will always be present in-

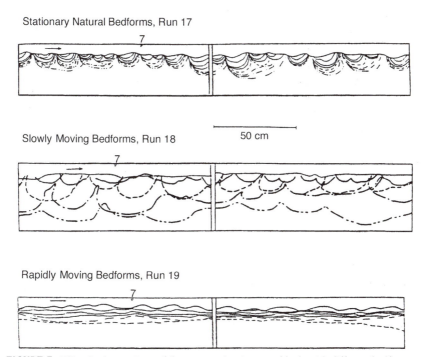

FIGURE 7 Elliott's observations of dye penetration into sand beds with different bedform velocities. The uppermost line in each figure is a typical bed surface profile, and each subsurface line is the dye front observed in the bed at a particular time. Stationary bedforms produce stable dye fronts obviously related to the streamlines depicted in Fig. 4, whereas slow-moving bedforms ($U_b^* = 0.35$) produce similar but more complicated and unstable fronts, and fast-moving bedforms ($U_b^* = 30$) result in horizontal fronts. Reprinted with permission from Elliott (1990). Copyright by A.H. Elliott.

stantaneously under each bedform and will thus be constant if viewed from the Lagrangian standpoint. However, the path taken by a parcel of water in the bed will be different when bedforms move because the position of the parcel will change relative to the pressure distribution at the bed surface. This can be accounted for in the porewater velocity distribution by adding a velocity component for the bed motion

$$u = u_{\text{pumping}} + u_{\text{u}} - U_{\text{b}}. \tag{31}$$

The effect of bedform motion is a reduction in the pumping penetration of water from the stream and a corresponding decrease in net exchange. This model was tested against experimental data and found to give good predictions of exchange for the intermediate case (Elliott and Brooks, 1997b). However, the basic pumping model does a better job of predicting exchange with very slow-moving bedforms.

Packman (1997) developed a model for the case of fast-moving bedforms, which is based on the porewater flow induced throughout the bed. The unsteady, periodic pressure disturbance from fast-moving bedforms causes porewater deep in the bed to move in small circular paths, producing downward mixing. Because the bedforms are regular and their velocity is always parallel to the bed surface, the exchange is simply one-dimensional— uniformly downward in the bed. This explains Elliott's observation that tracer penetration in the fast-moving case occurs as horizontal fronts, which was shown in Fig. 7. Because the bedforms are fast-moving, turnover still dominates the exchange at early times ($t < \lambda/U_{\text{b}}$), until the turnover zone becomes completely mixed. At later times, the rate of exchange is controlled by the downward front velocity, which can be determined by averaging the porewater velocity profile over the bedform wavelength. For a finite bed, this is

$$\bar{v}^* (y^*) = \frac{\tanh(d_b^*)\cosh(y^*) + \sinh(y^*)}{\pi}. \tag{32}$$

When tested against data from flume experiments, this model provided a reasonable prediction of exchange with fast-moving bedforms, but it underpredicted the exchange with lower bedform velocities (Packman, 1997).

C. Reactive Solute and Particle Exchange

The models described in the previous section can be applied to the exchange of reactive solutes by including chemical processes in the models. Because the flow of water in the bed is known, the transport of a reactive species can be determined by considering both its motion due to porewater flow and any reactions with the bed sediment. This can be difficult because the chemical behavior must be correctly parameterized at a scale appropriate for inclusion in the exchange model. Typically, the chemical interactions are analyzed to provide reaction rates or partition coefficients for use in the exchange model. This procedure will be illustrated in the examples that follow. To date, models have been developed for the exchange of metal ions and the consumption of dissolved oxygen. A similar approach has been used to predict the capture of suspended sediments by sand beds.

1. Metal Exchange

Eylers conducted an extensive study of the exchange of a variety of metal ions with a silica sand bed (Eylers, 1994; Eylers *et al.*, 1995). The experimental portion of this work included both flume experiments on net metal exchange and batch studies on metal sorption to the bed sediment. Both the stream-water composition and bed sediment surface chemistry were controlled in the flume experiments. The porewater velocity was sufficiently slow so that all metals investigated underwent equilibrium partitioning to the bed sediment even as they were being pumped through the bed (i.e., the characteristic pumping time was much slower than the timescale at which sorption occurred). As a result, the reactive solute transport could be modeled using the retardation coefficient approach, which is generally used for groundwater problems (Bear and Verrujit, 1987). The advection–dispersion equation for solutes with sorption to the bed sediment is

$$\theta\frac{\partial C}{\partial t} + \theta u\frac{\partial C}{\partial x} = \theta D\frac{\partial^2 C}{\partial x^2} - \rho\frac{\partial C_{ads}}{\partial t}, \tag{33}$$

where C is now the concentration of solute in the porewater, C_{ads} is the adsorbed mass per unit mass of bed sediment, ρ is the bulk density of the sediment, and D is the dispersion coefficient. At equilibrium, the sorbed mass of pollutant is related to the aqueous concentration by the partition coefficient:

$$C_{ads} = k_p C \tag{34}$$

Converting between mass and volume concentrations and then combining terms yields

$$\left(1 + \frac{\rho}{\theta}k_p\right)\frac{\partial C}{\partial t} + u\frac{\partial C}{\partial x} = D\frac{\partial^2 C}{\partial x^2}. \tag{35}$$

The retardation factor is then

$$R = 1 + \frac{\rho}{\theta}k_p. \tag{36}$$

Equilibrium sorption causes the reactive solute to travel through the bed with effective velocity R times less than the porewater velocity and to have concentration R times more than a conservative solute per unit of fully mixed volume of bed. The retardation coefficient can easily be included in the calculation of pumping exchange. Values of R for each metal were obtained by measuring k_p in batch sorption experiments. Simultaneous exchange of several metal ions could be modeled using this approach. Comparison between model predictions and the results of one flume experiment is given in Fig. 8.

2. Oxygen Exchange

The bedform-exchange models described earlier have been applied to oxygen consumption in natural rivers (Rutherford *et al.*, 1993, 1995; Rutherford, 1994). Knowledge of the exchange processes is combined with the rate of oxygen consumption by the sediments to yield the net oxygen consumption by the bed. Oxygen consumption due to turnover is calculated by determining the average residence time in the bed of a parcel of water buried

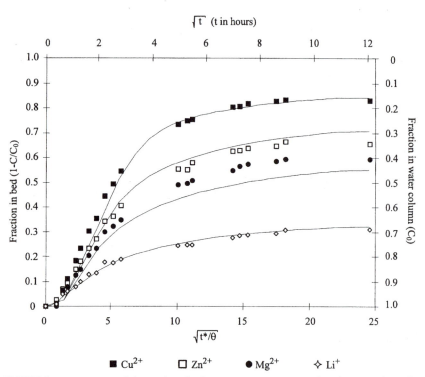

FIGURE 8 Comparison of the results of a flume experiment on metal exchange with predictions of the equilibrium sorption exchange model. The modeling curves include the same pumping exchange (based on the bedform geometry and the stream depth and velocity) and differ only due to the value of R, the retardation coefficient for each metal. $R = 15$ for copper, $R = 7$ for zinc, $R = 3$ for magnesium, and $R = 1$ for lithium (conservative). All metals were introduced to the flume simultaneously. Reprinted with permission from Eylers (1994). Copyright by H. Eylers.

by turnover. Assuming a given, constant rate of oxygen consumption in the bed, α, turnover results in an oxygen consumption rate of

$$O' = \frac{OHU_b}{\lambda}\left(1 - \frac{\theta OU_b}{2\alpha\rho\lambda}\right), \tag{37}$$

where O is the oxygen concentration in the stream water; all other terms are physical characteristics of the sediment bed, defined previously. Oxygen consumption due to pumping can be found by integrating the consumption rate over the pumping streamlines. Again assuming a constant rate of oxygen consumption, pumping results in

$$O_{out} = O_{in} - \alpha t \tag{38}$$

over each streamline, where t is the residence time for pumping along that streamline. Where the effluent is anoxic, the net consumption is given by the oxygen flux to the bed:

$$O' = u_m\, O_{in}\, /\, \pi. \tag{39}$$

In the case where oxygen consumption in the bed is limited by the input of stream biological oxygen demand (BOD), net consumption is given by the BOD flux to the bed (B_{in}):

$$O' = u_m\, B_{in}\, /\, \pi. \tag{40}$$

The pumping model can also be applied to first-order oxygen consumption kinetics, in which case

$$O_{out} = O_{in} - B_{in}\,(1 - e^{-\beta\, t}), \tag{41}$$

$$B_{out} = B_{in}\, e^{-\beta t} \tag{42}$$

along every streamline, where β is the first-order rate constant.

The zero-order model was applied to two streams in New Zealand that show significant deoxygenation—the Tarawera, a sand-bed river with large dunes, and the Waiotapu, a gravel-bed stream with irregular bed features. In the Tarawera, the relatively low hydraulic conductivity limits pumping flows and thus pumping-driven oxygen consumption. However, turnover causes a significant amount of oxygen consumption. In the Waiotapu, the more porous gravel bed allows for substantial pumping flows, and oxygen consumption due to pumping is predicted to be the dominant deoxygenation mechanism. However, in both cases a large fraction ($\sim 50\%$) of the actual measured oxygen consumption could not be accounted for by pumping and turnover. This discrepancy is probably due to processes not included in the model, either in the streambed or in the surface river water. Turbulent interactions between the stream and the upper layer of the bed likely enhance sediment–mediated oxygen consumption, especially in gravel-bed rivers. Thus, modeling of practical field problems is still hampered by our limited understanding of the fluid dynamics at the sediment-water interface. In addition, there could be field-scale processes beyond the scope of the bed-form models that also contribute to hyporheic exchange in these rivers. Clearly a considerable amount of research on exchange processes will have to be done before we can reliably model sediment-driven oxygen consumption in rivers.

3. Colloid Exchange

The exchange of colloidal particles has been modeled by adding colloidal processes to the fundamental hydraulic exchange models. Packman identified particle settling and filtration as the two mechanisms that cause colloid transport to differ from solute transport in sand streambeds (Packman and Brooks, 1995; Packman, 1997; Packman *et al.*, 1997, 1999a,b). Settling causes larger colloids ($d > 2\ \mu$) to deviate from the porewater flowpaths in the bed. Filtration is the attachment of colloids to stationary bed sediment grains due to chemical and electrostatic interactions. Straining (physical separation) is not an important trapping process because the colloids are assumed to be small enough (by definition, $d < 10\ \mu$) to pass through the pore spaces between sand grains. As for reactive solutes, the colloidal processes must be parameterized for inclusion in an exchange model. Particle paths in the bed can be determined by adding the particle settling velocity (v_s) to the advective porewater profile:

$$u_{\text{particle}} = u, \tag{43}$$

$$v_{\text{particle}} = v + v_s. \tag{44}$$

Particle settling is always downward, and thus can only result in higher penetration of particles into the bed (relative to solutes). The settling velocity is nondimensionalized with the maximum seepage velocity in the bed

$$v_s^* = \theta\, v_s\, /\, u_m. \tag{45}$$

Filtration can be modeled with the filtration coefficient, λ_f, which is an empirical description of the reduction in concentration of a colloidal suspension flowing through a porous media:

$$\partial C / \partial x = -\lambda_f C. \tag{46}$$

In this formulation, the filtration coefficient accounts for all the microscale processes (physical and chemical) that are responsible for the attachment of colloids to the bed sediment. In the colloid exchange model, net filtration is calculated numerically by applying Eq. (46) along the particle paths. This results in a residence time function for particles that includes both particle settling and the permanent trapping of particles in the bed due to filtration.

Packman also conducted flume experiments on the exchange of kaolinite clay with a sand bed and compared the results of these experiments with the colloid exchange model (Packman, 1997; Packman *et al.*, 1997, 1999b). The modeling value of the filtration coefficient was obtained from column experiments conducted with the same colloids, bed sediment, and water chemistry used in the flume, whereas the particle settling velocity was calculated from the size distribution of the colloids added to the flume. This model did an excellent job of predicting colloid exchange when all particle surface properties were well defined. A typical result is shown in Fig. 9.

Huettel *et al.* (1996) also conducted experimental studies on particle exchange and obtained three-dimensional profiles of the penetration of acrylic pigment particles into and around a single bedform. Their results demonstrated that colloids are carried into the bed at areas of high pumping inflow, where they end up trapped by the bed sediment.

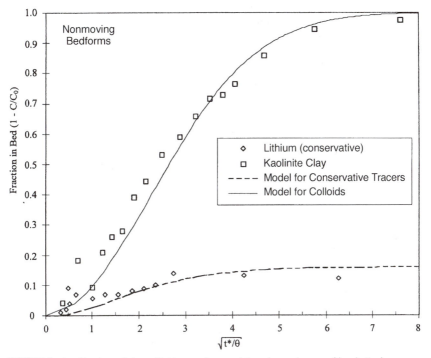

FIGURE 9 Application of the colloid pumping model to the exchange of kaolinite between a stream and sand bed in a laboratory flume. The model prediction is not fitted to the data but is an entirely independent estimate based on the average bedform geometry, the stream depth and velocity, the particle settling velocity, and the filtration coefficient measured in a column experiment with identical conditions. Data and model prediction for exchange of lithium (a conservative tracer) are given for comparison. Reprinted with permission from Packman (1997). Copyright by A.I. Packman.

D. Summary

Progress has been made in the last 10 years in identifying and modeling the processes responsible for localized exchange between a stream and the near-stream regions of the hyporheic zone. Laboratory flume experiments have allowed the study of these exchanges under controlled, easily observable conditions and aided the identification of exchange mechanisms. In addition, the application of fundamental fluid mechanics and hydraulics has resulted in the formulation of generalized exchange models. However, progress still needs to be made in applying these models to the field, which will require synthesizing models for competitive processes and including additional exchange processes that are beyond the scope of the current models.

It is now apparent that localized exchange through the stream–bed interface is primarily due to one of three main mechanisms: turbulent interaction between the stream and porewater flows, advective pumping induced by the pressure variation over bedforms, and turnover due to bedform motion. Gravel-bed streams will tend to have higher turbulent interactions resulting in a significant slip velocity at the bed surface and considerable non-Darcy flow in the bed. Pumping can also be an important exchange mechanism in gravel-bed streams. The low permeability of sand beds tends to severely restrict the turbulent interaction, making either pumping or turnover the dominant exchange process in sand-bed streams. A general model that analyzes the effect of all three processes operating simultaneously has not yet been developed. In addition, no model has been developed which includes the heterogeneities that can be found in natural streams.

The exchange of reactive solutes and colloids has also been modeled. The fundamental exchange processes described earlier obviously drive the exchange of reactive substances. Models for reactive solutes include pumping and turnover, as well as the chemical interactions unique to each individual solute. These models have been very successful in reproducing the results of laboratory experiments on the exchange of metal ions, but only accounted for about half of the observed oxygen consumption in natural streams. The model for colloid exchange adds particle settling and filtration to the fundamental hydraulic exchange. This model was successful in reproducing the results of laboratory experiments on clay exchange but has yet to be applied to the field.

IV. VIEWING THE INTERACTION FROM THE SUBSURFACE

Viewing hyporheic exchange from the subsurface is focused on determining the flow of water into, through, and out of the aquifer beneath the stream. Hyporheic exchange is considered purely as subsurface flow and determined based upon the mixing of stream-derived and aquifer-derived water. This approach is useful for the examination of hyporheic flowpaths that diverge significantly from the stream channel as well as the interaction of these flows with the larger-scale groundwater flow field.

The subsurface approach requires that groundwater motion be calculated throughout the stream system. Groundwater flows can be modeled at the field scale using a numerical package such as MODFLOW (McDonald and Harbaugh, 1988). This requires specifying the system geometry and calibrating to measured groundwater flows by the adjustment of modeled boundary conditions. Both the stream and the underlying aquifer must be described in an appropriate fashion to analyze the subsurface flows. In this approach, the stream itself is then viewed as a boundary condition that drives subsurface flow. Additional groundwater flow occurs due to other elements of the aquifer geometry (e.g., slope). Once the stream–aquifer system has been modeled, hyporheic exchange can be determined by analyzing the penetration of stream-derived water into the aquifer.

Because modeling is done at the field scale, all relevant aspects of the stream system can be included. Certainly the complete stream and aquifer geometry will be included in the model, and variations in aquifer composition can be modeled. This modeling approach requires extensive effort to obtain the information needed to characterize the study site. The surface and subsurface geometry and the aquifer flow properties (e.g., hydraulic conductivity) must be measured in the field in order to apply the model. In addition, groundwater heads and resulting flows must be measured in order to calibrate the model. The quality of the model simulation will depend on the quality of the data set used in preparing the model.

The use of MODFLOW to investigate hyporheic exchange will be demonstrated by considering specific case studies rather than the general functioning of the program.

A. Case Studies

Wroblicky applied this approach to hyporheic exchange in two streams—Aspen Creek and Rio Calveras (Wroblicky *et al.*, 1994, 1998; Wroblicky, 1995). These are both headwater (first-order) mountain streams but have distinctly different geological settings, bed compositions, and aquifer properties. Aspen Creek lies in sediment that is mostly fine sand with some gravel layers, whereas Rio Calveras lies in sediment that is poorly sorted coarse sand and gravel. Each of these sites was modeled over a 120-m stretch of stream, with the groundwater system represented by a two-dimensional finite-difference grid with 0.5-m node spacing. Each node was assigned properties to characterize the flow response of the water in the aquifer at that location. The study reaches were surveyed and instrumented to obtain the data necessary to implement the model. The hydraulic conductivity, porosity, and specific yield of the various aquifer materials were measured and assigned to each node. Forty wells were placed in five transects at each site and installed with piezometers to record groundwater elevations. Additional piezometers were installed in the stream itself, and staff gauges were added to measure stream stage. A variety of other data (e.g., precipitation, temperature) were also obtained to fully characterize the site.

With the study reach described in the model, the flow of ground water in the area can be predicted. However, because not all relevant parameters can be measured and there is uncertainty in the measured parameters, this type of model is generally calibrated to measured flow conditions in order to increase confidence in model predictions. Wroblicky adjusted the values of the inflow and outflow at the boundary of the study area, rates of recharge and evapo-transpiration, and distribution of stream and aquifer conductivity (within the uncertainty of the measured values) in order to fit measured groundwater heads. This type of calibration guarantees that modeled flow rates are consistent with measured groundwater profiles. However, because many parameters are used to obtain a model fit, it cannot be guaranteed that the values selected are representative of the actual conditions. In addition, these models are typically calibrated to obtain an overall fit to the conditions over the entire study area, and model predictions will typically fail to agree with measured flows in some regions. Thus, model simulations cannot be used to predict groundwater flow indiscriminately. Nonetheless, flowpaths can be identified given sufficient confidence in the calibration and functioning of the model.

In addition to identifying the hyporheic flows near the two streams with MODFLOW, Wroblicky used a particle tracking model (MODPATH, described in Pollock, 1989) to determine the extent to which stream-derived water penetrated into the aquifer. Model simulations indicated that hyporheic exchange in both streams was driven by local stream geometry, notably at stream meanders and regions of locally high slope. However, the locations where hyporheic exchange occurred changed with the stream-flow conditions. In Aspen Creek, there was stable (i.e., year-round) hyporheic exchange at four locations and flow-dependent hyporheic exchange at several other locations. Fig. 10 shows the hyporheic flowpaths present at intermediate groundwater levels, typical of winter conditions. The relative magnitude of the exchange at each location also changed depending on the flow conditions. At high stream discharge, the subsurface flow was mostly down-valley, but at low discharge there was relatively more flow into and out of the stream. Rio Calveras showed even more seasonal variability in hyporheic exchange. At high flow, the ground water generally moved either toward or away from the stream, and hyporheic flow (that diverged from and then returned to the stream) occurred only at two locations. At low flow, four additional near-stream hyporheic pathways were present. Thus, in both streams, the overall pattern of hyporheic exchange was found to vary considerably with the stream-flow conditions.

Wondzell used MODFLOW to analyze hyporheic exchange in McRae Creek, a fourth-order mountain stream with a poorly sorted sand/gravel bed (Wondzell, 1994; Wondzell and Swanson, 1996). The study site (100 m by 80 m in extent) featured a complex morphology with a flood plain, high ter-

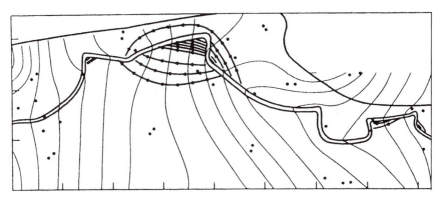

FIGURE 10 Simulated groundwater head contours and hyporheic flowpaths in Aspen Creek during intermediate flow conditions. Hyporheic flow (indicated by heavy lines with arrows) is predicted to occur at stream meanders and changes in slope. Reprinted from Wroblicky (1995).

races at the edge of the flood plain, and a secondary stream channel with a prominent gravel bar. As in the work of Wroblicky *et al.* (1998), a large number of wells and several river staff gauges were used to obtain contours of water head throughout the system, and the hydraulic conductivities of the various aquifer sediments were measured. Eight head-elevation data sets were used to calibrate the model.

Once the model was calibrated, it was used to predict the transient exchange during a storm and to examine the characteristics of the hyporheic exchange flow. Generally good agreement between predicted and measured heads in the transient simulation provided confidence that the groundwater flows were modeled correctly. The complex morphology of the study reach caused some difficulties: hillslope flow from the high terraces caused discrepancies between predicted and measured heads near the floodplain boundary. However, the model functioned well near the stream, allowing analysis of hyporheic flow patterns. There was substantial flow from the stream through the high-permeability gravel bar and secondary stream channel under all flow conditions, with the flow pattern shown in Fig. 11. The high subsurface flow through the gravel bar was confirmed by tracer studies (temperature and dye). Thus, hyporheic exchange in this reach was controlled by the gravel bar and secondary channel: changes in stream flow just affected the magnitude of the flow in these preferential pathways and did not result in the appearance of new zones of hyporheic exchange.

B. Summary

Numerical groundwater modeling packages such as MODFLOW are a useful tool for analyzing hyporheic flows. These models allow detailed description, analysis, and simulation of subsurface flow at a study site. Application of this type of model requires extensive site characterization and data

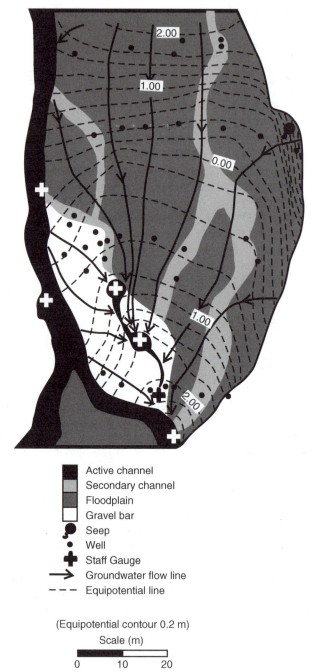

FIGURE 11   Simulated groundwater flowpaths in McRae Creek showing hyporheic flow through a gravel bar and secondary stream channel. Reprinted with permission from Wondzell and Swanson (1996) *Journal of the North American Benthological Society* 15: 3–19.

for model calibration. Once the model has been developed and calibrated, it can be used to distinguish hyporheic flowpaths by tracing the movement of water from the stream into and through the subsurface. Changes in these flowpaths due to seasonal variations in stream and groundwater flow can also be modeled. This model view of stream–subsurface interaction is site-specific and is thus a tool to determine the functioning of a specific reach of stream. By identifying the factors that tend to define or control hyporheic exchange at different scales in natural streams, these models will increase our overall understanding of hyporheic exchange and aid the development of more process-detailed models.

V. CHALLENGES

The existence of hyporheic flows is now well documented. Evidence of these flows comes, to a large extent, by implication from the numerous studies of biogeochemical hyporheic zone processes. Hydrometric study of these flows, independent of overarching biogeochemical motivations, has been limited to small physical scales both in the field (e.g., Harvey and Bencala, 1993) and in laboratory flumes (e.g., Elliot and Brooks, 1997b). The models developed from these studies are sufficient to empirically describe both hyporheic flowpaths and tracer exchange between the stream and subsurface and to predict exchange in controlled laboratory flumes based on fundamental hydraulic principles. However, the challenge for hydrometric study of surface–subsurface interactions still extends beyond the issue of "scaling-up." For even at the small scale, there is no clear methodology to generalize results beyond specific conditions at reach study sites in particular streams. Work remains to be done relating flume and extensive site-specific hydrometric parameters to "easily measurable" channel geomorphic and hydraulic parameters.

Finally, in the context of stream ecology and nutrient transport, it is important to remember that downstream solute transport is not solely determined by hydrologic processes. Allan (1995) summarizes this succinctly in commenting: "Rapid uptake and short transport distances are expected under strong nutrient limitation, whereas at the other extreme all solutes behave conservatively and simply are in transit downstream." The implication is that advances in the study of biogeochemical processes at substrate surfaces will increase the need for an understanding of the detailed fluid mechanics at the scale of these surfaces.

REFERENCES

Allan, J. D. 1995. "Stream Ecology: Structure and Function of Running Waters." Chapman & Hall, New York.

Bear, J., and A. Verrujit. 1987. "Modelling of Groundwater Flow and Pollution." Kluwer Academic Publishers, Norwell, MA.

Bencala, K. E. 1984. Interactions of solutes and streambed sediment. 2. a dynamic analysis of coupled hydrologic and chemical processes that determine solute transport. *Water Resources Research* **20**:1804–1814.

Bencala, K. E. 1993. A perspective on stream-catchment connections. *Journal of the North American Benthological Society* **12**:44–47.

Bencala K. E., and R. A. Walters. 1983. Simulation of solute transport in a mountain pool-and-riffle stream: A transient storage model. *Water Resources Research* **19**:718–724.

Bencala K. E., D. M. McKnight, and G. W. Zellweger. 1990. Characterization of transport in an acidic and metal-rich mountain stream based on a lithium tracer injection and simulations of transient storage. *Water Resources Research* **26**:989–1000.

Bencala K. E., J. H. Duff, J. W. Harvey, A. P. Jackman, and F. J. Triska. 1993. Modelling within the stream-catchment continuum. *In* "Modelling Change in Environmental Systems" (A. J. Jakeman, M. B. Beck, and M. J. McAleer, eds.), pp. 163–187. Wiley, New York.

Broshears R. E., K. E. Bencala, B. A. Kimball, and D. M. McKnight. 1993. Tracer-dilution experiments and solute-transport simulations for a mountain stream, Saint Kevin Gulch, Colorado. *U.S. Geological Survey Water Resources Internal Report* **92-4081.**

Castro, N. M., and G. M. Hornberger. 1991. Surface–subsurface interactions in an alluviated mountain stream channel. *Water Resources Research* **27**:1613–1621.

Constantz J., C. L. Thomas, and G. Zellweger. 1994. Influence of diurnal variations in stream temperature on streamflow loss and groundwater recharge. *Water Resources Research* **30**:3253–3264.

D'Angelo D. J., J. R. Webster, S. V. Gregory, and J. L. Meyer. 1993. Transient storage in Appalachian and Cascade mountain streams as related to hydraulic characteristics. *Journal of the North American Benthological Society* **12**:223–235.

Elliott, A. H. 1990. Transfer of solutes into and out of streambeds. Ph.D. Thesis, California Institute of Technology, Pasadena.

Elliott, A. H., and N. H. Brooks. 1997a. Transfer of nonsorbing solutes to a streambed with bed forms: Theory. *Water Resources Research* **33**:123–136.

Elliott, A. H., and N. H. Brooks. 1997b. Transfer of nonsorbing solutes to a streambed with bed forms: laboratory experiments. *Water Resources Research* **33**:137–151.

Eylers, H. 1994. Transport of adsorbing metal ions between stream water and sediment bed in a laboratory flume. Ph.D. Thesis, California Institute of Technology, Pasadena.

Eylers, H., N. H. Brooks, and J. J. Morgan. 1995. Transport of adsorbing metals from stream water to a stationary sand-bed in a laboratory flume. *Marine and Freshwater Research* **46**:209–214.

Fehlman, H. M. 1985. Resistance components and velocity distributions of open channel flows over bed forms. M.S. Thesis, Colorado State University, Ft. Collins.

Harvey, J. W., and K. E. Bencala. 1993. The effect of streambed topography on surface–subsurface water exchange in mountain catchments. *Water Resources Research* **29**:89–98.

Harvey, J. W., and C. C. Fuller. 1998. Effect of enhanced manganese oxidation in the hyporheic zone on basin-scale geochemical mass balance. *Water Resources Research* **34**:623–636.

Harvey, J. W., B. J. Wagner, and K. E. Bencala. 1996. Evaluating the reliability of the stream tracer approach to characterize stream-subsurface water exchange. *Water Resources Research* **32**:2441–2451.

Huettel, M., W. Ziebis, and S. Forster. 1996. Flow-induced uptake of particulate matter in permeable sediments. *Limnology and Oceanography* **41**:309–322.

Jackman, A. P., R. A. Walters, and V. C. Kennedy. 1984. Transport and concentration controls for chloride, strontium, potassium, and lead in Uvas Creek, a small cobble-bed stream in Santa Clara County, California, USA. 2. Mathematical modeling. *Journal of Hydrology* **75**:111–141.

Kim, B. K. A., A. P. Jackman, and F. J. Triska. 1992. Modeling biotic uptake by periphyton and transient storage of nitrate in a natural stream. *Water Resources Research* 28:2743–2752.

Larkin, R. G., and J. M. Sharp. 1992. On the relationship between river-basin geomorphology, aquifer hydraulics, and ground-water flow direction in alluvial aquifers. *Geological Society of America Bulletin* 104:1608–1620.

McDonald, M. G., and A. W. Harbaugh. 1988. A modular three-dimensional finite-difference ground-water flow model. *U.S. Geological Survey, Techniques of Water Resources Investigations.*

Mendoza, C. 1988. Comment on: Convective transport within stable river sediments, by S. A. Savant, D. D. Reible, and L. J. *Thibodeaux. Water Resources Research* 24(7):1206–1207.

Mendoza, C., and D. Zhou. 1992. Effects of porous bed on turbulent stream flow above bed. *Journal of Hydraulic Engineering* 118:1222.

Morrice, J. A., H. M. Valett, C. N. Dahm, and M. E. Campana. 1997. Alluvial characteristics, groundwater-surface water exchange and hydrologic retention in headwater streams. *Hydrological Processes* 11:253–267.

Mulholland, P. J., E. R. Marzolf, J. R. Webster, D. R. Hart, and S. P. Hendricks. 1997. Evidence that hyporheic zones increase hetrotrophic metabolism and phosphorus uptake in forest streams. *Limnology and Oceanography* 42:443–451.

Packman, A. I. 1997. Exchange of colloidal kaolinite between stream and sand bed in a laboratory flume. Ph.D. Thesis, California Institute of Technology, Pasadena.

Packman, A. I., and N. H. Brooks. 1995. Colloidal particle exchange between stream and stream bed in a laboratory flume. *Marine and Freshwater Research* 46:233–236.

Packman, A. I., N. H. Brooks, and J. J. Morgan. 1997. Experimental techniques for laboratory investigation of clay colloid transport and filtration in a stream with a sand bed. *Water, Air, and Soil Pollution* 99:113–122.

Packman, A. I., N. H. Brooks, and J. J. Morgan. 1999a. A physicochemical model for bedform-driven colloid exchange between a stream and a sand stream bed. *Water Resources Research* (submitted for publication).

Packman, A. I., N. H. Brooks, and J. J. Morgan. 1999b. Kaolinite exchange between a stream and stream bed—laboratory experiments and evaluation of a colloid transport model. *Water Resources Research* (submitted for publication).

Pinder, G. F., and S. P. Sauer. 1971. Numerical solution of flood-wave modification due to bank storage effects. *Water Resources Research* 7:63–70.

Pollock, D. W. 1989. Documentation of computer programs to compute and display pathlines using results from the U.S. Geological Survey modular three-dimensional finite-difference groundwater flow model. *Geological Survey Open File Report (U.S.)* 89-38.

Rutherford, J. C. 1994. "River Mixing." Wiley, New York.

Rutherford, J. C., G. J. Latimer, and R. K. Smith. 1993. Bedform mobility and benthic oxygen uptake. *Water Research* 27:1545–1558.

Rutherford, J. C., J. D. Boyle, A. H. Elliott, T. V. J. Hatherell, and T. W. Chiu. 1995. Modeling benthic oxygen uptake by pumping. *Journal of Environmental Engineering* 21:84–95.

Savant, S. A., D. B. Reible, and L. J. Thibodeaux. 1987. Convective transport within stable river sediments. *Water Resources Research* 23:1763–1768.

Sharp, J. M. 1988. Alluvial aquifers along major rivers. *In* "Hydrogeology, the Geology of North America" (W. Back, J. S. Rosenshein and P. R. Seaber, eds.), pp. 273–282. Geological Society of America, Boulder, CO.

Shen, H. W., H. M. Fehlman, and C. Mendoza. 1990. Bed form resistances in open channel flows. *Journal of Hydraulic Engineering* 116:799–815.

Shimizu, Y., T. Tsujimoto, and H. Nakagawa. 1990. Experiment and macroscopic modelling of flow in highly permeable porous medium under free-surface flow. *Journal of Hydrosciences and Hydrologic Engineering* 8:69–78.

Squillace, P. J., E. M. Thurman, and E. T. Furlong. 1993. Groundwater as a nonpoint source of atrazine and deethlatrazine in a river during base flow conditions. *Water Resources Research* **29**:1719–1729.

Tate, C. M., R. E. Broshears, and D. M. McKnight. 1995. Phosphate dynamics in an acidic mountain stream: Interactions involving algal uptake, sorption by iron oxide and photoreduction. *Limnology and Oceanography* **40**:938–946.

Thibodeaux, L. J., and J. D. Boyle. 1987. Bedform-generated convective transport in bottom sediment. *Nature (London)* **325**:341–343.

Valett, H. M., J. A. Morice, C. N. Dahm, and M. E. Campana. 1996. Parent lithology, surface-groundwater exchange, and nitrate retention in headwater streams. *Limnology and Oceanography* **41**:333–345.

Vanoni, V. A. 1975. "Sedimentation Engineering." American Society of Civil Engineering, New York.

Vittal, N., K. G. Ranga Raju and R. J. Garde. 1977. Resistance of two-dimensional triangular roughness. *Journal of Hydraulic Research* **15**:19–36.

Wagner, B. J., and J. W. Harvey. 1997. Experimental design for estimating parameters of rate-limited mass-transfer: Analysis of stream tracer studies. *Water Resources Research* **33**:1731–1741.

Winter, T. C. 1995. Recent advances in understanding the interaction of groundwater and surface water. *Reviews of Geophysics* **33**:985–994.

Wondzell, S. M. 1994. Flux of ground water and nitrogen through the floodplain of a fourth-order stream. Ph.D. Thesis, Oregon State University, Corvallis.

Wondzell, S. M., and F. J. Swanson. 1996. Seasonal and storm dynamics of the hyporheic zone of a 4th-order mountain stream. 1. Hydrologic processes. *Journal of the North American Benthological Society* **15**:3–19.

Wroblicky, G. J. 1995. Numerical modeling of stream-groundwater interactions, near-stream flowpaths, and hyporheic zone hydrodynamics of two first-order mountain stream-aquifer systems. M.S. Thesis, University of New Mexico, Albuquerque.

Wroblicky, G. J., M. E. Campana, C. N. Dahm, H. M. Valett, J. A. Morrice, K. S. Henry, and M. A. Baker. 1994. Simulation of stream-groundwater exchange and near stream flow paths of two first-order mountain streams using MODFLOW. *In* "Proceedings of the Second International Conference on Ground Water Ecology" (J. A. Stanford and H. M. Valett, eds.), pp. 187–196. American Water Resources Association, Bethesda, MD.

Wroblicky, G. J., M. E. Campana, H. M. Valett, and C. N. Dahm. 1998. Seasonal variation in surface–subsurface water exchange and lateral hyporheic area of two stream-aquifer systems. *Water Resources Research* **34**:317–328.

Zellweger, G. W. 1994. Testing and comparison of four ionic tracers to measure stream flow loss by multiple tracer injections. *Hydrological Processes* **8**:155–165.

Zhou, D. 1992. Turbulent channel flow over porous beds. Ph.D. Thesis, Columbia University, New York.

Zhou, D., and C. Mendoza. 1993. Flow through porous bed of turbulent stream. *Journal of Engineering Mechanics* **119**:365–383.

Zhou, D., and C. Mendoza. 1995. Pollutant transport beneath porous stream beds. *In* "Proceedings of the First International Conference on Water Resources Engineering," pp. 219–223. American Society of Civil Engineering, New York.

SECTION *TWO*

BIOGEOCHEMISTRY

Subsystem Interactions

with Stream Surface Water

Stream Chemistry and Riparian Zones

Alan R. Hill

Department of Geography
York University
North York, Ontario, Canada

I. Introduction   83
II. Riparian Zone Hydrological–Chemical Interactions:
     An Overview   84
III. Riparian Influences on Stream Chemistry   87
     A. Nitrogen   87
     B. Dissolved Organic Carbon   93
     C. Other Elements   94
IV. Riparian Zone Influence on Stream Chemistry in Relation to
     Watershed Hydrogeology: A Conceptual Framework   100
V. Future Research Directions   106
     References   107

I. INTRODUCTION

Research on stream chemistry has given increased attention in recent years to the analysis of watershed processes that define the overall supply and availability of mineral elements to streams. How water chemistry is altered along various hydrologic pathways from uplands to streams is increasingly recognized as critical to understanding stream biogeochemistry (Hynes, 1983; Likens, 1984). This broader perspective of stream ecosystems which emphasizes land–water interactions has focused attention on the stream riparian zone as a critical interface. Because of their streamside location, riparian zones have a significant potential to regulate the chemistry of overland and subsurface flows moving from uplands to streams.

The riparian zone is composed of the area of land adjacent to streams and rivers. This zone is often a narrow strip between the channel and hill slope in headwater areas but also occurs as a more extensive floodplain adjacent to larger rivers (Naiman and Decamps, 1997). Riparian areas are zones of transition between terrestrial and aquatic environments. They can often be distinguished from upland areas of watersheds by vegetation, soils, and topography. Many riparian zones can be identified as wetlands where the land surface is saturated with water for a long enough time to be the dominant influence on soil and vegetation. However, some riparian areas do not flood frequently and may not have a high water table because of channel incision.

The purpose of this chapter is to review the literature on riparian zone controls and influences on stream chemistry during stream baseflow and storm conditions. Baseflows often dominate the annual hydrograph, and element concentrations during baseflow have a major influence on stream biological processes. Storms, depending on their magnitude and timing, may serve as an element subsidy by providing additional inputs and elevated concentrations in stream water. The initial section of the chapter discusses the interactions between hydrology and chemistry that are essential for an understanding of riparian dynamics. This is followed by a review of current research on riparian influences on stream chemistry and the presentation of a conceptual model which describes how riparian zone controls on stream chemistry vary in the context of watershed hydrogeologic setting. The chapter concludes with a discussion of future research priorities.

Attention is focused mainly on riparian zones of headwater streams in humid landscapes (orders 1–3). These low-order streams comprise most of the total stream length and are where the most extensive interaction between land surfaces and aquatic systems occurs. The role of riparian zones may be restricted in some headwater landscapes. In areas of mountain terrain, small streams are commonly incised in steep valleys with minimal riparian zones. The widespread use of ditches and the extension of cultivation to the stream bank in some agricultural landscapes has reduced the effect of riparian areas on streams. In this chapter, I review studies of headwater riparian zones where hydrology is dominated by groundwater flow from uplands to the stream. However, stream water may penetrate laterally to create an extensive hyporheic zone beneath the riparian area in some landscapes where groundwater inputs are small and alluvial sediments have high hydraulic conductivities (Triska *et al.*, 1993; Valett *et al.*, 1996).

II. RIPARIAN ZONE HYDROLOGICAL–CHEMICAL INTERACTIONS: AN OVERVIEW

Most studies of riparian zone chemistry have used an input–output perspective involving the comparison of element concentrations of water in the riparian zone with the adjacent upland and stream. Mass balance studies of small valley bottom wetlands have also frequently been used to quantify the retention, transformation, or export of elements to downstream reaches. These studies provide little insight into the hydrologic and chemical processes that produce different riparian influences on stream chemistry. An integrated process-based approach which links hydrologic flowpaths with chemical and biological transformation processes is needed to provide a better understanding of riparian zone regulation of stream chemistry.

Recently, researchers have begun to analyze relationships between riparian zone hydrology and chemical transformation processes (Hill, 1990; McDowell *et al.*, 1992; Cirmo and McDonnell, 1997; Devito and Hill, 1997). A hydrological–chemical interaction perspective focuses attention on the external hydrologic links between riparian zones and the watershed, as well as on the internal hydrology of the riparian area. Element inputs from adjacent uplands are influenced by the magnitude and seasonality of groundwater flows to the riparian zone (Devito *et al.*, 1996). Retention efficiency (outputs divided by inputs) often increases with the magnitude of input up to a level where efficiency either reaches a plateau or decreases (Brinson, 1993). The flowpath of water from upslope areas influences chemical inputs, as well as the degree and location of interaction within the riparian zone (Phillips *et al.*, 1993). The size and seasonality of upland hydrologic connections controls riparian zone water table fluctuations and the extent of surface saturation (Roulet, 1990). These hydrologic factors affect riparian zone soil redox, vegetation dynamics, chemistry, and microbial processes. Hydraulic gradient, water table position, and water retention capacity of sediments also exert a strong control on the magnitude, rate, and type of hydrologic flowpath within the riparian zone.

Riparian zones function as the immediate source of ground water, which sustains stream baseflows in interstorm periods. Riparian zones and adjacent uplands in some landscapes are underlain at a shallow depth by impermeable materials (bedrock, clay, dense till) which form an aquitard restricting the downward movement of water. This hydrogeologic setting produces a thin surficial aquifer that discharges seasonally variable amounts of ground water locally to the adjacent riparian zone. Where riparian zone sediments are permeable above the shallow aquitard, ground water flows mainly in a shallow subsurface direction toward the stream, resulting in considerable interaction with vegetation and soils.

In headwater areas where shallow aquitards are absent, a thicker surficial aquifer develops beneath the upland, and ground water follows deeper longer flowpaths toward the riparian zone. In larger watersheds, thick extensive aquifers may contain discrete flow systems at several scales (Toth, 1963). Riparian zones in the downstream areas of these watersheds may receive groundwater inputs from local systems which recharge on adjacent slopes, from intermediate systems that include at least one local flow system between their recharge and discharge points, and regional flow systems that begin at regional topographic divides. Riparian zones connected to thick aquifers receive large constant groundwater inputs in comparison to landscapes with shallow impermeable layers. If permeable layers are present at depth, ground water may flow rapidly under the riparian zone and discharge directly to the stream channel. Ground water can also flow upward to the surface of the riparian zone at the upland perimeter because of the abrupt flattening of the water table slope or the presence of low permeability alluvial or organic sediments near the surface of the riparian zone (Brusch and Nilsson, 1993). This upward discharge of ground water can maintain areas of permanent surface saturation and a stable shallow water table in riparian zones.

During storms, riparian zones can be an important source of runoff as well as function as a zone of interaction with drainage water from adjacent slopes. Recent hydrology research in humid landscapes emphasizes the importance of variable source areas of runoff generation within drainage basins (Pearce et al., 1986; Anderson and Burt, 1990). Runoff-contributing zones expand and contract seasonally and during individual storms. These areas usually occupy a limited area of the drainage basin and are frequently located in riparian areas adjacent to streams. Although several different storm runoff-generating mechanisms have been identified in humid headwater landscapes, tracer studies using stable isotopes indicate that water stored in soils prior to storms (pre-event water) usually dominates storm discharge in streams (Sklash, 1990).

Subsurface storm flow on hill slopes contributes significantly to runoff in humid forest landscapes with permeable soils and steep slopes. Recent research indicates that preferred flowpaths through macropores in forest soils transmit water and solutes rapidly downslope to the streamside zone (McDonnell, 1990; Peters et al., 1995). Infiltration-excess overland flow occurs where rainfall exceeds the soil infiltration rate. This runoff process rarely occurs in undisturbed forest landscapes but is often important on cropland. Sediments and sediment-bound elements carried in surface runoff from agricultural fields are often deposited effectively in adjacent forest and grass riparian areas.

Saturation-excess overland flow from riparian zones is often the dominant mechanism of storm runoff production in landscapes with thin soils, high water tables, and gentle concave slopes. Saturation overland flow develops in areas of saturated soils where no infiltration can occur. Direct runoff from rain onto the saturated surface can be augmented by return flow where upslope water emerges from the soil. Runoff generated by saturation overland flow is frequently composed mainly of water stored in the saturated soils prior to the storm, although occasional storms of higher intensity or longer duration produce a larger precipitation (event water) contribution (Hill and Waddington, 1993; Eshleman et al., 1994). Rapid mixing of small volumes of storm water with a larger pool of pre-event water stored in saturated areas is responsible for the dominance of pre-event water in surface-generated storm flows (Waddington et al., 1993).

Subsurface storm flow composed of pre-event water may be generated in streamside zones by increased transmission if the water table rises into highly conductive surface soils (Rodhe, 1989). Subsurface storm flow can also be produced by groundwater ridging which results in a rapid increase in the riparian water table hydraulic gradient toward the stream because of the conversion of the capillary fringe into phreatic water by a small amount of infiltrating rainfall (Gillham, 1984). This hydrologic research on storm runoff processes suggests that variations in water residence time, limited interaction of overland flow with soils, and the extent of mixing of ground water, unsaturated zone soil water, and event water may influence chemical transport from riparian zones to streams (Cirmo and McDonnell, 1997).

Hydrologic flowpaths and water table fluctuations within riparian zones are linked to temporal and spatial patterns of microbial redox reactions (see also Baker *et al.*, Chapter 11 in this book). A mechanistic and predictive understanding of stream riparian zone biogeochemistry can be enhanced by considering how microbial communities interact with variations in the supply of electron donors and acceptors (Hedin *et al.*, 1998). Various chemical and biological transformations occur in a predictable sequence within narrow redox ranges. Under oxidized conditions, aerobic bacteria use oxygen (O_2) as a terminal electron acceptor during the oxidation of organic matter. As the environment becomes increasingly reduced, anaerobic bacteria are capable of using alternative electron acceptors. Nitrate (NO_3^-) is reduced to nitrous oxide (N_2O) or dinitrogen (N_2), manganese is transformed from manganic (Mn^{4+}) to manganous (Mn^{2+}) compounds, and ferric iron (Fe^{3+}) is reduced to ferrous iron (Fe^{2+}). Under permanently anaerobic conditions, obligate anaerobic bacteria are involved in sulfate (SO_4^{2-}) reduction and methanogenesis.

In riparian zones where ground water flows in a shallow subsurface path toward the stream, a sequence of lateral zones dominated by aerobic respiration, denitrification, and sulfate reduction may occur between the upland perimeter and the stream (Correll and Weller, 1989). Similar sequences of redox reactions can be associated with ground water discharging upward into riparian sediments and streambeds (Hedin *et al.*, 1998). Vertical patterns of increased reduction and oxidation are also linked to water table fluctuations in riparian zones. Increasing oxidation of soils because of seasonal water table declines may produce a vertical sequence of methane (CH_4) loss, accumulation of sulfate (SO_4^{2-}), and increased nitrate concentrations due to nitrification.

III. RIPARIAN INFLUENCES ON STREAM CHEMISTRY

A. Nitrogen

During stream baseflows, considerable differences in nitrate and ammonium (NH_4^+) in subsurface water frequently occur between upland, riparian, and stream environments in forest watersheds. McClain *et al.* (1994) found that the nitrate concentration of oxygenated upland ground water in a central Amazon watershed decreased from 300–650 μgN L^{-1} to <50 μgN L^{-1} in the riparian zone, whereas ammonium increased (Fig. 1). These spatial patterns of nitrate and ammonium are influenced by differences in hydrologic flowpaths and geomorphology. Nitrate concentration declined from 500 to 9 μgN L^{-1} and ammonium increased from <30 to >500 μgN L^{-1} as upland ground water flowed in a subsurface path through a highly reduced sand layer below the riparian zone of a Puerto Rico rain forest watershed (McDowell *et al.*, 1992). No clear trends were evident in a second watershed where clays restricted subsurface flow to variably oxidized riparian surface soils.

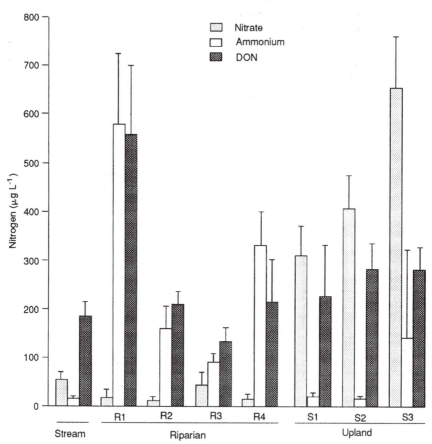

FIGURE 1   Mean nitrogen concentrations in stream and ground water at piezometer locations in the riparian zone and upland of the Barro Branco watershed in the Central Amazon Basin. Bars denote standard deviations. Modified from McClain *et al.* (1994) *Biogeochemistry* 27:113–127, with kind permission from Kluwer Academic Publishers.

Declines in nitrate in anoxic riparian zone subsurface water have been attributed to denitrification, whereas low nitrification and continued ammonification increase ammonium concentration. However, the extent to which these riparian zone processes regulate stream nitrogen chemistry is unclear in some forest watersheds. McClain *et al.* (1994) noted that stream ammonium concentration was considerably lower than riparian groundwater in a small Amazon watershed (Fig. 1). Large decreases in ammonium and small increases in nitrate between the riparian zone and the stream have been reported in other studies (McDowell *et al.*, 1992; Mulholland, 1993). These data suggest that riparian zone inorganic nitrogen concentrations may be further altered by processes occurring in hyporheic or in-stream environments.

Headwater riparian zones in forest watersheds may differ in their effect on downstream nitrate chemistry. Small conifer swamps on the Canadian Shield effectively retain nitrate during baseflows (Devito *et al.*, 1989; Devito and Dillon, 1993). Elevated groundwater nitrate concentration in a partly clear-cut Swedish watershed were depleted by a small fen (Jacks *et al.*, 1994). During the winter, nitrate concentration in the wetland outlet stream showed only a minor increase despite a large seasonal increase in groundwater nitrate inputs (Fig. 2). Greater winter retention of nitrate was attributed to a rise in water table that favored denitrification as ground water interacted with more labile carbon in surface organic soils.

A positive correlation between stream nitrate and riparian soil nitrate concentrations in the riparian zone of forested watersheds in Japan suggests that the streamside area was an important source of nitrate to streams (Ohrui and Mitchell, 1998). The steep slopes of these watersheds resulted in small moist riparian areas, which exhibited a high soil nitrification rate. Stream riparian zones dominated by nitrogen-fixing plant communities may also act as a source of nitrate to streams. High stream nitrate concentration with seasonal peaks in autumn and winter coincided with nitrate release from riparian stands of mountain alder (*Alnus tenuifolia*) in a Lake Tahoe watershed (Leonard *et al.*, 1979).

Some riparian zones in forest landscapes have little effect on nitrate concentration. For example, nitrate concentration in stream baseflow was controlled by the mixing of local and regional ground water and not by transformation processes in the Glen Major riparian swamp in southern Ontario. Local oxygenated ground water with nitrate concentration of 110–180 μgN L^{-1} emerged at the upland edge of the riparian zone to produce small surface streamlets which crossed the swamp to the stream (Fig. 3a). The upward flow within the swamp of regional ground water, which had trace nitrate and lower chloride concentrations than the local ground water, produced a decline in nitrate and chloride in streamlet water entering the stream. Water in the streamlets had a high oxygen concentration, limited interaction with surface organic soils and short residence times of <1 h in the swamp, which restrict-

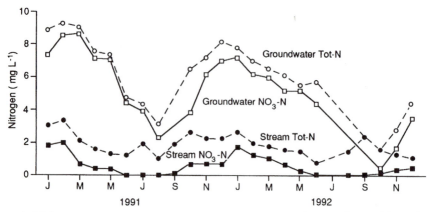

FIGURE 2   Nitrate-N and total nitrogen concentrations in groundwater inflow and the outlet stream of an oligotrophic fen in the Tonnerjoheden watershed, southwest Sweden. Modified from Jacks *et al.* (1994), with permission.

ed nitrate loss by denitrification (Hill, 1991). A progressive downstream de-cline in nitrate and chloride concentrations in the stream resulted from dilu-tion by a larger proportion of regional ground water discharging to the stream with distance from the source (Fig. 3b). Although the riparian swamp did not remove nitrate, it did influence stream baseflow ammonium concentration (Hill and Warwick, 1987). Microbial immobilization of ammonium in stream-let substrates resulted in a low ammonium concentrations of 5 μgN L^{-1} in swamp surface streamlets entering the stream in contrast to ammonium con-centrations of 30–60 μgN L^{-1} in local and regional ground water (Fig. 3a).

Many studies have examined the effects of forest and grass riparian zones on the nitrate concentration of subsurface flow from agricultural land. Rapid declines in nitrate of up to 80–95% over short distances have been noted at many sites (Lowrance *et al.*, 1984; Peterjohn and Correll, 1984; Haycock and Pinay, 1993). Loss of nitrate was not associated with large increases in ammonium and organic-N in subsurface water entering streams. However, most riparian zones studied in agricultural landscapes occur in wa-tersheds with impermeable layers near the land surface (Hill, 1996). In this setting, ground water follows shallow subsurface horizontal pathways that increase water residence times and contact with vegetation roots and organ-ic-rich riparian soils, facilitating rapid nitrate removal by plant uptake and denitrification.

A minority of studies has noted limited nitrate depletion by riparian zones in agricultural landscapes. A riparian forest in a Maryland catchment had little effect on ground water nitrate flow to the stream, which had a ni-trate concentration of >9 mgN L^{-1} during baseflows (Bohlke and Denver, 1995). Ground water at this site flowed at depth, effectively bypassing the riparian zone and discharging upward to the streambed. Effectiveness of ni-

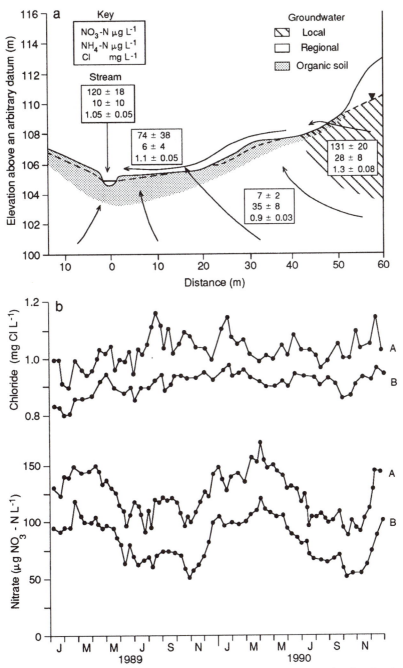

FIGURE 3 (a) Mean nitrate, ammonium, and chloride concentrations in local and regional groundwater inputs, swamp streamlets and the outlet stream of the Glen Major riparian swamp, southern Ontario. (b) Longitudinal variations in nitrate and chloride concentrations in Glen Major stream 400 m (site A) and 650 m (site B) downstream from the stream source.

trate removal can also vary within individual riparian zones. During low flow in a New Zealand pasture, most ground water flowed through patches of riparian organic soil, where nitrate was denitrified (Cooper, 1990). Large decreases or increases in stream nitrate concentration occurred over short distances as flow increased seven times between the spring source and the watershed outlet (Fig. 4). Nitrate increase was usually associated with reaches that received small lateral inputs from mineral soils that removed little nitrate. A decrease in nitrate concentration often occurred in short 10 to 20-m reaches where large lateral flow inputs from organic soil patches were depleted in nitrate (Fig. 4).

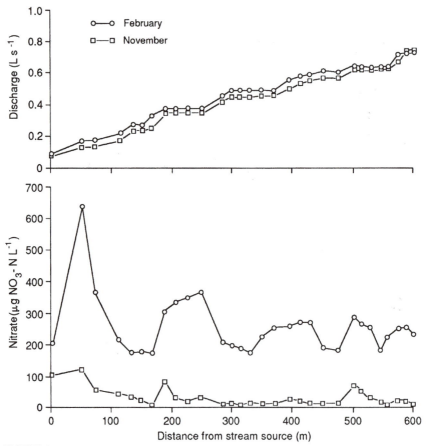

FIGURE 4 Longitudinal variation in stream discharge and nitrate concentration along a small New Zealand headwater stream. Decreases in nitrate concentration occur in short reaches where large lateral flow inputs from organic riparian soil patches are depleted in nitrate. Reproduced from Cooper (1990) *Hydrobiologia* **202**:13–26, with kind permission from Kluwer Academic Publishers.

Rainstorms or periods of snowmelt frequently produce riparian zone effects on stream nitrogen chemistry that differ from baseflow conditions. In many northern landscapes, >50% of the annual stream runoff occurs during spring snowmelt. Dominance of saturation overland flow in combination with short water residence times and low temperatures during snowmelt strongly influence element transport and transformations in riparian zones. Nitrate concentration in the outlet stream of a Canadian Shield conifer swamp was often >100 μgN L^{-1}, and the swamp retained only 52 and 28% of nitrate inputs in winter and spring, respectively (Devito and Dillon, 1993). In contrast, the swamp was a large source of organic nitrogen and phosphorus, which was released from sediments and senescent vegetation during these seasons.

Saturation overland flow from riparian wetlands may also influence stream nitrogen chemistry during larger summer and autumn storms (Hill, 1993a). In Glen Major swamp, maximum stream nitrate concentration was <2–6 times higher than the baseflow level of 130 μgN L^{-1}, whereas stream ammonium concentration increased from baseflow values of 5–10 μgN L^{-1} to 20–50 μgN L^{-1} during storms. These high inorganic-N concentrations preceded or coincided with the peak stream discharge. The stream nitrate concentration was controlled mainly by the mixture of rainfall and ground water transported by saturation overland flow from the swamp rather than by riparian zone transformation processes. Although maximum instantaneous contributions of rain were <25% in most storms, this component strongly influenced stream nitrate because of high nitrate concentration in rain. In contrast, stream ammonium concentration was much lower than concentration predicted from the proportion of rain and ground water present in stream flow. The rapid loss of ammonium from mixtures of surface storm flow and swamp soils in laboratory experiments and the absence of uptake in sterilized substrates suggested that ammonium in saturation overland flow was removed by microbial uptake in the riparian zone (Hill, 1993a).

B. Dissolved Organic Carbon

Streams with riparian wetlands in undisturbed watersheds often have high baseflow dissolved organic carbon (DOC) concentrations. In the Reedy Creek watershed, Virginia, DOC concentration increased from 2–3 mgC L^{-1} in the ground water at the hillslope base to 6–18 mgC L^{-1} in the riparian wetland, and ranged from 5 to 11 mgC L^{-1} in the stream (O'Brien *et al.*, 1994). Leaching of DOC from wetland soils was the dominant pathway of organic transport to a small blackwater stream in South Carolina (Dosskey and Bertsch, 1994). Although the 50- to 100-m-wide riparian swamp only occupied 6% of the forest watershed, it contributed 14,600 kgC yr^{-1} during baseflow conditions which represented 75% of the annual stream DOC export. High mean DOC concentration of 9.4 mgC L^{-1} was also measured in the riparian soil water of a forest watershed in Wales (Fiebig *et al.*, 1990). The mean baseflow DOC concentration in the adjacent stream, however, was only 2.1 mgC L^{-1}, suggesting that immobilization of DOC occurred in the streambed. Riparian wetland soils are a significant source of DOC because of high organic matter content and absence of strong adsorption processes (e.g., adsorption to iron and aluminum-rich clays). The chemical nature of DOC inputs to streams is also influenced by oxidation-reduction reactions in riparian zones. Labile products such as volatile fatty acids and carbohydrates occur in higher concentrations under anaerobic conditions (Meyer, 1990).

Dissolved organic carbon concentration is often positively correlated with stream discharge (Meyer and Tate, 1983; Sedell and Dahm, 1990). Nevertheless, few studies have directly measured riparian zone influences on stream DOC concentration during storms. Dalva and Moore (1991) showed that stream DOC increased during high flows because of inputs of DOC-rich ground water from a small valley swamp in Quebec (Fig. 5a). After major rainfall or spring snowmelt, low DOC concentration (<3 mgC L^{-1}) in headwaters of a stream fed by a hill slope spring at the upper weir increased rapidly to 20 mgC L^{-1} as the stream flowed through the swamp where mean porewater DOC concentration was 58.6 mgC L^{-1} (Fig. 5b). Concentration declined toward the outlet because of dilution by two tributaries that only flowed through a small area of the swamp.

The importance of riparian soils as a source of DOC during storms was influenced by hydrologic flowpaths in two Precambrian Shield watersheds (Hinton *et al.*, 1998). In Harp 4-21 watershed, the dominant riparian flowpath was through organic-rich upper soil horizons; consequently, riparian soils contributed 73–84% of the stream DOC export during an autumn storm. In Harp 3A watershed, where riparian flowpaths were mainly through lower mineral B horizons, near-stream areas contributed $<50\%$ of the stream DOC export.

C. Other Elements

Research in agricultural and forest watersheds suggests that some riparian zones function as a sink for phosphorus during baseflow (Peterjohn and Correll, 1984; Devito and Dillon, 1993), whereas others are seasonal sources (Mulholland, 1993). High seasonal concentrations of soluble reactive phosphorus (SRP) in riparian zone ground water may result from mobilization of phosphorus as iron and manganese oxides are reduced under anoxic conditions. Some riparian zones may not regulate phosphorus transport during baseflows. Differences in background SRP concentrations between two adjacent tributary streams draining riparian swamps in Costa Rica were not related to riparian zone transformations (Pringle and Triska,

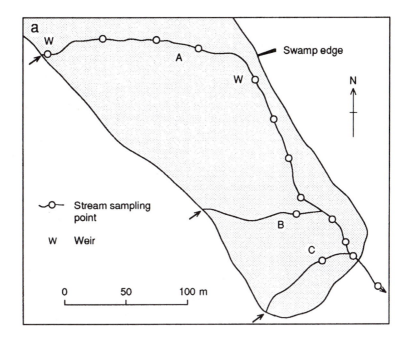

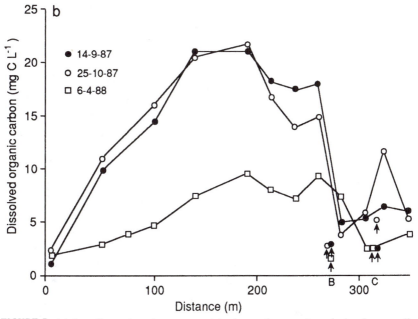

FIGURE 5 (a) Sampling points along stream A in the southern section of a headwater valley swamp, Mont St. Hilare, Quebec. (b) DOC concentration in stream A on 14 September 1987, 25 October 1987, and 6 April 1988. Arrows indicate the DOC concentration in tributaries B and C. Modified from Dalva and Moore (1991) *Biogeochemistry* **15**:1–19, with kind permission from Kluwer Academic Publishers.

1991). The Salto River had a high baseflow SRP concentration of 50–200 μgP L^{-1} because it received inputs from phosphorus-rich geothermal water from a regional ground water flow system, as well as local ground water from adjacent hill slopes. The neighboring Pantano stream, where regional ground water inputs were absent, had a low SRP concentration (<10 μgP L^{-1}).

The influence of riparian wetlands on stream sulfate concentration has received attention, particularly in watersheds with high atmospheric sulfate deposition. Studies have documented decreases in sulfate concentration as subsurface water passes through riparian zones. In Reedy Creek, Virginia, O'Brien *et al.* (1994) noted that a high sulfate concentration of 8–31 $mgSO_4$ L^{-1} in ground water declined progressively from the hill slope across the riparian wetland to levels of 2–5 $mgSO_4$ L^{-1} in the creek bank and 2–4 $mgSO_4$ L^{-1} in stream water (Fig. 6). During low flow periods, concentration of sulfate in outlet streams from small headwater conifer swamps on the southern Canadian shield declined to <3 $mgSO_4$ L^{-1} compared to 8–12 $mgSO_4$ L^{-1} in stream and hillslope ground water inputs (Devito and Hill, 1997). During baseflows, sulfate reduction is favored by lower water table elevation that increases interaction of subsurface water with anoxic riparian soils.

Several studies have noted high stream sulfate concentration and little retention in riparian wetlands during storms (Bayley *et al.*, 1986; O'Brien *et al.*, 1994). During high runoff in autumn and winter in two Canadian Shield conifer swamps, maximum water table elevation and dominance of saturation overland flow resulted in water largely bypassing deeper anoxic peat (Devito and Hill, 1997). Consequently, sulfate concentration of 8–10 $mgSO_4$ L^{-1} in outlet streams was similar to upland inputs. Significant net export of sulfate during snowmelt in another Canadian Shield Basin has been linked to flushing from organic and upper mineral soil layers in a small valley wetland (Steele and Buttle, 1994).

Large episodic releases of previously stored sulfate have been observed in some riparian wetlands during the first autumn storms following water table drawdown and oxidation of reduced sulfur in dry summers (Bayley *et al.*, 1986; Lazerte, 1993; O'Brien *et al.*, 1994). A comparative study of two valley bottom conifer swamps on the southern Canadian Shield provides more detailed data on the connection between hydrological and biogeochemical processes that produce cyclic sulfate release to streams from some riparian wetlands but not others (Devito *et al.*, 1996; Devito and Hill, 1997). During extended dry summers such as 1990, absence of hillslope ground water inputs to Plastic swamp, where watershed mean till depths are <1 m, resulted in a large water table drawdown of >60 cm. Aeration of surface peat was associated with the oxidation of reduced sulfur compounds and enhanced mineralization of sulfur which produced pulses of sulfate with maximum concentration up to 35 $mgSO_4$ L^{-1} during the initial autumn runoff

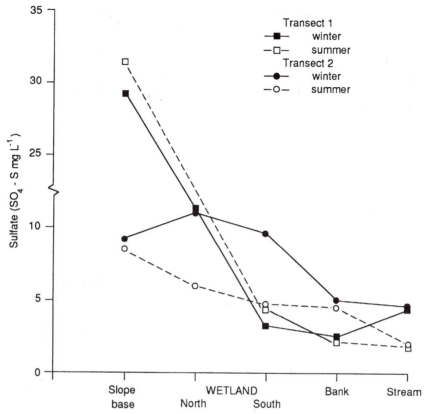

FIGURE 6 Down-gradient variability in sulfate concentration along groundwater-monitoring transects on winter (22 January 1991) and summer (13 June 1991) collection dates in Reedy Creek watershed, Virginia. Modified from O'Brien *et al.* (1994), *Hydrological Processes* **8**: 411–427. Copyright John Wiley & Sons Limited. Used with permission.

(Fig. 7a). During the wet summer of 1992, there was only a small water table decline and no sulfate pulse. In Harp swamp, where watershed till depths are >2–3 m, a small but continuous ground water input from the uplands maintained a high water table and efficient sulfate retention during both wet and dry summers (Fig. 7b). Higher sulfate concentration in the outlet stream from Plastic swamp during the autumn and winter of 1990–1991 relative to 1991–1992 indicate that sulfate transformations in wetlands during extended droughts can influence stream sulfate chemistry through the following winter (Fig. 7a).

Recent research indicates that organic-rich soils in riparian areas are important sources of methylmercury (MeHg) in regions of the United States, Canada, and Sweden where mercury contamination of surface waters is a major problem. Bishop *et al.* (1995) hypothesized that stream baseflow

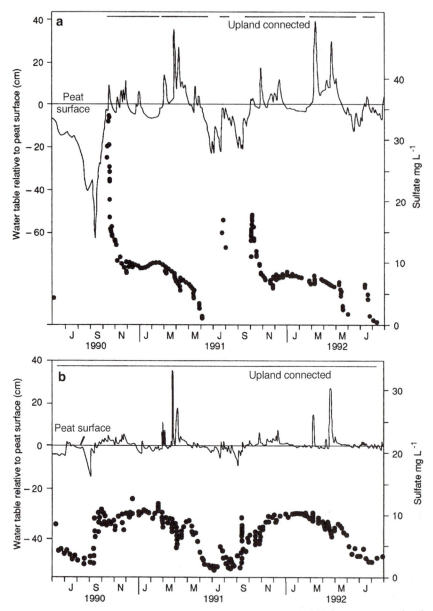

FIGURE 7   Sulfate concentration of the outlet stream and water table elevation in two head-water valley conifer swamps on the southern Canadian Shield, Ontario. (a) Plastic Swamp; (b) Harp Swamp. Water table elevation (centimeters above outlet weir, continuous line), sulfate concentration (solid circles). The horizontal bars at the top of the graphs show periods when the upland is hydrologically connected to the riparian swamps. Modified from Devito and Hill (1997), *Hydrological Processes* **11**: 485–500. Copyright John Wiley & Sons Limited. Used with permission.

methylmercury concentration was controlled by riparian zone biogeochemical processes in Svalberget watershed in northern Sweden. Concentration of methylmercury in podzols, which covered 70% of the forest watershed, was too low to sustain stream exports. Subsurface water from the upslope podzols flowed laterally through a 10- to 20-m zone of organic-rich soils and mosses on the stream bank that contained a very high methylmercury level. The role of the riparian zone as a methylmercury source has also been studied in a headwater stream which flows through a riparian peatland to a pond in the Experimental Lakes Area, northwest Ontario (Branfireun *et al.*, 1996). Under summer baseflows, methylmercury concentration at the stream origin adjacent to the hillslope–peatland contact was 0.03–0.09 ngMeHg L^{-1}, whereas at the pond outlet of the stream concentration increased by seven to eight times to 0.25–0.65 ngMeHg L^{-1} (Fig. 8).

Some riparian peatlands are a large source of methylmercury to streams

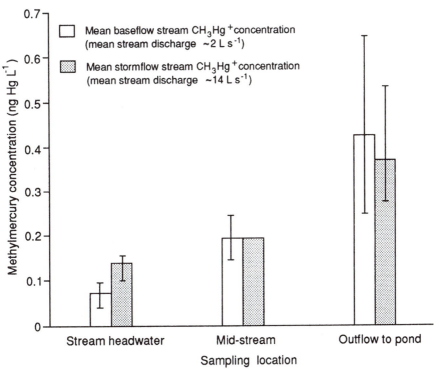

FIGURE 8   Mean methylmercury concentration from the seep origin to the pond outflow of a stream draining a headwater peatland in northwest Ontario. Error bars indicate the range of concentration. Only one set of samples was taken at the midstream location under stormflow conditions. Distance from the stream headwater to the outflow is approximately 100 m. Reproduced with permission from Branfireun *et al.* (1996), *Water Resources Research* **32:** 1785–1794, copyright by the American Geophysical Union.

during storms. Branfireun *et al.* (1996) noted that the relative increase in methylmercury concentration was maintained as a headwater stream flowed through a riparian wetland even though discharge increased up to ten times during a summer storm (Fig. 8). During the storm, the water table rose to the surface, and runoff occurred as shallow subsurface and saturation overland flow. This runoff transported near-surface peat porewater that was laden with methylmercury from areas of groundwater discharge in the peatland. During the June–August period, 58% of the methylmercury was transported from the peatland by stormflows, which occurred only 16% of the time.

Considerable element flushing may occur in riparian zones which function as variable source areas that only contribute runoff to streams during major storms or spring snowmelt. For example, in the Telford basin in southern Ontario, a small swamp isolated from regional ground water by a clay layer is only a major source of runoff to an ephemeral stream in mid-March to early May (Pierson and Taylor, 1994). After a large rainstorm and increased spring snowmelt, potassium concentration increased rapidly to 2.3–3.1 mgK L^{-1} in the ephemeral stream and remained high for 10 days until potassium-rich standing water was displaced from the swamp.

In some riparian zones, stream cation dynamics may be influenced by the mixing of event water and ground water, as well as by interactions with riparian substrates during storms. Increased stream flow during summer and autumn storms in the Glen Major swamp was associated with declines in calcium, magnesium, and sodium concentrations, whereas potassium concentration often remained constant or increased (Hill, 1993b). Comparisons of chemical and isotope hydrograph separations indicated that stream calcium, magnesium, and sodium variations resulted from the proportion of rain and ground water transported by riparian zone saturation overland flow in storms, with a small event water contribution of <25% of discharge. These cations were desorbed from organic soils due to H^+ addition when storm runoff contained >40% rain. For storms where rainfall potassium concentrations were <1 mgK L^{-1}, potassium was released as saturated overland flow interacted with the swamp surface soils.

IV. RIPARIAN ZONE INFLUENCE ON STREAM CHEMISTRY IN RELATION TO WATERSHED HYDROGEOLOGY: A CONCEPTUAL FRAMEWORK

Analysis of the literature suggests that riparian zones exhibit a considerable range of effects on stream chemistry. It is therefore important to develop a conceptual framework to explain why riparian zones differ in their influence on stream chemistry. This understanding is necessary if we are to develop generalizations that can be applied from one riparian zone to others. A conceptual model based on watershed hydrogeology has been developed to describe the range of sulfate dynamics in Canadian Shield valley bottom swamps (Devito and Hill, 1997; Devito *et al.*, 1999). This model can be expanded to provide an initial framework for considering riparian zone controls on headwater stream chemistry in humid landscapes. Riparian zones are located along a hydrogeologic gradient that extends from landscapes with small to large upland water storage. The model suggests how interactions between both catchment and internal riparian zone hydrology and biogeochemistry regulate riparian zone effects on stream chemistry during base and stormflow conditions (Table I).

Riparian zones at one end of the hydrogeologic gradient are located in landscapes where upland water storage is small because of climate or thin unconsolidated soils overlying impermeable materials at a shallow depth. Headwater wetlands linked to perched water tables function as variable source areas which only discharge water to ephemeral streams for periods of days to a few months in response to very large rainfalls or spring snowmelt in northern regions (Fig. 9A). Perched aquifer riparian zones do not influence perennial stream baseflow chemistry but can produce large flushes of elements to ephemeral streams during stormflows. This riparian zone function is exemplified by elevated potassium concentration in snowmelt runoff from the Telford swamp in Ontario (Pierson and Taylor, 1994).

In landscapes with more extensive but thin aquifers, ground water inputs to the riparian zone are small and seasonally variable (Table I). Thin aquifer riparian zones are hydrologically disconnected from uplands during the summer in most years and are characterized by large seasonal water table fluctuations (Fig. 9B). Because of frequent water table fluctuations and small ground water inputs, these riparian zones usually lack well-developed wetland characteristics and thick organic soils. This category of riparian zone, which has been studied extensively in agricultural landscapes, influences stream baseflow chemistry by rapid depletion of nitrate as ground water flows in a shallow subsurface path above impermeable layers to the stream. Riparian redox potentials favor aerobic respiration and denitrification but are often not low enough for significant sulfate reduction.

The effect of this riparian zone type on stream chemistry during storms has received little attention. Storm runoff may occur mainly as shallow subsurface flow if the water table rises into more conductive surface soils, whereas saturation overland flow is limited by the absence of extensive surface saturation. These riparian areas are a potential element source to the stream if runoff flushes solutes that accumulate in surface soils by evapotranspiration and mineralization processes during water table drawdowns.

In some landscapes with a thin upland aquifer, higher rainfall and lower evapotranspiration rate in summer maintain the water table near the riparian zone surface in most years despite a seasonal absence of a ground water connection to upland aquifers (Fig. 9C). However, during extended dry summers, these riparian zones have large water table drawdowns. It is

TABLE 1 Conceptual Model Relating Catchment Hydrogeology to Variations in the Influence of Headwaters Riparian Zones on Stream Chemistry in Humid Temperate Landscapes

	Hydrogeologic gradient				
		Riparian zone			
Variable	Perched aquifer, intermittent input to riparian zone				Thick aquifer, large constant input to riparian zone
	Perched aquifer	Thin aquifer	Thin aquifer-rain dependent	Intermediate aquifer	Thick aquifer
Water table variation	Large each year	Large each year	Large only in dry summers	Moderate to small	Small
Hydrologic link to stream	Discontinuous; no stream flow except in large storms	Discontinuous; no stream flow days to weeks each year	Discontinuous; no stream flow during extended dry periods	Continuous; stream baseflow varies	Continuous; large constant stream baseflow
Flowpaths during baseflows	None	Shallow subsurface	Shallow subsurface	Shallow and deeper subsurface	Various paths (a) surface springfed and (b) deep bypass flow
Flowpaths during storms	SOF	SSSF; limited SOF	SOF and SSSF	SOF and SSSF	Flow SOF if (a) above; SSSF is (b) above

Example of effects on stream baseflow chemistry	None	Retention of NO$_3$	Retention of NO$_3$ and SO$_4$; source of DOC	Retention of NO$_3$ and SO$_4$; source of DOC	Element retention often imited; can be influenced by mixing of local and regional ground water
Example of effects on stream storm chemistry	Large element source	Element source?	Less retention of NO$_3$ and SO$_4$, source of DOC, organic N and P; Large episodic source (SO$_4$) after dry summers	Less retention of NO$_3$ and SO$_4$, source of DOC, organic N and P	Variable effects (see text)
Example and reference	Telford Swamp Ontario[a]	Rhode River, Maryland[b]	Plastic Swamp, Ontario[c], Reedy Creek, Virginia[d]	Harp Swamp, Ontario[c]	Overland flow[e] deep bypass flow[f] local and regional ground water effects, Glen Major swamp, Ontario[g] Salto River, Costa Rica[h]

SSSF, subsurface storm flow. SOF, saturation-excess overland flow.

[a]Pierson and Taylor (1994).
[b]Peterjohn and Correll (1984).
[c]Devito and Hill (1997).
[d]O'Brien et al. (1994).
[e]Brusch and Nilsson (1993).
[f]Bohlke and Denver (1995).
[g]Hill (1990).
[h]Pringle and Triska (1991).

103

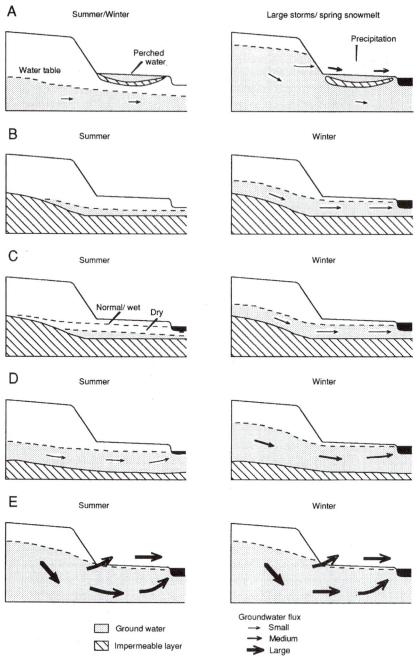

FIGURE 9 Schematic groundwater flow systems that might be expected for a generalized land-scape consisting of an upland adjacent to the riparian zone of headwater streams along a hydrogeologic gradient of increasing upland water storage and magnitude of groundwater connection to the riparian zone. (A) Perched aquifer riparian zone. (B) Thin aquifer riparian zone. (C) Thin aquifer-rain dependent riparian zone. (D) Intermediate aquifer riparian zone. (E) Thick aquifer riparian zone.

important to distinguish this category of rainfall-dependent riparian zone from the thin aquifer riparian zone type where the seasonal absence of a ground water connection in combination with low rainfall or high evapotranspiration in summer produces large water table fluctuations each year. Riparian zones in the former category are often wetlands with thick organic soils and redox potentials that favor the storage of reduced sulfur. These riparian zones remove nitrate and sulfate from subsurface ground water during baseflows but retain elements less effectively during seasons when saturation overland flow is dominant. During storm periods, these riparian zones may be a source of DOC, organic nitrogen, and phosphorus to streams via mixing and entrainment by saturation overland flow. An important characteristic of these riparian zones is the episodic release of sulfate and possibly other elements to streams during storms following periods when the water table declines in occasional years with extended dry periods.

Farther along the hydrogeologic gradient, riparian zones connected to an upland aquifer with intermediate storage have seasonal variations in ground water inputs but never become disconnected from the aquifer (Fig. 9D). Although intermediate aquifer riparian zones have small to moderate water table fluctuations, the presence of small continuous ground water inputs prevents large water table drawdowns even in droughts. The threshold between riparian zones that are seasonally and permanently connected to uplands may be the result of small differences in average depth of storage in some landscapes (Devito *et al.*, 1996). Apart from an absence of episodic element export after droughts, intermediate aquifer riparian zones often function as element sinks and sources in a manner similar to riparian zones which experience occasional large water table fluctuations.

At the large upland water storage end of the hydrogeologic gradient, riparian zones connected to thick upland aquifers have large seasonally constant ground water inputs, which maintain stable riparian water tables and relatively constant spatial and temporal redox patterns (Table I). Several ground water flowpaths besides shallow lateral subsurface flow may dominate in these thick aquifer riparian zones (Fig. 9E). Extensive areas of permanent surface saturation and spring-fed zones of overland flow are important in riparian zones where ground water discharges upward to the surface. In riparian zones with deep permeable sediments, ground water may bypass surface soils and vegetation at depth and discharge upward to the channel. Riparian zones linked to thick local aquifers, which are dominated by deep bypass flow and overland flow, have been studied by Bohlke and Denver (1995) and Brusch and Nilsson (1993), respectively. Riparian zones with these flowpaths often have limited effects on the chemistry of water entering streams during baseflows. Riparian zones with large constant ground water inputs that maintain areas of surface saturation are important sites of saturation overland flow during storms. Stream chemistry during storms may be influenced mainly by variations in the proportion of rain and ground water transported by saturation overland flow.

In landscapes with extensive thick ground water aquifers, riparian zones in the downstream areas of second- and third-order streams may receive inputs from adjacent hill slopes and from larger-scale flow systems that recharge in more distant parts of the catchment. The chemistry of these streams is often controlled by the mixing of local and regional ground water that differ in element composition rather than by riparian zone processes. An example is the Salto River in Costa Rica studied by Pringle and Triska (1991).

V. FUTURE RESEARCH DIRECTIONS

There is considerable evidence that riparian zones play an important role in regulating the flux of elements to streams in many headwater landscapes. Nevertheless, many aspects of riparian zone biogeochemistry are inadequately understood. Current knowledge of riparian zone effects on stream chemistry is limited by a failure to understand the hydrogeologic setting and integrate hydrology with the chemistry and biology of these complex environments. Although several recent studies have focused on hydrologic flow-path-chemistry linkages, most riparian studies still rely on transects of single wells with wide slot zones which provide limited data on ground water flowpaths and three-dimensional patterns of redox status and solute chemistry. There is a clear need for a more appropriate approach involving the use of parallel transects of piezometer nests consisting of pipes with slotted tips installed to various depths and a water table well at each location across the riparian zone.

Studies showing high ammonium and DOC concentrations in riparian zone and stream bank ground water and low concentration of these solutes in stream water suggest that hyporheic zones may strongly regulate the movement of some elements from the riparian zone to the stream. Even though many researchers have examined how stream-water chemistry is modified by interaction with the hyporheic zone, few have studied the role of the hyporheic zone in regulating chemical fluxes from riparian zone ground water to streams. Uncertainty concerning hydrologic linkages between the riparian zone and the stream hyporheic zone underscores the fact that stream chemistry is influenced by the interaction of hydrologic and biogeochemical processes at various locations within the catchment. Although the general concentration levels of nitrogen and phosphorus in Walker Branch, Tennessee, are controlled by catchment soil and bedrock processes, seasonal variations are regulated by stream processes (Mulholland, 1993). In contrast, the role of in-stream processes is considerably less important than the riparian zone in regulating baseflow nitrate flux in a small New

Zealand pasture catchment (Cooper, 1990). With the exception of these studies, comparative research on the role of upland, riparian zone, hyporheic, and in-stream environments in regulating stream chemistry is limited and should be conducted in a range of different watersheds.

Most studies of riparian zone element transport and transformations have analyzed streamside areas that can be considered wetlands. This bias toward wetland environments dominated by organic soils is particularly evident with respect to the role of riparian zones as DOC sources and sulfate sinks. There is a need for studies of the water chemistry of nonwetland riparian zones where higher redox conditions and the presence of mineral soils may result in the dominance of different water quality processes.

The biogeochemical response of riparian zones to increased element loading and climate change is also deserving of attention. The long-term effect of high-nitrogen loading on stream riparian zones is not well understood, and there is concern that nitrogen saturation of plant and microbial pools will occur leading to increased stream-water nitrate concentrations (Hanson *et al.*, 1994). Global climate change may result in warmer and drier summers in some regions which will influence riparian zone chemistry. Riparian zones in landscapes with limited upland water storage are likely to be particularly sensitive to climate change. Many valley bottom swamps on the southern Canadian Shield contain large stores of reduced sulfur and effectively retain sulfate even in years with dry summers because a small continuous ground water connection to uplands maintains surface saturation. The absence of this upland hydrologic link because of an increase in catchment evapotranspiration rates may result in a large water table drawdown during dry periods and the widespread episodic release of sulfate to streams when the water table rises again (Devito *et al.*, 1999).

Knowledge gained from detailed studies of specific riparian zones must be generalized to understand differences in riparian zone influence on stream chemistry within and among watersheds. This requires a detailed understanding of hydrology. We need to recognize that both external hydrological connections with the watershed and internal hydrologic pathways regulate riparian zone biogeochemistry. Thus the biogeochemical function of the riparian zone is influenced by the riparian zone position in the landscape and by watershed hydrology and will vary with differences in hydrogeologic setting and climate.

REFERENCES

Anderson, M. G., and T. P. Burt. 1990. *In* "Process Studies in Hillslope Hydrology" (M. G. Anderson T.P. Burt, eds.), pp. 365–400. Wiley, Chichester.

Bayley, S. E., R. S. Behr, and C. A. Kelly. 1986. Retention and release of S from a freshwater wetland. *Water, Air, and Soil Pollution* 31:101–114.

Bishop, K., Y. H. Lee, C. Pettersson, and B. Allard. 1995. Terrestrial sources of methylmercury in surface waters: The importance of the riparian zone on the Svartberget catchment. *Water, Air, and Soil Pollution* 80:435–444.

Bohlke, J. K., and J. M. Denver. 1995. Combined use of ground water dating, chemical and isotopic analysis to resolve the history and fate of nitrate contamination in two agricultural watersheds, Atlantic coastal plain, Maryland. *Water Resources Research* 31:2319–2339.

Branfireun, B. A., A. Heyes, and N. T. Roulet. 1996. The hydrology and methylmercury dynamics of a Precambrian Shield headwater peatland. *Water Resources Research* 32:1785–1794.

Brinson, M. M. 1993. Changes in the functioning of wetlands along environmental gradients. *Wetlands* 13:65–74.

Brusch, W., and B. Nilsson. 1993. Nitrate transformation and water movement in a wetland area. *Hydrobiologia* 251:103–111.

Cirmo, C. P., and J. J. McDonnell. 1997. Linking the hydrologic and biogeochemical controls of nitrogen transport in near-stream zones of temperate-forested catchments: A review. *Journal of Hydrology* 199:88–120.

Cooper, A. B. 1990. Nitrate depletion in the riparian zone and stream channel of a small headwater catchment. *Hydrobiologia* 202:13–26.

Correll, D. L., and D. E. Weller. 1989. Factors limiting processes in freshwater wetlands: An agricultural primary stream riparian forest. *In* "Freshwater Wetlands and Wildlife" (R. R. Sharitz and J. W. Gibbons, eds.), DOE Symposium Series, pp. 9–23. U.S. Department of Energy, Washington, DC.

Dalva, M., and T. R. Moore. 1991. Sources and sinks of dissolved organic carbon in a forested swamp catchment. *Biogeochemistry* 15:1–19.

Devito, K. J., and P. J. Dillon. 1993. The influence of hydrologic condition and peat oxia on the phosphorus and nitrogen dynamics of a conifer swamp. *Water Resources Research* 29:2675–2685.

Devito, K. J., and A. R. Hill. 1997. Sulphate dynamics in relation to ground water-surface water interactions in headwater wetlands of the southern Canadian Shield. *Hydrological Processes* 11:485–500.

Devito, K.J., P. J. Dillon, and B. D. Lazerte. 1989. Phosphorus and nitrogen retention in five Precambrian Shield wetlands. *Biogeochemistry* 8:185–204.

Devito, K. J., A. R. Hill, and N. T. Roulet. 1996. Ground water-surface water interactions in headwater forested wetlands of the Canadian Shield. *Journal of Hydrology* 181:127–147.

Devito, K. J., A. R. Hill, and P. J. Dillon. 1999. Episodic sulphate export from wetlands in acidified headwater catchments: Prediction at the landscape scale. *Biogeochemistry* 44:187–203.

Dosskey, M. G., and P. M. Bertsch. 1994. Forest sources and pathways of organic matter transport to a blackwater stream: A hydrological approach. *Biogeochemistry* 24:1–19.

Eshleman, K. N., J. S. Pollard, and A. K. O'Brien. 1994. Interactions between ground water and surface water in a Virginia coastal plain watershed. 1. Hydrological flow paths. *Hydrological Processes* 8:389–410.

Fiebig, D. M., M. A. Lock, and C. Neal. 1990. Soil water in the riparian zone as a source of carbon for a headwater stream. *Journal of Hydrology* 116:217–237.

Gillham, R. W. 1984. The capillary fringe and it's effect on water table response. *Journal of Hydrology* 67:307–324.

Hanson, G. G., P. M. Groffman, and A. J. Gold. 1994. Symptoms of nitrogen saturation in a riparian wetland. *Ecological Applications* 4:750–756.

Haycock, N. E., and G. Pinay. 1993. Ground water nitrate dynamics in grass and poplar vegetated riparian buffer strips during the winter. *Journal of Environmental Quality* 22:273–278.

Hedin, L. O., J. C. von Fischer, N. E. Ostrom, B. P. Kennedy, M. G. Brown, and G. P. Robertson. 1998. Thermodynamic constraints on nitrogen transformations and other biogeochemical processes at soil-stream interfaces. *Ecology* 79:684–703.

Hill, A. R. 1990. Ground water flow paths in relation to nitrogen chemistry in the near-stream zone. *Hydrobiologia* **206**:39–52.

Hill, A. R. 1991. A ground water nitrogen budget for a headwater swamp in an area of permanent ground water discharge. *Biogeochemistry* **14**:209–224.

Hill, A. R. 1993a. Nitrogen dynamics of storm runoff in the riparian zone of a forested watershed. *Biogeochemistry* **20**:19–44.

Hill, A. R. 1993b. Base cation chemistry of storm runoff in a forested headwater wetland. *Water Resources Research* **29**:2663–2673.

Hill, A. R. 1996. Nitrate removal in stream riparian zones. *Journal of Environmental Quality* **25**:743–755.

Hill, A. R., and J. M. Waddington. 1993. Analysis of storm runoff sources using oxygen-18 in a headwater swamp. *Hydrological Processes* **7**:305–316.

Hill, A. R., and J. Warwick. 1987. Ammonium transformations in springwater within the riparian zone of a small woodland stream. *Canadian Journal of Fisheries and Aquatic Sciences* **44**:1948–1956.

Hinton, M. J., S. L. Schiff, and M. C. English. 1998. Sources and flowpaths of dissolved organic carbon during storms in two forested watersheds of the Precambrian Shield. *Biogeochemistry* **41**:175–197.

Hynes, H. B. N. 1983. Ground water and stream ecology. *Hydrobiologia* **100**:93–99.

Jacks, G., A. Joelsson, and S. Fleischer. 1994. Nitrogen retention in forested wetlands. *Ambio* **23**:358–362.

Lazerte, B. D. 1993. The impact of drought and acidification on the chemical exports from a minerotrophic conifer swamp. *Biogeochemistry* **18**:153–175.

Leonard, R.L., L. A. Kaplan, J. F. Elder, R. N. Coats, and C. R. Goldman. 1979. Nutrient transport in surface runoff from a subalpine watershed, Lake Tahoe basin, California. *Ecological Monographs* **49**:281–310.

Likens, G. E. 1984. Beyond the shoreline: a watershed ecosystem approach. *Verhandlungen Internationale Vereinigung für Theoretische und Angewandte Limnologie* **22**:1–22.

Lowrance, R. R., R. L. Todd, and L. E. Asmussen. 1984. Nutrient cycling in an agricultural watershed: 1. Phreatic movement. *Journal of Environmental Quality* **13**:22–27.

McClain, M. E., J. E. Richey, and J. P. Pimentel. 1994. Ground water nitrogen dynamics at the terrestrial-lotic interface of a small catchment in the Central Amazon Basin. *Biogeochemistry* **27**:113–127.

McDonnell, J. J. 1990. A rationale for old water discharge through macropores in a steep humid catchment. *Water Resources Research* **26**:2821–2832.

McDowell, W. H., W. H. Bowden, and C. E. Ashbury. 1992. Riparian nitrogen dynamics in two geomorphologically distinct tropical rain forest watersheds: Subsurface solute patterns. *Biogeochemistry* **18**:53–75.

Meyer, J. L. 1990. Production and utilization of dissolved organic carbon in riverine ecosystems. *In* "Organic Acids in Aquatic Ecosystems" (E. M. Perdue and E. T. Gjessing, eds.), pp. 281–299. Wiley, New York

Meyer, J. L., and C. M. Tate. 1983. The effects of watershed disturbance on dissolved organic carbon dynamics of a stream. *Ecology* **64**:33–44.

Mulholland, P. J. 1993. Regulation of nutrient concentrations in a temperate forest stream: roles of upland, riparian and instream processes. *Limnology and Oceanography* **37**:1512–1526.

Naiman, R. J., and H. Decamps. 1997. The ecology of interfaces. *Annual Review of Ecology and Systematics* **28**:621–658.

O'Brien, A. K., K. N. Eshleman, and J. S. Pollard. 1994. Interactions between ground water and surface water in a Virginia coastal plain watershed. 2. Acid-base chemistry. *Hydrological Processes* **8**:41–427.

Ohrui, K., and M. J. Mitchell. 1998. Spatial patterns of soil nitrate in Japanese forested watersheds: Importance of the near-stream zone as a source of nitrate in stream water. *Hydrological Processes* **12**:1433–1445.

Pearce, A.J., M. K. Stewart, and M. G. Sklash. 1986. Storm runoff generation in humid headwater catchments. 1. Where does the water come from? *Water Resources Research* **22**:1263–1272.

Peterjohn, W. T., and D. L. Correll. 1984. Nutrient dynamics in an agricultural watershed: Observations on the role of the riparian forest. *Ecology* **65**:1466–1475.

Peters, D. L., J. M. Buttle, C. H. Taylor, and B. D. Lazerte. 1995. Runoff production in a forested, shallow soil, Canadian Shield basin. *Water Resources Research* **31**:1291–1304.

Phillips, P. J., J. M. Denver, R. J. Shedlock, and P. A. Hamilton. 1993. Effect of forested wetlands on nitrate concentrations in ground water and surface water on the Delmarva peninsula. *Wetlands* **13**:75–83.

Pierson, D. C., and C. H. Taylor. 1994. The role of surface and subsurface runoff processes in controlling cation export from a wetland watershed. *Aquatic Sciences* **56**:80–96.

Pringle, C. M., and F. J. Triska. 1991. Effects of geothermal ground water on nutrient dynamics of a lowland Costa Rican stream. *Ecology* **72**:951–965.

Rodhe, A. 1989. On the generation of stream runoff in till soils. *Nordic Hydrology* **20**:1–8.

Roulet, N. T. 1990. Hydrology of a headwater basin wetland. 1. ground water discharge and wetland maintenance. *Hydrological Processes* **4**:387–400.

Sedell, J. R., and C. N. Dahm. 1990. Spatial and temporal scales of dissolved organic carbon in streams and rivers. *In* "Organic Acids in Aquatic Ecosystems" (E. M. Perdue and E. T. Gjessing, eds.), pp. 261–279. Wiley, New York.

Sklash, M. G. 1990. Environmental isotope studies of storm and snowmelt runoff generation. *In* "Process Studies in Hillslope Hydrology" (M. G. Anderson and T. P. Burt, eds.), pp. 401–435. Wiley, Chichester.

Steele, D. W., and J. M. Buttle. 1994. Sulphate dynamics in a northern wetland catchment during snowmelt. *Biogeochemistry* **27**:187–211.

Toth, J. 1963. A theoretical analysis of groundwater flow in small drainage basins. *Journal of Geophysical Research* **68**:4795–4812.

Triska, F. J., J. H. Duff, and R. J. Avanzino. 1993. The role of water exchange between a stream channel and its hyporheic zone in nitrogen cycling at the terrestrial-aquatic interface. *Hydrobiologia* **251**:167–184.

Valett, H. M., J. A. Morrice, C. N. Dahm, and M. E. Campana. 1996. Parent lithology, surface-ground water exchange, and nitrate retention in headwater streams. *Limnology and Oceanography* **41**:333–345.

Waddington, J. M., N. T. Roulet, and A. R. Hill. 1993. Runoff mechanisms in a forested ground water discharge swamp. *Journal of Hydrology* **147**:36–60.

4

Flood Frequency and Stream–Riparian Linkages in Arid Lands

Eugènia Martí,[1] Stuart G. Fisher,
John D. Schade, and Nancy B. Grimm

Department of Biology
Arizona State University
Tempe, Arizona

I. Introduction   111
II. Riparian Zones in Arid Catchments   112
III. Hydrological Linkages in Mesic and Arid Catchments   113
IV. Conceptual Model   115
V. Case Study: Sycamore Creek   117
   A. Site Description for Empirical Studies of Stream–Riparian
      Interactions   117
   B. Results of Empirical Studies   118
VI. Synthesis   123
   A. Hydrological Linkages   123
   B. Implications for the Riparian Zone   128
VII. Conclusions: Intermediate Disturbance and Nutrient
     Retention   131
   References   133

I. INTRODUCTION

Riparian zones are important landscape elements by virtue of their spatial location in catchments. Because riparian zones lie at the interface between terrestrial and aquatic ecosystems, they can strongly influence flows of energy and matter between them (Naiman *et al.*, 1988; Gregory *et al.*,

---

[1] Present address: Centre d'Estudis Avançats de Blanes (C.S.I.C.), 17300 Blanes, Giroma, Spain.

Streams and Ground Waters

1991; Decamps, 1996). Over the last decade, stream ecologists have recognized the tight linkage between riparian zones and surface streams and have incorporated the riparian zone as a key component of the stream–riparian ecosystem (Junk *et al.*, 1989; Pinay *et al.*, 1990; Bencala *et al.*, 1993; Stanford and Ward, 1993; Fisher *et al.*, 1998a). Alan Hill (Chapter 3 in this book) correctly stresses the importance of the riparian zone to chemistry of the surface stream; however, most of the work reviewed in that chapter was done in streams in which the main direction of water flow is from upland, through the riparian zone, and into the stream channel. In arid and semiarid regions, in contrast, the predominant direction of flow is from upland, directly into the surface stream via overland flow, and then into the riparian zone.

This hydrologic difference, described in more detail later, provides an opportunity to consider in general terms how upland, riparian, and surface stream linkages can influence riparian processes, surface stream chemistry, and material retention by the stream–riparian ecosystem. In this chapter, we focus on stream–riparian exchanges during flood and low flow periods and develop a simple model proposing that material exchange and retention are greatest when regimes of flooding are of intermediate frequency. This model mainly applies to streams of arid and semiarid regions and is based on empirical studies of Sycamore Creek, a Sonoran Desert stream in central Arizona. The model is similar to the Intermediate Disturbance Hypothesis (Connell, 1978) used widely to explain the relationship between disturbance and species diversity; however, our model is different in that it deals with the effect of several disturbance attributes (frequency, intensity, evenness) on *nutrient processing*, an ecosystem property, rather than on species diversity of communities.

II. RIPARIAN ZONES IN ARID CATCHMENTS

Vegetation in arid catchments is water-limited, and large trees are restricted to areas of high water availability such as streams and their adjacent floodplains. In these catchments, the riparian zone can be easily distinguished as a green swath within the gray-brown landscape, as well as by differences in physiognomy of the vegetation (i.e., scrub or grassland in upland regions compared to forest in the riparian zone). Riparian zones are clearly important components of arid landscapes, contributing significantly to the overall production of the watershed, provided that water and nutrients are ample. They often are not. Several studies have shown that distribution, species composition, and plant growth of the riparian vegetation in arid land streams is strongly influenced by depth to the water table in the riparian zone (Richter, 1993; Stromberg *et al.*, 1993; 1996; Busch and Smith, 1995). Similar results have been reported in mesic regions, such as the upper Rhone basin in France (Patou and Décamps, 1985). In all these studies, fluctuations in ground water beneath the riparian zone are correlated with stream channel flow, indicating that these zones are hydrologically connected. Therefore, maintenance of riparian vegetation depends upon floods, even when flow does not inundate floodplains. Even though tolerance of flooding seems to be an important determinant of structure and composition of riparian vegetation everywhere, in arid and semiarid regions some minimal flood magnitude and duration is necessary if riparian trees are to persist.

In contrast to the extensive literature on the influence of water on riparian vegetation in arid land streams, much less is known about nutrient availability and its effect on vegetation. Although it is well known that absence of water limits plant growth, nutrient limitation owing to decreasing nutrient influx rate (discharge times concentration) to the riparian zone may restrict growth well before the riparian water table drops below the rooting zone. This contention is plausible but speculative in the absence of data. This information is important for appropriate management of arid catchments where water scarcity has such a major influence on the structure and function of riparian zones. This chapter represents an initial step in understanding the relationships between nutrient dynamics and hydrological variables in arid land riparian zones.

III. HYDROLOGICAL LINKAGES IN MESIC AND ARID CATCHMENTS

In all streams during storms, water moves across the upland landscape, into the stream channel, and then into and out of riparian and hyporheic subsystems (Fig. 1). As water moves along complex flowpaths, water chemistry changes as a function of subsystem-specific biological and chemical processes and the residence time of water in a given subsystem. Surface–subsurface hydrological linkages among different subsystems, either in the watershed or in the stream–riparian ecosystem, can influence structure and function of both a subsystem (e.g., distribution of algae in surface stream) and the entire watershed ecosystem (e.g., nutrient retention; Fisher *et al.*, 1998b). Effects are often most pronounced at patch edges; for example, high nitrate concentration where water emerges from sandbars stimulates algal growth along the stream edge (Fisher *et al.*, 1998a). Edge effects are dependent on direction of flow; for example, hyporheic sediments below downwelling zones are metabolically more active than those below upwelling zones (Jones *et al.*, 1995a). After disturbance, epibenthic algal growth above upwelling zones is faster than above downwelling zones (Stanley and Valett, 1992; Valett *et al.*, 1994).

These generalizations probably hold regardless of the climatic setting of the stream under consideration; however, streams in mesic and arid regions may differ greatly in the routing of water through the watershed (i.e., the structure of flowpaths). In temperate-mesic regions, the main direction of water flow is from upland, vertically through the soil profile, horizontally (subsurface) across the riparian zone, and then into the stream channel (Fig.

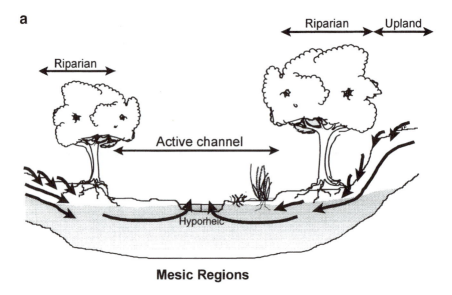

Mesic Regions

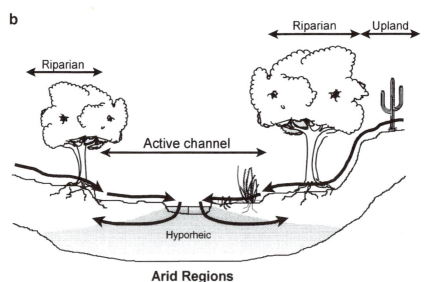

Arid Regions

FIGURE 1   Contrasting routes of hydrologic linkage between upland, stream (active channel), and riparian zone. Curved arrows show routing of water (i.e., flowpaths). (a) An upland–riparian–stream route characterizes stream–riparian linkage in mesic zones, where storm water infiltrates soils before entering the riparian zone and then the stream. (b) In arid regions, water moves across the surface (overland flowpaths) from the uplands into the surface stream, where it then enters the riparian zone via subsurface flowpaths.

1a). In these regions, upland soils are highly permeable, and overland flow is uncommon, occurring only under heavy precipitation when amount of rain exceeds soil infiltration rate. Thus, most of the water reaches the surface stream via subsurface flowpaths (Hill, 1990; Fetter, 1994).

In contrast, in arid-climate catchments, upland vegetation is sparse, distinct riparian vegetation is restricted to higher-order channels with ample water, and soils are less permeable. Rainwater moves rapidly by sheetflow into rills and low-order tributaries without first moving through the soil profile and without traversing the below-ground component of the riparian zone. In these catchments, overland flow is the main contributor to the flood peak (Fetter, 1994). This is accentuated when storms are intense. In terms of dissolved inorganic nitrogen, desert floodwater is chemically very similar to rainwater (Fisher and Grimm, 1985). Once water is in channels, it moves freely from channel to riparian zone and back, mainly by subsurface routes (Fig. 1b). Overbank flooding is not required for hydrologic recharge of riparian subsurface water.

These arid-mesic differences lead us to hypothesize that the structure and function of the riparian zone in arid regions is more dependent on processes in the active channel than in the adjacent terrestrial ecosystem. The situation in mesic streams is likely to be the reverse, whereas highly altered streams (e.g., in urban areas) may be somewhere in between. The importance of this lateral linkage has also been recognized in large river ecosystems where there is two-way exchange between the main channel and the adjacent floodplain ("flood-pulse concept"; Junk *et al.*, 1989). Connections in arid land streams involve linkages between subsurface and surface components of the stream–riparian ecosystem and their reciprocal effect—the theme of this book.

IV. CONCEPTUAL MODEL

So far we have described response to individual events; however, at the scale of a year or more, several floods occur. Furthermore, flood frequency and distribution vary among years. Grimm (1993) reported that the number of floods ranged from 2 to 18 per year over a 30-year period in Sycamore Creek, Arizona. The capacity for riparian sediments to absorb floodwater from the stream channel is partially a function of the position of the riparian water table relative to the surface stream, which in turn is related to elapsed time since the last flood. If the interval between floods is long, the riparian subsystem has a substantial capacity to absorb floodwater because the riparian water table will be low. The second of two closely spaced floods may exit the catchment without entering the riparian zone at all. Given this pattern of hydrologic connectance between surface stream and riparian zones, we propose that in arid land streams the effect of this reciprocal interaction, in terms of water and nutrient delivery to the riparian zone, is greatest when floods occur at an intermediate frequency.

This conceptual model has two main components: (1) hydrologic connection and flowpath-related processing and (2) the magnitude and frequency of connection as a function of hydrologic regime. Among the different attributes defining the hydrologic regime (see Poff et al., 1997), we argue that flood frequency is a key factor determining the magnitude of the linkage between stream and riparian zone, especially in arid land stream ecosystems. When flood frequency is very low, such as during an extended drought, delivery of water and nutrients to riparian biota approaches zero, and riparian and surface stream subsystems become disjunct. At the other extreme, frequent flooding, the riparian water table is fully charged, and the hydraulic head differential between the active channel and the riparian zone is small, thereby limiting exchange of water and nutrients. Under intermediate flood frequency, the riparian water table can fluctuate greatly. Floods occurring when the riparian water table is depressed recharge riparian "ground water" rapidly via subsurface flow across steep hydraulic gradients. Under these conditions, the total supply of materials to riparian biota is highest (Fig. 2). This

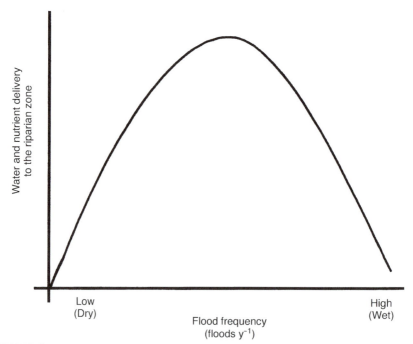

FIGURE 2 Hypothesized rate or amount of water and nutrient delivery to the riparian zone as a function of one attribute of the hydrologic regime—flood frequency.

is due to (a) enhanced movement of nutrient-rich floodwaters across steep hydraulic gradients and (b) short residence time in metabolically active hyporheic zones that are interposed between surface stream and riparian subsystems and that can act as nutrient filters of floodwater.

The model predicts that at intermediate flood frequencies water delivery to the riparian zone from the stream will be greater, annual nutrient flux through the riparian zone will be higher, annual riparian primary production and rates of microbial processes will be higher, potential for nutrient retention in the riparian zone will be greater, and potential for a riparian effect on the chemistry of water returning to the stream will be enhanced. This model applies to streams in arid and semiarid regions where the lateral subsurface hydrological connection from the active channel to the riparian zone is most likely to occur. In the sections that follow, we empirically evaluate four specific predictions of this conceptual model using data from an 18-month study of riparian dynamics in Sycamore Creek, Arizona. If our model holds, the following will be true:

1. The water table of the riparian zone will rise quickly during floods.
2. The rise will be greater after a long interflood period than after a short one. The magnitude of increase will be dependent on riparian groundwater levels and will be largely unrelated to flood magnitude.
3. High nutrient concentrations of floodwaters will be reflected in subsurface water of the riparian zone.
4. The influence of an individual flood on nutrient concentration in riparian subsurface water will decrease as flood frequency increases.

V. CASE STUDY: SYCAMORE CREEK

A. Site Description for Empirical Studies of Stream–Riparian Interactions

Sycamore Creek is a Sonoran Desert stream with headwaters in the Mazatzal mountains northeast of Phoenix, Arizona. Elevation ranges from 2164 m in the headwaters to 427 m at the confluence of Sycamore Creek with the Verde River. Catchment area is 505 km^2, and stream length is approximately 65 km. Annual precipitation is about 51 and 30 cm yr^{-1} at higher and lower elevations, respectively (Thomsen and Schumann, 1968). Precipitation is greatest during winter and summer; however, rain is irregularly distributed among years. Heavy storms may cause floods; flash floods during summer recede much faster than those in winter, and discharge can return to baseflow in <1 day. The intensity and frequency of floods greatly influence the morphology of the stream channel, temporal variation in nutrient concentration, and stream community structure and function (Fisher *et al.*, 1982). Evapotranspiration is high (>300 cm yr^{-1}), especially in summer. High evapotranspiration and low precipitation (or intense and short rain storms) during summer cause intermittency in surface stream flow along some reaches of this stream (Stanley *et al.*, 1997).

A study site consisting of a 400-m run located at an elevation of approximately 700 m in the middle reaches of Sycamore Creek was established for examining stream–riparian interactions. Stream substrata in the study reach consist primarily of sand and fine gravel, and depth of alluvial sediments averages >1.5 m (Holmes *et al.*, 1994). Surface flow occurs during most of the year; however, at baseflow the surface stream typically occupies <25% of the active channel with the remainder being dominated by extensive gravel bars with subsurface flow. The riparian zone is dominated by large trees such as Gooding's willow (*Salix goodingii*), sycamore (*Platanus wrightii*), cottonwood (*Populus fremontii*), mesquite (*Prosopis glandulosa*), ash (*Fraxinus pennsylvanica velutina*), and woody shrubs (<2 m high) such as seep willow (*Baccharis salicifolia*) and burro bush (*Hymenoclea monogyra*). These plants are distributed in a narrow strip floodplain bounded on one side by upland Sonoran Desert scrub and restricted on the other side by high flow stream margins that experience severe flash floods, preventing the establishment of vegetation (particularly large trees) in the active channel.

We installed a network of piezometers in the riparian zone of this reach. Piezometers were located approximately every 20 m along the reach and, on average, 4 m lateral to the active stream channel. The piezometers consisted of PVC pipes (16 mm diameter and 1.5–2 m long) slotted over the lower 5 cm. To examine the temporal pattern of change in water table in the riparian zone, we measured the height of the water in each piezometer (hereafter referred to as wells) at least biweekly. The frequency of measurement was increased during periods of marked hydrologic change (e.g., after floods).

We sampled water for chemical analysis from a subset of wells once a month and within the first week after floods. Additionally, we placed 6 wells in riparian bank sediments immediately adjacent to the active channel and 15 in gravel bars within the active channel (in 3 transects across the channel). These wells, and surface water, were sampled at the same time as riparian wells for comparison of chemical characteristics between subsystems. Water samples were analyzed for nitrate (NO_3^-), ammonium (NH_4^+), dissolved organic nitrogen (DON), soluble reactive phosphorus (SRP), dissolved organic carbon (DOC), conductivity, and oxygen using methods described elsewhere (Holmes *et al.*, 1994).

B. Results of Empirical Studies

1. Flood Effects

During an 18-month study period, four floods occurred (22 August 1996, 5 September 1996, 14 January 1997, and 28 February 1997) which provided an excellent opportunity to examine the hydrological linkages between riparian zone and the active channel under flood conditions. Only during the February 1997 flood did water overflow the active channel and inundate part of the riparian zone (approximately 2–3 m from the banks) via surface flow. A comparison of stream surface water chemistry during these floods with average chemistry during baseflow (i.e., >30 days after a flood) strongly suggests that floods represent an input of water and solutes (mainly nutrients) to the stream ecosystem (Table I).

2. Patterns of Water Table Fluctuation in the Riparian Zone

a. Temporal Variation.   The water table in the riparian zone fluctuated at least 80 cm during the study period (Fig. 3a). The fluctuation was likely greater than 80 cm, because the water table probably continued to drop after the wells dried (zero values in Fig. 3a). Temporal variation in water table height showed seasonality, being lowest during summer and highest during fall and winter. The sudden rises in water table superimposed on this seasonal pattern coincided with floods (Figs. 3a and 3b) and occurred very quickly after the flood peak. For instance, after the September 5th event, 75% of the increase measured one day after the flood occurred within just two hours of the passage of the flood wave. These data support the prediction that the water table will rise fast during floods (prediction 1). The magnitude of increase in the riparian water table was not proportional to flood magnitude, but instead was related to the degree of saturation of riparian soils and sediments (i.e., position of riparian water table) prior to flooding, supporting prediction 2. Overall, the temporal pattern suggests that, in Sycamore Creek, hydrologic linkage between active channel and riparian zone is episodic and mainly controlled by floods.

TABLE I   Chemical and Physical Characteristics of Surface Water during Baseflow (Average for Data Collected from 1978 to 1996) and for the Four Floods Occurring during the Study Period

		Floods			
	Baseflow	*22 Aug 1996*	*5 Sep 1996*	*14 Jan 1997*	*28 Feb 1997*
Days since flood	>30	368	14	131	45
Discharge ($m^3 s^{-1}$)	0.098	0.44	0.33	14.25	20.0
Conductivity ($\mu S\ cm^{-1}$)	431	215	117	160	128
DOC ($mgC\ L^{-1}$)	2–5	17.1	15.8	17.3	16.6
DIN ($\mu gN\ L^{-1}$)	60	1850	1920	2800	2250
SRP ($\mu gP\ L^{-1}$)	40	600	390	180	160

Note. DOC, dissolved organic carbon; DIN, dissolved inorganic nitrogen; and SRP, soluble reactive phosphorus.

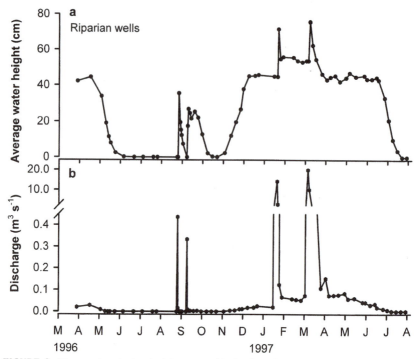

FIGURE 3 Temporal variation in (a) water table level in the riparian zone and (b) surface stream discharge of Sycamore Creek, Arizona, from March 1996 to August 1997. Temporal variation in water table was examined using the average height of water measured in all wells on each date. Zero values denote that wells were dry. Note gradual decrease during dry periods (April–June 1996, June–Aug 1997), gradual increase during cooler months (October–December 1996), and abrupt changes during floods (corresponding to discharge spikes).

b. Spatial Variation: Longitudinal and Lateral Flowpaths. After each flood, all wells along the reach experienced a simultaneous rise in water table, although the magnitude of increase varied slightly (Fig. 4). The rise in water table coincided with a 30–60% decrease (depending on the date) in surface discharge as flood water moved through the 400-m reach. Based on data from riparian area and soil porosity, we estimated that the decline in surface discharge was large enough to account for the rise in water table in the riparian zone during these events. These results suggest that as floodwater moves downstream, some fraction moves laterally into subsurface zones (hyporheic, gravel bars, and riparian zones), causing an increase in the height of the water table in these zones and a concomitant decrease in surface flow.

In contrast, during the October–December 1996 period when no floods occurred, the rise in water table was not simultaneous for all wells but occurred gradually along the reach over time. Water table level in upstream wells rose 25 days before increases were observed in downstream wells (Fig.

5). We also observed a gradual increase in surface water flow in the study reach; from entirely dry in October to continuous flow by December. The gradual increase in water availability in the reach was probably caused by a decline in evapotranspiration, since no floods and little rain occurred during this period. We suggest that the decline in evapotranspiration increased the input of subsurface water from upstream, and the increase in water table was due mainly to longitudinal flow in all lateral subsystems of the reach. Based on the time-lag between the increase in water table in wells located at the head and at the end of the reach, we estimated the average velocity of sub-surface water in the riparian zone to be 0.7 m h^{-1}. This velocity is slightly slower than that of the subsurface water in the active channel (1–2 m h^{-1}; Holmes *et al.,* 1994) and two orders of magnitude slower than the average velocity in the surface stream at baseflow. These two contrasting patterns of change in water table together suggest that the direction of movement of wa-ter (lateral or longitudinal) depends on hydrologic conditions in the whole riparian–stream ecosystem (i.e., flood and interflood periods).

3. Patterns of Solute Variation in the Stream–Riparian Corridor

a. Temporal Variation. Solute concentrations in the riparian zone do not show any clear seasonal pattern for any of the solutes measured; instead, ma-

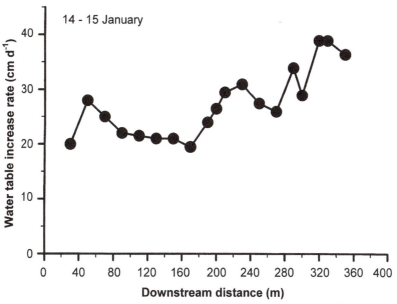

FIGURE 4 Rate of increase in water table level of riparian wells located along the 400-m study reach, during a one-day period. This water level increase occurred in response to a single flood on 14 January 1997.

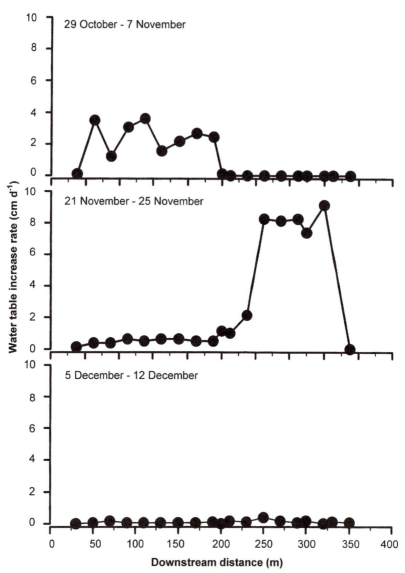

FIGURE 5 Rate of increase in water table level of riparian wells located along the 400-m study reach, during three time periods. In contrast to data in Fig. 4, increases in water table were not simultaneous in all wells; rather, upstream wells increased first, and downstream wells increased 25 days later. Changes in water table level during this period were not associated with any floods. For simplification, the graphs only show data from the right bank of the reach, but similar results were found on the left bank.

jor changes in water chemistry were associated with floods (Fig. 6). Concentrations of solutes were significantly higher after floods than during baseflow, except for SRP and DON, supporting prediction 3 (Table II). Dissolved oxygen was slightly higher after floods, but concentrations never exceeded 4 $mgO_2 L^{-1}$. After floods, dissolved inorganic nitrogen (DIN, $NO_3^- + NH_4^+$) and DOC concentrations decreased more rapidly than the concentrations of major ions (expressed as conductivity; Fig. 6). We also observed qualitative changes associated with floods, particularly for DIN. During baseflow, ammonium was the dominant form, and concentrations of nitrate were mostly below detection levels. After floods, concentrations of both nitrate and ammonium significantly increased; however, there was a shift from dominance of ammonium (reduced DIN form) to nitrate (oxidized DIN form; Fig. 7). The largest increases in nutrient concentrations were observed after the 22 August flood (>300 days after a flood), and minor changes were observed on the 5 September flood only 14 days after a flood, supporting prediction 4.

b. Spatial Variation, Comparison among Subsystems.   Water chemistry in the riparian zone differed significantly from both subsurface and surface water in the active channel (Table III). Riparian zone chemistry, however, did not differ from water sampled from wells located in the stream bank. Sharp differences in subsurface water chemistry occurred across the interface between the active channel and the riparian zone at all times. During baseflow, water in the riparian zone was characterized by lower DIN, SRP, and dissolved oxygen concentrations relative to the active channel (Table II). Conductivity was significantly higher in the riparian zone. Similar to the riparian pattern, nutrients in the active channel increased after floods (Table II), but differences between riparian zone and active channel persisted for nutrients and were accentuated for conductivity (Table II). Subsystems also differed in the relative abundance of nitrate and ammonium (Table IV). Nitrate dominated in the active channel, and ammonium dominated in the riparian zone. Overall, both hydrology and subsystem type had significant effects on both the amount and relative abundance of nutrients and ions in Sycamore Creek (Table III).

VI. SYNTHESIS

A. Hydrological Linkages

The temporal pattern of water table fluctuation observed in the riparian zone in Sycamore Creek appears to be caused by a combination of hydrology and plant activity. Change in position of the water table can be either gradual or abrupt. Gradual change reflects evapotranspiration-driven seasonal variation in water availability in the stream–riparian ecosystem.

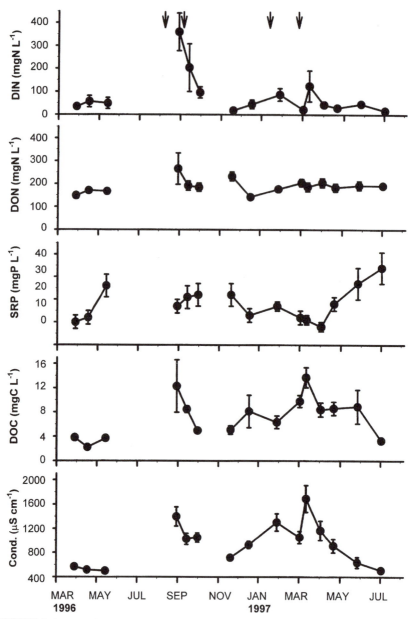

FIGURE 6   Temporal variation in conductivity and concentrations of selected solutes in riparian subsurface water during the study period. Points are means ± standard error of 16 wells sampled; vertical arrows show the timing of flash floods. Gaps correspond to periods when wells were dry.

TABLE II Comparison of Water Chemistry among Different Stream–Riparian Subsystems under Baseflow and Flood Conditions

Parameter	Riparian zone			Bank			Hyporheic zone			Surface stream		
	Baseflow	Flood		Baseflow	Flood		Baseflow	Flood		Baseflow	Flood	
NO_3 ($\mu gN\ L^{-1}$)	8±2	110±27	*	7±2	159±60	*	156±12	1145±113	*	28±6	985±161	*
NH_4 ($\mu gN\ L^{-1}$)	33±5	84±31	*	28±6	78±24	*	17±3	30±7		6±2	98±34	*
DON ($\mu gN\ L^{-1}$)	185±5	204±17		180±8	167±13		166±5	175±22		207±21	218±26	
SRP ($\mu gP\ L^{-1}$)	18±1	16±2	*	23±3	15±2		36±2	68±7	*	19±2	201±61	*
DOC ($mgC\ L^{-1}$)	6.2±0.4	10.2±1.2	*	5.5±0.3	10.4±0.8	*	4.3±0.1	14.9±2.4	*	4.4±0.3	9.8±1.3	*
DO ($mgO_2\ L^{-1}$)	1.6±0.1	2.0±0.1	*	1.6±0.1	1.7±0.2		5.4±0.3	3.8±0.5	*	>8	>8	
Conductivity ($\mu S\ cm^{-1}$)	784±30	1354±84	*	626±26	1028±96	*	525±8	545±19		503±13	369±46	*

Note. DON, dissolved organic nitrogen; SRP, soluble reactive phosphorus; DOC, dissolved organic carbon; and DO, dissolved oxygen. Average values (±1 SEM) for the 1996–1997 period. Asterisks denote that concentrations at baseflow and at flood are significantly different (*$p < 0.05$) within each subsystem (*t*-test, hydrologic conditions as a factor).

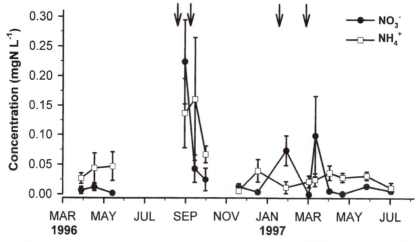

FIGURE 7   Temporal variation in concentration of the two major DIN species, nitrate (filled symbols) and ammonium (open symbols) in riparian subsurface water during the study period. Points are means ± standard error of 16 wells sampled. Note the switch in dominance from ammonium to nitrate associated with flash floods (vertical arrows).

Seasonal variation in evapotranspiration rates is mainly caused by seasonal changes in temperature and activity of the riparian trees, which, in Sycamore Creek, are mostly deciduous. Patou and Décamps (1985) also reported that a significant proportion of the annual change in water table in the upper Rhone River was due to biological activity (i.e., variation in evapotranspiration). Evapotranspiration may also generate daily variation in water table depth in riparian zones, especially on warm days (Fetter, 1994).

Abrupt rises in riparian zone water table are related to floods, supporting prediction 1. The fact that floods rise quickly during storms in Sycamore Creek suggests that floodwaters are mainly generated by overland flow. This is in contrast to mesic streams where floodwater mostly originates from subsurface water derived from upland areas (Sklash and Farvolden, 1979; Mulholland, 1993). Floods in arid regions may suddenly produce large hydraulic gradients between surface water in the active channel and subsurface water in the riparian zone, thus forcing water from the stream channel to the riparian zone. Results suggest that hydrological linkage between surface streams and riparian zones is strongest during floods, at least at the reach scale (i.e., < 500 m). The hydrologic regime in arid land streams is characterized by extreme variation in discharge (i.e., flash floods and droughts) that drive episodic linkages between streams and riparian zones. Although linkage between the two subsystems is mainly mediated by floods, biological activity (evapotranspiration) may enhance the connection by increasing the hydraulic gradient between the active channel and the riparian zone during interflood periods.

TABLE III Results from Two-Way ANOVAs Testing the Influence of Subsystem and Hydrology on Different Water Chemistry Parameters ($n = 498$)

	Subsystem		Hydrology		Subs * Hydr		Main effects		Interaction		
	F	p	F	p	F	p	Subsys.	Hydrol.	Baseflow	Flood	
NO_3^-	146.7	0.000	327.4	0.000	89.5	0.000	NA	NA	E=R<H	E=R<S=H
NH_4^+	3.0	0.032	18.3	0.000	1.7	0.159	R>H	BF>FL	NA	NA
DON	3.1	0.025	0.4	0.546	0.6	0.637	S>H	NA	NA	NA
SRP	53.6	0.000	93.2	0.000	48.7	0.000	NA	NA	R<H	E=R<H<S
DOC	1.9	0.127	61.8	0.000	5.9	0.001	NA	NA	R=E=H=S	R<H
Conductivity	69.7	0.000	26.3	0.000	19.9	0.000	NA	NA	S=H<R	W=H<E<R

Note. R, riparian zone; E, bank; H, hyporheic zone; S, surface stream; BF, baseflow ; FL, flood. Cases not shown in main effects or interaction columns were not statistically significant ($p > 0.05$) from any other cases. NA, not applicable.

TABLE IV Percentage of Total Dissolved Nitrogen as Nitrate, Ammonium, and Dissolved Organic Nitrogen: Comparison among Different Stream–Riparian Subsystems under Baseflow and Flood Conditions

	Baseflow			Flood		
	NO_3	NH_4	DON	NO_3	NH_4	DON
Riparian zone	3±1	13±2	84±2	22±5	12±3	66±6
Bank	4±1	12±3	84±3	32±8	14±4	54±8
Hyporheic zone	42±4	6±1	52±4	82±6	4±3	14±3
Surface stream	13±3	4±1	83±3	53±19	3±2	44±18

Note. Values are the average (±1 SEM) for samples collected during the 1996–1997 period.

The temporal pattern described earlier, together with the spatial pattern in water table fluctuation in the riparian zone, support the hypothesis that, during floods, water moves laterally from the active channel into the riparian zone in Sycamore Creek (prediction 1). This direction of water movement is typical of streams of arid areas (Hughes, 1990) but is not unique to them. For example, in the Flathead River (Montana), a tight relationship exists between increases in river discharge and a rise in water level of floodplain wells located up to 650 m from active channel (Stanford and Ward, 1993). Groundwater pumping near channels also can cause movement of water from the stream to riparian zones, by steepening the hydraulic gradient. The direction of this linkage is opposite to that typical of mesic stream ecosystems (Hill, 1990; McDowell *et al.,* 1992; Triska *et al.,* 1993). For this reason, riparian zones in arid land streams are probably more dependent on water supplied from the stream than from the upland.

B. Implications for the Riparian Zone

During baseflow, water chemistry in the riparian zone differs from the chemistry of surface and subsurface water in the active channel. Dissolved inorganic nitrogen and SRP are lower and DOC is slightly higher in the riparian zone than beneath the active channel. Differences in the chemical form of nitrogen also occur with ammonium being dominant in the riparian zone and nitrate being higher beneath the active channel. During baseflow, when hydrological linkages are weaker, nutrient concentrations are likely to be more strongly influenced by biological or chemical processes occurring within each subsystem. The differences in chemistry among subsystems suggest that different processes are controlling nutrient dynamics in each subsystem. Nutrient dynamics in hyporheic zones are mainly controlled by microbial processes such as aerobic respiration (Jones, 1995), nitrification (Jones *et al.,* 1995b), and denitrification (Holmes *et al.,* 1996). In the riparian zone, however, plant activity may also directly or indirectly influence nutrient concentrations. Differences in redox conditions may also be important, as has been shown by several other studies of streams and their riparian zones (Dahm *et al.*, 1987; 1998; Hedin *et al.*, 1998). Our results show that stream channel and riparian subsystems differ significantly in dissolved oxygen concentration. In Sycamore Creek, rates of anaerobic processes (e.g., denitrification, methanogenesis) were higher in sediments in the banks of the active channel (and presumably in riparian zones beyond the stream banks) than in the hyporheic zone, whereas rates of aerobic processes (e.g., nitrification) were higher in the hyporheic zone (Jones *et al.*, 1994; 1995c). These metabolic differences among subsystems probably account for the differences observed in DIN forms.

In Sycamore Creek, most of the temporal variation in solute concentrations in the riparian zone, as well as in the hyporheic zone and the surface stream, is associated with floods. The lack of seasonality in nutrient concentrations in the riparian zone has also been observed for other streams (McDowell *et al.*, 1992; McClain *et al.*, 1992), but in those studies temporal variation was unrelated to floods as well. Sycamore Creek flood waters are commonly rich in DIN, DOC, and SRP relative to baseflow; thus, floods represent an input of nutrients to the stream. Within one week after floods, nutrients in hyporheic (DIN, DOC, and SRP) and riparian zones (only DIN and DOC) also increased, supporting prediction 3. In the riparian zone, we also found an increase in major ions (expressed as conductivity) and a shift from reduced to oxidized DIN. Thus, floods not only represent a pulse of water, but also of solutes, to the riparian zone. Our data show that during a single flood (22 August 1996), 60% of the floodwater was retained in subsurface subsystems (the riparian and hyporheic zones). This movement of water represented a pulse of 3 kg $NO_3-N/$ ha to these subsystems, twice the amount that annually enters the Sycamore Creek watershed via precipitation (M.S. Holland, Arizona State University, unpublished data). This pulse can represent from 15 to 100% of the annual nitrogen input via precipitation in temperate watersheds (Likens and Bormann, 1995). Therefore, the effective retention of water also represents a large supply of nutrients (in particular nitrogen) to the riparian zone in this arid land stream.

The flood-pulse concept (Junk *et al.*, 1989) was first developed for large rivers and mainly focused on surface interactions between river and adjacent floodplain as water overflowed stream banks. The flood-pulse increases the area of "active zones" in the river ecosystem by connecting the main channel with the floodplain from which it is usually disconnected. Floods in Sycamore Creek act similarly, but because floods often do not overflow banks, the stream–riparian linkage occurs via subsurface flowpaths.

The increase in solutes in riparian zone water may result from direct input of nutrient-rich floodwater from the stream channel, but other mechanisms may also be important. For example, changes in chemistry along stream to riparian flowpaths during small floods that do not fill the active channel may be the result of processes occurring in the intervening hyporheic zone. In addition, rewetting of dry riparian or gravel bar sediments may trigger redissolution of salts (evaporites) and increase rates of riparian biological processes such as mineralization or nitrification. Redissolution of evaporites is supported by observed increases in ion concentrations in the riparian zone after floods because floodwater from the stream channel is ion-poor. Increases in rates of riparian biological processes after floods may explain the persistent differences in subsystem water chemistry despite overall increases in solute concentrations. The relative importance of these processes may vary with season, flood magnitude, and solute and may depend on how water is routed from the surface stream to the riparian zone. Processes are not necessarily simultaneous. Redissolved sediment evaporites may enter the riparian zone during the first hour or two of flooding; however, microbial responses may lag the flood peak by a day or more. Thus, increases in riparian nutrient concentrations after floods may be caused not just by direct solute delivery but also by indirect effects of increases in water table.

After floodwater enters the riparian zone, it moves downstream slowly, allowing longer interaction with riparian sediments and biota. Change in riparian solute concentrations after floods will depend on hydrologic residence time, and rates and types of biochemical processes in the riparian zone (Triska *et al.*, 1993). For three of the four floods we studied, concentrations of nutrients in riparian zones returned to preflood values one month after flood. Conservative solutes, expressed as conductivity, responded more slowly, suggesting that biological processes partially control nutrient reduction. In riparian zones elsewhere, nitrogen removal mechanisms include plant uptake and denitrification (McDowell *et al.*, 1992; Triska *et al.*, 1993; Hedin *et al.*, 1998; Groffman *et al.*, 1992; Pinay *et al.*, 1994; Hill, 1996). Less work has been done to assess mechanisms regulating dynamics of other nutrients (carbon and phosporus; but see Pinay *et al.*, 1992). Some of the water and nutrients entering the riparian zone during the flood seep slowly back into the surface stream during the interflood period as hydraulic gradients reverse. Depending on relative concentrations, returning water may enrich or dilute stream water with respect to nitrogen, the limiting nutrient.

Our study has focused on water and nutrient delivery to the riparian zone rather than mechanisms controlling nutrient dynamics within the riparian zone. Studies in riparian zones of other arid land streams have emphasized effects of water availability on vegetation production and composition (Stromberg *et al.*, 1996). Jacobson *et al.* (1995) showed that in ephemeral rivers of the Namib Desert the flood-pulse is a key ecological process structuring and maintaining water-limited riparian ecosystems. Presumably, in Sycamore Creek the episodic inputs of water and nutrients are frequent enough to support the requirements of riparian vegetation and other organisms in these zones.

VII. CONCLUSIONS: INTERMEDIATE DISTURBANCE AND NUTRIENT RETENTION

We tested four predictions of our intermediate disturbance conceptual model with studies in Sycamore Creek, an arid land stream: (1) the water table of the riparian zone will rise quickly during floods via subsurface flow from the adjacent stream channel; (2) the rise will be greater after a long interflood period than after a short one, and the magnitude of increase will be unrelated to flood magnitude; (3) high nutrient concentrations of floodwaters will be reflected in subsurface water of the riparian zone; and (4) the effect of an individual flood on nutrient concentrations in riparian subsurface water will decrease as flood frequency increases. All these predictions deal with individual floods, or combinations of a few events, and all were supported by data from an 18-month period encompassing four floods in Sycamore Creek. We anticipate that long-term study across several years with differing hydrologic regimes will present an opportunity to test additional predictions of the model.

Nutrient-rich floodwaters enter the riparian zone rapidly. Both rate and amount of recharge depend upon available storage, which is a function of time since the last flood (or the time over which drying occurred). From an ecosystem perspective, lateral linkage between stream and riparian zones during floods enhances the capacity of the riparian subsystem to retain episodic water and nutrient inputs. Retention includes both storage *in situ* and losses to the atmosphere (e.g., of water and nitrogen via transpiration and denitrification). In this sense, riparian zones in arid lands act as a large transient storage zone effectively retaining a fraction of floodwater and, therefore, decreasing the rapid downstream export of water and solutes. This effect has been acknowledged in large rivers with extensive floodplains. Channel bed, banks, and hydrologically linked floodplains together attenuate flow volume, which would otherwise be routed downstream (see Wilby and Gibert, 1996, for a review).

On a long-term basis, large floods recur, and several statistics could be used to describe these disturbance regimes (Connell, 1978; Grimm, 1993; Sousa, 1984; Richter *et al.*, 1996). One of the most basic of these is disturbance frequency (number of events per year). In our conceptual model, we proposed that water and nutrient subsidy and potential nutrient retention in the riparian zone should respond to disturbance frequency parabolically. At low flood frequencies (few floods per year), delivery of water and nutrients over a year is low and may lead to limitation of riparian processes by water, nutrients, or both. At high flood frequencies, hydraulic heads are low, and nutrient-rich water moves only slowly into the riparian zone. At an intermediate flood frequency, intermittent recharge followed by discharge against steep hydraulic gradients should maximize annual nutrient delivery and potential retention response (Fig. 8).

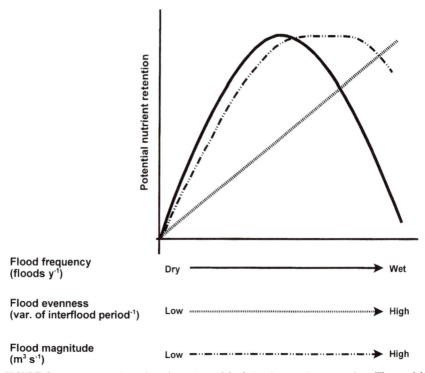

FIGURE 8 An "intermediate disturbance" model of riparian nutrient retention. The model proposes that the potential for nutrient retention by riparian zones in arid lands is controlled by disturbance (hydrologic) regime. Three different attributes of hydrologic regime are plotted on the abscissa from low to high values: flood frequency (dry to wet), flood evenness (low to high; measured as the inverse of the variance in interflood interval length), and flood magnitude (low to high). See text for explanation of hypothesized patterns.

Joseph Connell (1978) developed the Intermediate Disturbance Hypothesis as a parabolic response of species diversity to disturbance frequency. Even though species diversity is a community characteristic and is quite different from nutrient delivery, an ecosystem property, they both have a parabolic response to disturbance frequency. Connell's model also plotted disturbance intensity and time after the event and proposed that species diversity would respond as to frequency. Species are eliminated at high (or frequent) disturbance by biophysical means and at low disturbance frequencies by competitive exclusion. This is not the case with our model, which is predicated on physical–chemical relationships only. We propose that, all else being equal, an increase in flood intensity would generate an asymptotic response with a sharp break point associated with overbank flow (Fig. 8). At some point, exceedingly severe floods may erode banks and remove riparian vegetation, thereby decreasing both storage volume and uptake capacity.

Odum *et al.* (1979) described a "subsidy-stress gradient model" which showed how imposed disturbances such as flooding or nutrient enrichment can benefit an ecosystem in such a way that a "humpbacked" performance curve is generated. According to this model, a subsidy to the ecosystem is evident at some intermediate level along a stress gradient. Megonigal *et al.* (1997) evaluated the subsidy-stress hypothesis with respect to primary production of southeastern floodplain forests along a flooding gradient and showed that these forests are stressed by any additional flooding above extant rates, leading them to reject Odum's hypothesis. Our expectation for desert stream riparian zones would be that low frequencies should depress primary production. This prediction contrasts with findings of Megonigal *et al.* (1997); however, in their study, the deprivation of subsidy (e.g., drought) was not considered.

Time since last flood can be expressed as variance in interflood interval during a given time period. If floods are clustered in time, variance will be high, whereas if they are evenly spaced, variance will be low. We propose that, all else being equal, nutrient delivery and processing will be highest when floods are moderately frequent and evenly spaced but will decline as variance increases (that is, as 1/var decreases, Fig. 8). In summary, our model proposes that potential nutrient retention is greatest at intermediate flooding frequency, increases with evenness of flood recurrence intervals, and decreases with individual flood magnitude. These relationships hold when these attributes of disturbance regime are considered individually. In reality, this relationship is multivariate, and the net effect of the interaction of all three variables determines whole system response in a manner that is not intuitively obvious without more empirical data than we currently have. We suspect, however, that within a reasonable range of hydrologic events, flood frequency will explain most of the variance in nutrient retention capacity.

REFERENCES

Bencala, K. E., J. H. Duff, J. W. Harvey, A. P. Jackman, and F. J. Triska. 1993. Modelling within the stream-catchment continuum. *In* "Modelling Change in Environmental Systems" (A. J. Jakeman, M. B. Beck, and M. J. McAleer, eds.), pp. 163–187. Wiley, Chichester.

Busch, D. E., and S. D. Smith. 1995. Mechanisms associated with decline of woody species in riparian ecosystems of the southwestern U.S. *Ecological Monographs* 65:347–370.

Connell, J. H. 1978. Diversity in tropical rain forests and coral reefs. *Science* 199:1302–1310.

Dahm, C. N., E. H. Trotter, and J. R. Sedell. 1987. Role of anaerobic zones and processes in stream ecosystem productivity. *In* "Chemical Quality of Water and the Hydrologic Cycle" (R. C. Averett and D. M. McKnight, eds.), pp. 157–178. Lewis Publishers, Chelsea, MI.

Dahm, C. N., N. B. Grimm, P. Marmonier, H. M. Valett, and P. Vervier. 1998. Nutrient dynamics at the interface between surface waters and ground waters. *Freshwater Biology* 40:427–451.

Décamps, H. 1996. The renewal of floodplain forests along rivers: A landscape perspective. *Verhandlungen Internationale Vereinigung für Theoretische und Angewandte Limnologie* 26:35–59.

Fetter, C. W. 1994. "Applied Hydrogeology." Macmillan College Publishing Company, New York.

Fisher, S. G., and N. B. Grimm. 1985. Hydrologic and material budgets for small Sonoran Desert watershed during three consecutive cloudburst floods. *Journal of Arid Environments* 9:105–118.

Fisher, S. G., L. J. Gray, N. B. Grimm, and D. E. Busch. 1982. Temporal succession in a desert stream ecosystem following flash flooding. *Ecological Monographs* 52:93–110.

Fisher, S. G., N. B. Grimm, E. Martí, and R. Gomez. 1998a. Hierarchy, spatial configuration, and nutrient cycling in streams. *Australian Journal of Ecology* 23:41–52.

Fisher, S. G., N. B. Grimm, E. Martí, R. M. Holmes, and J. B. Jones. 1998b. Material spiralling in river corridors: A telescoping ecosystem model. *Ecosystems* 1:19–34.

Gregory, S. V., F. J. Swanson, W. A. McKee, and K. W. Cummins. 1991. An ecosystem perspective of riparian zones. *BioScience* 41:540–551.

Grimm, N. B. 1993. Implications of climate change for stream communities. *In* "Biotic Interactions and Global Chanage" (P. Kareiva, J. Kingsolver, and R. Huey, eds.), pp. 293–314. Sinauer Associates, Sunderland, MA.

Groffman, P. M., A. J. Gold, and R. C. Simmons. 1992. Nitrate dynamics in riparian forests: Microbial studies. *Journal of Environmental Quality* 21:666–671.

Hedin, L. O., J. C. Fischer, N. E. Ostrom, B. P. Kennedy, M. G. Brown, and G. Philip Robertson. 1998. Thermodynamic constrains on biogeochemical structure and transformations of nitrogen at terrestrial-lotic interfaces. *Ecology* 79:684–703.

Hill, A. R. 1990. Ground water flow paths in relation to nitrogen chemistry in the near-stream zone. *Hydrobiologia* 206:39–52.

Hill, A. R. 1996. Nitrate removal in stream riparian zones. *Journal of Environmental Quality* 25:743–755.

Holmes, R. M., S. G. Fisher, and N. B. Grimm. 1994. Parafluvial nitrogen dynamics in a desert stream ecosystem. *Journal of the North American Benthological Society* 13:468–478.

Holmes, R. M., J. B. Jones, S. G. Fisher, and N. B. Grimm. 1996. Denitrification in a nitrogen-limited stream ecosystem. *Biogeochemistry* 33:125–146.

Hughes, F. M. R. 1990. The influence of flooding regimes on forest distribution and composition in the Tana River floodplain, Kenya. *Journal of Applied Ecology* 27:475–491.

Jacobson, P. J., K. M. Jacobson, and M. K. Seely. 1995. "Ephemeral Rivers and their Catchments: Sustaining People and Development in Western Namibia." Desert Research Foundation of Namibia, Windhoek.

Jones, J. B. 1995. Factors controlling hyporheic respiration in a desert stream. *Freshwater Biology* 34:91–101.

Jones, J. B., R. M. Holmes, S. G. Fisher, and N. B. Grimm. 1994. Chemoautotrophic production and respiration in the hyporheic zone of a Sonoran Desert stream. *In* "Proceedings of the Second International Conference on Groundwater Ecology" (J. A. Stanford and H. M. Valett, eds.), pp. 329–338. American Water Resources Association, Herndon, VA.

Jones, J. B., S. G. Fisher, and N. B. Grimm. 1995a. Vertical hydrologic exchange and ecosystem metabolism in a Sonoran Desert stream. *Ecology* 76:942–952.

Jones, J. B., S. G. Fisher, and N. B. Grimm. 1995b. Nitrification in the hyporheic zone of a desert stream ecosystem. *Journal of the North American Benthological Society* 14:249–258.

Jones, J. B., R. M. Holmes, S. G. Fisher, N. B. Grimm, and D. M. Greene. 1995c. Methanogenesis in Arizona, USA, dryland streams. *Biogeochemistry* 31:155–173.

Junk, W. J., P. B. Bayley, and R. E. Sparks. 1989. The flood pulse concept in river-flood plain systems. *In* "Proceedings of the International large River Symposium (LARS)" (D. P. Dodge, ed.), Special Publication, pp. 110–127. Canadian Fisheries and Aquatic Sciences, Ottawa, ON.

Likens, G. E., and F. H. Bormann. 1995. "Biogeochemistry of a Forested Ecosystem," 2nd ed. Springer-Verlag, New York.

McClain, M. E., J. E. Richey, and T. P. Pimentel. 1992. Groundwater nitrogen dynamics at the terrestrial-lotic interface of a small catchment in the Central Amazon Basin. *Biogeochemistry* **27**:113–127.

McDowell, W. H., W. B. Bowden, and C. E. Asbury. 1992. Riparian nitrogen dynamics in two geomorphologically distinct tropical rain forest watersheds: Subsurface solute patterns. *Biogeochemistry* **18**:53–75.

Megonigal, J. P., W. H. Connor, S. Kroeger, and R. R. Sharitz. 1997. Above ground production in southeastern floodplain forests: A test of the subsidy-stress hypothesis. *Ecology* **78**:370–384.

Mulholland, P. J. 1993. Hydrometric and stream chemistry evidence of three storm flowpaths in Walker Branch Watershed. *Journal of Hydrology* **151**:291–316.

Naiman, R. J., H. Décamps, J. Pastor, and C. A. Johnston. 1988. The potential importance of boundaries to fluvial ecosystems. *Journal of the North American Benthological Society* **7**:289–306.

Odum, E. P., J. T. Finn, and E. H. Franz. 1979. Perturbation theory and the subsidy-stress gradient. *BioScience* **29**:349–352.

Patou, G., and H. Décamps. 1985. Ecological interactions between the alluvial forest and hydrology of the upper Rhone. *Archiv für Hydrobiologie* **104**:13–37.

Pinay, G., H. Décamps, E. Chauvet, and E. Fustec. 1990. Functions of ecotones in fluvial systems. *In* "The Ecology and Management of Aquatic-Terrestrial Ecotones" (R. J. Naiman and H. Décamps, eds.), pp. 141–169. UNESCO and Parthenon Publishing Group, Paris.

Pinay, G., A. Fabre, P. Vervier, and F. Gazelle. 1992. Control of C, N, P distribution in soils of riparian forests. *Landscape Ecology* **6**:121–132.

Pinay, G., N. E. Haycock, and C. Ruffinoni. 1994. The role of denitrification in nitrogen removal in river corridors. *In* "Global Wetlands, Old World and New" (W. J. Mitsch, ed.), pp. 107–116. Elsevier, Amsterdam.

Poff, N. L., J. D. Allan, M. B. Bain, J. R. Karr, K. L. Prestegaard, B. D. Richter, R. E. Sparks, and J. C. Stromberg. 1997. The natural flow regime. *BioScience* **47**:769–784.

Richter, B. D., J. V. Baumgartner, J. Powell, and D. P. Braun. 1996. A method for assessing hydrologic alteration within ecosystems. *Conservation Biology* **10**:1163–1174.

Richter, H. E. 1993. Development of a conceptual model for floodplain restoration in a desert riparian system. *Arid Lands Newsletter* **32**:13–17.

Sklash, M. G., and R. N. Farvolden. 1979. The role of groundwater in storm runoff. *Journal of Hydrology* **43**:45–65.

Sousa, W. P. 1984. The role of disturbance in natural communities. *Annual Review of Ecology and Systematics* **15**:353–391.

Stanford, J. A., and J. V. Ward. 1993. An ecosystem perspective of alluvial rivers: Connectivity and the hyporheic corridor. *Journal of the North American Benthological Society* **12**:48–60.

Stanley, E. H., and H. M. Valett. 1992. Interaction between drying and the hyporheic zone of a desert stream ecosystem. *In* "Climate Change and Freshwater Ecosystems" (P. Firth and S. G. Fisher, eds.), pp. 234–249. Springer-Verlag, New York.

Stanley, E. H., S. G. Fisher, and N. B. Grimm. 1997. Ecosystem expansion and contraction in streams. *BioScience* **47**:427–435.

Stromberg, J. C., S. D. Wilkins, and J. A. Tress. 1993. Vegetation-hydrology models: Implications for management of Prosopis Velutina (Velvet Mesquite) riparian ecosystems. *Ecological Applications* **3**:307–314.

Stromberg, J. C., R. Tiller, and B. Richter. 1996. Effects of groundwater decline on riparian vegetation of semiarid regions: The San Pedro, Arizona. *Ecological Applications* **6**:113–131.

Thomsen, B. W., and H. H. Schumann. 1968. Water resources of the Sycamore Creek watershed, Maricopa County, Arizona. *Geological Survey Water-Supply Paper (U.S.)* **1861**.

Triska, F. J., J. H. Duff, and R. J. Avanzino. 1993. Patterns of hydrological exchange and nutrient transformation in the hyporheic zone of a gravel-bottom stream: Examining terrestrial-aquatic linkages. *Freshwater Biology* **29**:259–274.

Valett, H. M., S. G. Fisher, N. B. Grimm, and P. Camill. 1994. Vertical hydrologic exchange and ecological stability of a desert stream ecosystem. *Ecology* **75**:548–560.

Wilby, R., and J. Gibert, 1996. Hydrological and hydrochemical dynamics. *In* "Fluvial Hydrosystems" (G. E. Petts and C. Amoros, eds.), pp. 37–67. Chapman & Hall, London.

5

The Importance of Ground Water to Stream Ecosystem Function

Robert M. Holmes

The Ecosystems Center
Marine Biological Laboratory
Woods Hole, Massachusetts

I. Introduction   137
II. Influence of Ground Water on Stream Functioning   139
    A. Hydrology and Stream-Flow Generation   139
    B. Nutrients   139
    C. Dissolved Organic Matter   141
    D. Dissolved Inorganic Carbon   142
    E. Temperature   143
III. Summary   144
    References   145

I. INTRODUCTION

The objective of this chapter is to highlight some of the ways ground water influences stream ecosystems. Hynes had a similar objective more than a decade ago (Hynes, 1983), and much progress has been made since then. For example, our understanding of the role of hyporheic, parafluvial, and riparian zones to stream functioning has greatly increased (Triska *et al.*, 1989; Ward, 1989; Jones and Holmes, 1996). As a result, our view of what constitutes a stream ecosystem has expanded spatially, and we now have a more holistic understanding of stream structure and functioning. As noted by

Brunke and Gonser (1997), "the boundaries between river and ground water ecological research are dissolving, and both fields are beginning to merge
towards a comprehensive understanding of the hydrological continuum."

Even though much progress has been made, most of the advances have
dealt with the interactive hyporheic or parafluvial zones (subsurface zones
with appreciable content of surface water) as opposed to "true" ground water. This may be because ground water is difficult to study or because it influences streams on scales greater than most stream studies are conducted.
However, a review of the literature, both within and outside of the field of
stream ecology, reveals many ways in which ground water influences stream
functioning. I will summarize results of these studies and suggest ways in
which explicit consideration of ground water may improve our understanding of stream ecosystems. The emphasis will be on how ground water inputs
affect biogeochemical characteristics of the surface stream, but I will also
consider the impact of ground water on stream biota.

"Ground water" means different things to different people (Triska *et al.*,
1989; Gibert *et al.*, 1990; Vervier *et al.*, 1992; Fig. 1). Some investigators
consider all subsurface water to be ground water, whereas others define
ground water as noninteractive or as subsurface water entering a stream corridor for the first time. In this chapter, I will use the more restricted definition. Specifically, I will consider ground water to be any subsurface water

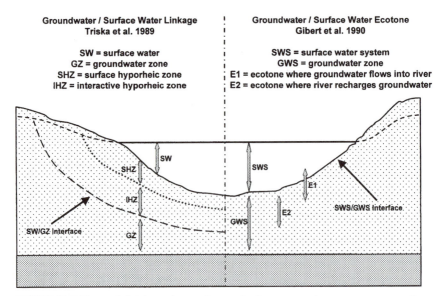

| Groundwater / Surface Water Linkage | Groundwater / Surface Water Ecotone |
| Triska et al. 1989 | Gibert et al. 1990 |

SW = surface water
GZ = groundwater zone
SHZ = surface hyporheic zone
IHZ = interactive hyporheic zone

SWS = surface water system
GWS = groundwater zone
E1 = ecotone where groundwater flows into river
E2 = ecotone where river recharges groundwater

FIGURE 1 Different definitions of ground water in the context of stream ecosystems. From
Vervier *et al.* (1992); reprinted with permission of the Journal of the North American Benthological Society.

that has not yet exchanged with surface water. This definition is similar to that used by Triska *et al.* (1989) and includes interflow, shallow ground water, and deep aquifers but excludes subsurface water in the stream corridor that interacts with surface water, such as hyporheic and parafluvial water (Holmes *et al.*, 1994).

II. INFLUENCE OF GROUND WATER ON STREAM FUNCTIONING

A. Hydrology and Stream-Flow Generation

The most obvious impact of ground water on stream ecosystems is in creating them. That is, flow in most streams is dominated by ground water inputs. Therefore, the baseline chemistry and hydrology of streams is a function of processes that occurred as precipitation percolated through soil horizons and moved along ground water flowpaths. Even during storms when stream discharge is greatly elevated, the majority of water in stream channels is typically recently discharged ground water displaced from soils and bedrock by incoming precipitation (Sklash and Farvolden, 1979; Neal *et al.*, 1992; Ogunkoya and Jenkins, 1993; Waddington *et al.*, 1993; Buttle 1994).

Depending on catchment characteristics, precipitation takes different routes from upland to stream ecosystems, and these different flowpaths influence material fluxes entering stream corridors (Mulholland, 1993; Hill, 1996; Fisher *et al.*, 1998). In forested or agricultural catchments with deep, well-drained soils, precipitation percolates below rooting depth and does not again interact with vegetation until reaching the riparian zone. When soils are shallow, vegetation throughout the catchment may intercept ground water nutrients. Finally, desert catchments with hydrophobic soils transport precipitation to stream channels as overland flow or in washes and rills, effectively bypassing the lateral "riparian filter" (Holmes *et al.*, 1996; Fisher *et al.*, 1998). In arid land streams, therefore, extensive hyporheic, parafluvial, and riparian sediments are recharged by storm water that entered the channel as overland flow, and these zones supply stream flow during interstorm periods (see Marti *et al.*, Chapter 4 in this book).

B. Nutrients

Ground water contributes not only water to streams but also dissolved materials. I will focus on ground water inputs of nitrogen and phosphorus, although ground water also significantly influences dynamics of other elements as well. The emphasis on nitrogen and phosphorus is due to their importance in limiting primary productivity in aquatic ecosystems and also reflects the more extensive research that has been done on nitrogen and phosphorus cycling by stream ecologists.

Nitrate contamination of ground water is widespread and is increasingly leading to elevated surface-water nitrate concentration (Valiela *et al.*, 1990, 1997; Smith *et al.*, 1991; Spaulding and Exner, 1993; McMahon and Bohlke, 1996; Fig. 2). Nitrate contamination of ground water results most commonly from leaching of agricultural fertilizers, but sewage and industrial inputs are also significant (Valiela *et al.*, 1997). In many regions, ground water nitrate concentration exceeds drinking water standards, and eutrophication of surface waters is increasing; consequently, research is being directed at improving our understanding of nitrogen transport and retention in ground water.

A number of studies have shown that riparian processes, chiefly denitrification and plant uptake, can greatly reduce ground water nitrate concentration and fluxes from upland to stream ecosystems (Peterjohn and Correll, 1984; Pinay and Decamps, 1988; Lowrance, 1992; Pinay *et al.*, 1994; Groffman *et al.*, 1996; Hill, 1996). This topic is treated more fully in this book in

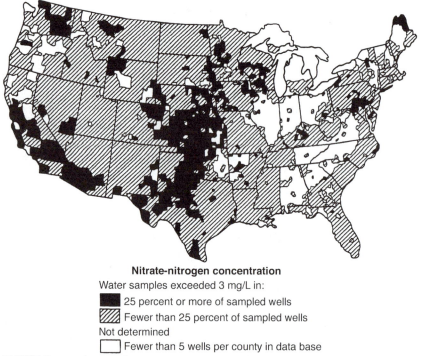

Nitrate-nitrogen concentration

Water samples exceeded 3 mg/L in:

■ 25 percent or more of sampled wells

▨ Fewer than 25 percent of sampled wells

Not determined

☐ Fewer than 5 wells per county in data base

FIGURE 2 Distribution of nitrate in ground water of the United States. The solid black fill shows regions were nitrate exceeded 3 mgN/L in greater than 25% of wells sampled and the striped pattern indicates that less than 25% of wells had nitrate exceeding 3 mgN/L. No fill indicates insufficient sample size. Reprinted with permission from Spaulding and Exner (1993); originally from Madison and Brunett (1985).

Chapter 3 by Hill. In addition to processing in the riparian zone, substantial nitrate retention may occur along ground water flowpaths before entering the riparian zone (Mariotti *et al.*, 1988; Smith *et al.*, 1991, 1996; McMahon and Bohlke, 1996). A challenge in all these studies is to differentiate among the various processes and their spatial distribution that could result in apparent nitrate retention. One promising approach is the use of natural abundance stable isotope ratios. For example, the natural abundance distribution of ^{15}N-nitrate has been used to distinguish between denitrification and dilution by low nitrate ground water along ground water flowpaths (Mariotti *et al.*, 1988; McMahon and Böhlke, 1996). Similarly, the natural abundance of ^{15}N-nitrate was used to investigate nitrogen cycling in the riparian zone of an Amazonian stream (Brandes *et al.*, 1996), and improved methods of measuring ^{15}N-DIN are making this powerful technique feasible in a greater range of studies (Sigman *et al.*, 1997; Holmes *et al.*, 1998).

Nitrogen has received the most attention as a nutrient contaminate of aquifers, but ground water also carries phosphorus to streams. For example, in a series of geothermal streams at La Selva Biological Station in Costa Rica, phosphorus concentration in streams in adjacent catchments was very different depending on the source of ground water (Pringle *et al.*, 1986, 1990, 1993; Pringle and Triska, 1991). Streams with relatively large inputs of geothermally impacted ground water had a high phosphorous concentration (up to 400 μgP/L), whereas nearby streams with lower inputs of geothermal ground water had a much lower phosphorus concentration (average equals 8.9 μgP/L). Nutrient limitation assays in light-gaps in these streams indicated that phosphorus was limiting in low-phosphorus streams but that micronutrients limited algal growth in streams with high inputs of geothermal ground water. This study is an excellent example of how explicit consideration of ground water processes and inputs control many aspects of surface stream chemistry and biology. Similarly, in Walker Branch, a temperate forest stream in Tennessee, phosphorus inputs to the stream were predominately from weathering of parent dolomite in the upland (Mulholland 1992, 1993). The riparian zone was also a potential source of inorganic phosphorus (and ammonium) when dissolved oxygen in the riparian zone was low. Even though ground water processes largely regulated the input of nutrients to the stream, the spatial and temporal variability of stream-water nutrient concentrations was primarily a function of biotic processing in the stream. As with the Costa Rican streams, research on Walker Branch clearly illustrates the connection between ground water hydrology and chemistry, and surface-stream chemistry and productivity.

C. Dissolved Organic Matter

Ground water also influences the metabolism of stream ecosystems through inputs of dissolved organic matter (DOM; Fisher and Likens, 1973;

Wallis *et al.,* 1981; Hynes, 1983; Rutherford and Hynes, 1987; Ford and Naiman, 1989; Kaplan and Newbold, 1993; Nelson *et al.,* 1993; Dosskey and Bertsch, 1994; Mulholland, 1997). In Bear Brook, New Hampshire, about 25% of the annual energy inputs to the stream were from subsurface DOM (Fisher and Likens, 1973). Surface and groundwater DOM concentrations in Bear Brook were similar. More commonly, ground water is elevated in DOM relative to the surface stream, and substantial immobilization occurs in hyporheic sediments as ground water enters the stream ecosystem (Wallis *et al.,* 1981; Rutherford and Hynes, 1987; Fiebig *et al.,* 1990; Fiebig, 1995). In general, 45–80% of ground water DOM appears to be immobilized in hyporheic sediments, which provides an important energy source for hyporheic microbes and subsequently for higher trophic levels.

The transport of DOM via ground water to stream ecosystems is regulated by soil and catchment properties (Nelson *et al.,* 1993; Dosskey and Bertsch, 1994; Currie *et al.,* 1996). In a comparison of streams in two catchments with differing soil characteristics but otherwise similar properties, higher DOM adsorption capacity in soils in one catchment led to streamwater DOM concentration nearly an order of magnitude lower than the stream in the adjacent catchment (Nelson *et al.,* 1993). In a blackwater stream on the Atlantic Coastal Plain of South Carolina, upland detrital sources were unimportant to stream DOM inputs, but riparian wetlands accounted for 63% of all organic matter entering the stream (Dosskey and Bertsch, 1994). In summary, catchment and ground water processes control the supply of DOM to streams, where substantial processing may occur during initial entry through the hyporheic zone or after entering the surface channel.

D. Dissolved Inorganic Carbon

Another way that ground water can influence stream ecosystems is through processes associated with dissolved inorganic carbon (DIC) inputs. This topic has received substantial attention over the past two decades in part because of its association with the alkalinity and acid neutralizing capacity of surface waters, and hence susceptibility of surface waters to acid precipitation (Bailey *et al.,* 1987; Piñol and Avila, 1992; Herlihy *et al.,* 1993). More recently, streamwater CO_2 has been used as an integrative measure of catchment processes such as soil respiration (Jones and Mulholland, 1998a,b) and to improve terrestrial carbon budgets by accounting for carbon loss via gas exchange across the surface-water–atmosphere interface (Kling *et al.,* 1991).

Processes occurring in terrestrial ecosystems influence the partial pressure of CO_2 of ground and surface waters. Precipitation is in equilibrium with atmospheric CO_2 ($CO_2 = 360$ ppmv), but as precipitation percolates through soils, it becomes supersaturated with CO_2 due to soil and root respiration. This CO_2 and carbonic acid-enriched water makes its way to ground water aquifers, dissolving carbonates such as calcite and dolomite along the way. The ground water eventually discharges into streams, which is one of the reasons stream water is generally supersaturated in CO_2 with respect to the atmosphere (Hope *et al.*, 1994; Jones and Mulholland, 1998a). In order to reach equilibrium with atmospheric CO_2, stream-water CO_2 degasses to the atmosphere, and this process is enhanced at turbulent locations in the stream channel (Herman and Lorah, 1987; Lorah and Herman, 1988). This CO_2 degassing can represent a significant carbon flux in some ecosystems (Kling *et al.*, 1992; Cole *et al.*, 1994). In addition to CO_2 degassing, biotic uptake during photosynthesis may be an important CO_2 sink in some stream systems (Hoffer-French and Herman, 1989).

In catchments underlain by limestone, dissolution of calcite by CO_2-enriched ground water leads to high Ca^{2+} and HCO_3^- levels in ground water and, consequently, elevated surface-water concentrations (Herman and Lorah, 1987; Lorah and Herman, 1988; Pentecost, 1995). As CO_2 degasses, the stream water can become supersaturated with respect to calcite, and calcite precipitation can occur after some minimum energy barrier is passed and when suitable nucleation sites are present. Precipitates of calcite (travertine) can form important geomorphological features of streams in karstic regions (Hoffer-French and Herman, 1989).

In addition, degassing of CO_2 in the surface stream leads to consumption of H^+ and, consequently, increases in pH according to the equation

$$CO_{2\,(g)} + H_2O \rightleftharpoons H_2CO_{3\,(aq)} \rightleftharpoons H^+ + HCO_3^-{}_{(aq)}$$

For example, in a small stream in a limestone region in Virginia (Falling Springs Creek), springwater entering the stream had a pH of 7.13 but increased to 8.24 approximately 2 km downstream (Hoffer-French and Herman, 1989). Over the same distance, HCO_3^- dropped from 352 to 313 mg L^{-1}, Ca^{2+} decreased from 72.7 to 63.3 mg L^{-1}, and the partial pressure of CO_2 declined from 25,120 to 1778 ppmv. Although this and other similar studies (e.g., Choi *et al.*, 1998) focused on geochemical aspects rather than their biological significance, CO_2-induced variation in pH and other factors associated with the carbonate systems are relevant to understanding biotic processes in streams.

E. Temperature

Ground water also influences stream functioning by affecting water temperature, which in turn influences rates of many processes (Ward and Stanford, 1982; White *et al.*, 1987). Ground water temperatures tend to be relatively constant at about the mean annual air temperature, whereas surface-water temperatures often vary greatly daily and seasonally (Brunke and Gonser, 1997). During summer, ground water inputs tend to be cooler than surface water, whereas in winter ground water is usually warmer than surface water. Therefore, as ground water contributions increase, temperature variations in the surface stream tend to be moderated, as do rates of temperature-dependent processes.

Although in general ground water inputs have a moderating effect on stream temperature, in some cases subsurface inputs may cause extreme spatial variation in stream temperature that may strongly influence the ecology of the ecosystem. For example, the Firehole River in Yellowstone National Park, Wyoming, receives substantial inputs of geothermal ground water, resulting in an approximate 12°C stream-water temperature increase in the region of geothermal influence (Boylen and Brock, 1973). This temperature increase causes changes in bacterial (Zeikus and Brock, 1972) and algal (Boylen and Brock, 1973) productivity, as well as in the growth and ecology of stream insects (Armitage, 1958) and fishes (Kaeding and Kaya, 1978; Kaeding, 1996). Although this is an extreme example of the impact of thermal ground water on stream functioning, even a moderate temperature change of a few degrees at locations of large ground water input may strongly influence stream processes.

Temperature also influences hydrology. Mid-afternoon reduction in surface-stream flow is a fairly common observation and has typically been attributed to increased evapotranspiration. However, a recent study at sites in New Mexico and Colorado demonstrated that diel fluctuations in water temperature caused changes in hydraulic conductivity, and, consequently, in recharge of subsurface waters, and these changes explained the majority of diel stream-flow variation (Constantz *et al.*, 1994). Thus, temperature changes generated by ground water inputs can influence the magnitude and possibly even direction of surface–subsurface exchange in downstream reaches. Because ground water inputs are often localized and patchy, they are one factor influencing the high degree of spatial heterogeneity that characterizes stream ecosystems.

III. SUMMARY

A visceral way to appreciate the influence of ground water on streams is to walk barefoot through a small stream. Before long, locations where water temperature changes markedly can be detected, reflecting a spot of ground water to surface stream hydrologic exchange. Even though a change in temperature is the most obvious difference, the water likely also differs from the surrounding stream water in nutrient concentrations and N:P ratio, degree of CO_2 saturation, and composition of other solutes. These differences can result in changes in community composition and rates of processes such as primary productivity and bacterial production.

On one level, stream functioning can be better understood by describing the spatial and temporal variability generated by factors such as ground water inputs and determining their influence on some process of interest. However, to really understand the stream, we must also fully understand catchment processes, recognizing that it is all part of the same hydrologic continuum. Currently aquatic, terrestrial, and ground water ecologists are meeting on the middle ground of riparian forests (Fisher *et al.*, 1998), and if Brunke and Gonser (1997) are correct, at some point the distinction between stream and ground water ecology will disappear, and we will fully recognize that we are all interested in the same continuum. At this point, the fundamental unit of study will be the watershed, which is the appropriate scale to truly understand "stream" processes such as nutrient transport.

REFERENCES

Armitage, K. B. 1958. Ecology of riffle insects of the Firehole River, Wyoming. *Ecology* **39:**571–580.

Bailey, S. W., J. W. Hornbeck, C. W. Martin, and D. C. Buso. 1987. Watershed factors affecting stream acidification in the White Mountains of New Hampshire, USA. *Environmental Management* **11:**53–60.

Boylen, C. C., and T. D. Brock. 1973. Effects of thermal additions from the Yellowstone geyser basins on benthic algae of the Firehole River. *Ecology* **54:**1282–1291.

Brandes, J. A., M. E. McClain, and T. P. Pimentel. 1996. 15N evidence for the origin and cycling of inorganic nitrogen in a small Amazonian catchment. *Biogeochemistry* **34:**45–56.

Brunke, M., and T. Gonser. 1997. The ecological significance of exchange processes between rivers and ground water. *Freshwater Biology* **37:**1–33.

Buttle, J. M. 1994. Isotope hydrograph separations and rapid delivery of pre-event water from drainage basins. *Progress in Physical Geography* **18:**16–41.

Choi, J., S. M. Hulseapple, M. H. Conklin, and J. W. Harvey. 1998. Modelilng CO_2 degassing and pH in a stream-aquifer system. *Journal of Hydrology* **209:**297–310.

Cole, J. J., N. F. Caraco, G. W. Kling, and T. K. Kratz. 1994. Carbon dioxide supersaturation in the surface waters of lakes. *Science* **265:**1568–1570.

Constantz, J., C. L. Thomas, and G. Zellweger. 1994. Influence of diurnal variations in stream temperature on streamflow loss and ground water recharge. *Water Resources Research* **30:**3253–3264.

Currie, W. S., J. D. Aber, W. H. McDowell, R. D. Boone, and A. H. Magill. 1996. Vertical transport of dissolved organic C and N under long-term N amendments in pine and hardwood forests. *Biogeochemistry* **35:**471–505.

Dosskey, M. G., and P. M. Bertsch. 1994. Forest sources and pathways of organic matter transport to a blackwater stream: A hydrologic approach. *Biogeochemistry* **24:**1–19.

Fiebig, D. M. 1995. Ground water discharge and its contribution of dissolved organic carbon to an upland stream. *Archiv für Hydrobiologie* **134:**129–155.

Fiebig, D. M., M. A. Lock, and C. Neal. 1990. Soil water in the riparian zone as a source of carbon for a headwater stream. *Journal of Hydrology* **116:**217–237.

Fisher, S. G., and G. E. Likens. 1973. Energy flow in Bear Brook, New Hampshire: An integrative approach to stream ecosystem metabolism. *Ecological Monographs* **43:**421–439.

Fisher, S. G., N. B. Grimm, E. Martí, R. M. Holmes, and J. B. Jones. 1998. Material spiraling in stream corridors: A telescoping ecosystem model. *Ecosystems* **1:**19–34.

Ford, T. E., and R. J. Naiman. 1989. Ground water-surface water relationships in boreal forest watersheds: Dissolved organic carbon and inorganic nutrient dynamics. *Canadian Journal of Fisheries and Aquatic Sciences* **46:**41–49.

Gibert, J., M. J. Dole-Olivier, P. Marmonier, and P. Vervier. 1990. Surface water/groundwater eco-tones. *In* "Ecology and Management of Aquatic-Terrestrial Ecotones" (R. J. Naiman and H. Décamps, eds.), Vol. 4, pp. 199–225. UNESCO and Parthenon Publishing Group, Paris.

Groffman, P. M., G. Howard, A. J. Gold, and W. M. Nelson. 1996. Microbial nitrate process-ing in shallow ground water in a riparian forest. *Journal of Environmental Quality* **25:**1309–1316.

Herlihy, A. T., P. R. Kaufmann, M. R. Church, P. J. Wigington, J. R. Webb, and M. J. Sale. 1993. The effects of acidic deposition on streams in the Appalachian mountain and piedmont re-gion of the mid-Atlantic United States. *Water Resources Research* **29:**2687–2703.

Herman, J. S., and M. M. Lorah. 1987. CO_2 outgassing and calcite precipitation in Falling Spring Creek, Virginia, U.S.A. *Chemical Geology* **62:**251–262.

Hill, A. R. 1996. Nitrate removal in stream riparian zones. *Journal of Environmental Quality* **25:**743–755.

Hoffer-French, K. J., and J. S. Herman. 1989. Evaluation of hydrological and biological influ-ences on CO_2 fluxes from a karst stream. *Journal of Hydrology* **108:**189–212.

Holmes, R. M., S. G. Fisher, and N. B. Grimm. 1994. Parafluvial nitrogen dynamics in a desert stream ecosystem. *Journal of the North American Benthological Society* **13:**468–478.

Holmes, R. M., J. B. Jones, S. G. Fisher, and N. B. Grimm. 1996. Denitrification in a nitrogen-limited stream ecosystem. *Biogeochemistry* **33:**125–146.

Holmes, R. M., J. W. McClelland, D. M. Sigman, B. Fry, and B. J. Peterson. 1998. Measuring ^{15}N-NH_4^+ in marine, estuarine, and fresh waters: An adaptation of the ammonia diffu-sion method for samples with low ammonium concentrations. *Marine Chemistry* **60:**235–243.

Hope, D., M. F. Billet, and M. S. Cresser. 1994. A review of the export of carbon in river wa-ter: Fluxes and process. *Environmental Pollution* **84:**301–324.

Hynes, H. B. N. 1983. Ground water and stream ecology. *Hydrobiologia* **100:**93–99.

Jones, J. B., and R. M. Holmes. 1996. Surface-subsurface interactions in stream ecosystems. *Trends in Ecology and Evolution* **11:**239–242.

Jones, J. B., and P. J. Mulholland. 1998a. Carbon dioxide variation in a hardwood stream: An integrative measure of whole catchment soil respiration. *Ecosystems* **1:**183–196.

Jones, J. B., and P. J. Mulholland. 1998b. Influence of drainage basin topography and elevation on carbon dioxide and methane supersaturation of stream water. *Biogeochemistry* **40:**57–72.

Kaeding, L. R. 1996. Summer use of coolwater tributaries of a geothermally heated stream by rainbow and brown trout, *Oncorhynchus mykiss* and *Salmo trutta*. *American Midland Naturalist* **135:**283–292.

Kaeding, L. R., and C. M. Kaya. 1978. Growth and diets of trout from contrasting environ-ments in a geothermally heated stream: The Firehole River of Yellowstone National Park. *Transactions of the American Fisheries Society* **107:**432–438.

Kaplan, L. A., and J. D. Newbold. 1993. Biogeochemistry of dissolved organic carbon entering streams. *In* "Aquatic Microbiology: An Ecological Approach" (T. E. Ford, ed.), pp. 139–165. Blackwell, Oxford.

Kling, G. W., G. W. Kipphut, and M. C. Miller. 1991. Arctic lakes and streams as gas conduits to the atmosphere: Implications for tundra carbon budgets. *Science* **251:**298–301.

Kling, G. W., G. W. Kipphut, and M. C. Miller. 1992. The flux of CO_2 and CH_4 from lakes and rivers in arctic Alaska. *Hydrobiologia* **240:**23–36.

Lorah, M. M., and J. S. Herman. 1988. The chemical evolution of a travertine-depositing stream: Geochemical processes and mass transfer reactions. *Water Resources Research* **24:**1541–1552.

Lowrance, R. 1992. Ground water nitrate and denitrification in a coastal plain riparian forest. *Journal of Environmental Quality* **21:**401–405.

Madison, R. J., and J. O. Brunett. 1985. Overview of the occurrence of nitrate in ground wa-ter of the United States. *Geological Survey Water-Supply Paper (U.S.)* **2275.**

Mariotti, A., A. Landreau, and B. Simon. 1988. ^{15}N isotope biogeochemistry and natural denitrification process in ground water: Application to the chalk aquifer of northern France. *Geochimica et Cosmochimica Acta* **52**:1869–1878.

McMahon, P. B., and J. B. Bohlke. 1996. Denitrificatation and mixing in a stream—aquifer system: Effects on nitrate loading to surface water. *Journal of Hydrology* **186**:105–128.

Mulholland, P. J. 1992. Regulation of nutrient concentrations in a temperate forest stream: Roles of upland, riparian, and instream processes. *Limnology and Oceanography* **37**:1512–1526.

Mulholland, P. J. 1993. Hydrometric and stream chemistry evidence of three storm flowpaths in Walker Branch watershed. *Journal of Hydrology* **151**:291–316.

Mulholland, P. J. 1997. Dissolved organic matter concentration and flux is streams. *Journal of the North American Benthological Society* **16**:131–141.

Neal, C., M. Neal, A. Warrington, A. Avila, J. Piñol, and F. Roda. 1992. Stable hydrogen and oxygen isotope studies of rainfall and streamwaters for two contrasting holm oak areas of Catalonia, northeastern Spain. *Journal of Hydrology* **140**:163–178.

Nelson, P. N., J. A. Baldock, and J. M. Oades. 1993. Concentration and composition of dissolved organic carbon in streams in relation to catchment soil properties. *Biogeochemistry* **19**:27–50.

Ogunkoya, O., and A. Jenkins. 1993. Analysis of storm hydrograph and flow pathways using a three-component hydrograph separation model. *Journal of Hydrology* **142**:71–88.

Pentecost, A. 1995. Geochemistry of carbon dioxide in six travertine-depositing waters of Italy. *Journal of Hydrology* **167**:263–278.

Peterjohn, W. T., and D. L. Correll. 1984. Nutrient dynamics in an agricultural watershed: Observations on the role of a riparian forest. *Ecology* **65**:1466–1475.

Pinay, G., and H. Decamps. 1988. The role of riparian woods in regulating nitrogen fluxes between the alluvial aquifer and surface water: A conceptual model. *Regulated Rivers* **2**:507–516.

Pinay, G., N. E. Haycock, C. Ruffinoni, and R. M. Holmes. 1994. The role of denitrification in nitrogen removal in river corridors. *In* "Global Wetlands: Old World and New" (W. J. Mitsch, ed.), pp. 107–116. Elsevier, Amsterdam.

Piñol, J., and A. Avila. 1992. Streamwater pH, alkalinity, pCO_2 and discharge relationships in some forested Mediterranean catchments. *Journal of Hydrology* **131**:205–225.

Pringle, C. M., and F. J. Triska. 1991. Effects of geothermal ground water on nutrient dynamics of a lowland Costa Rican stream. *Ecology* **72**:951–965.

Pringle, C. M., P. Paaby-Hansen, P. D. Vaux, and C. R. Goldman. 1986. In situ nutrient assays of periphyton growth in a lowland Costa Rican stream. *Hydrobiologia* **134**:207–213.

Pringle, C. M., F. J. Triska, and G. Browder. 1990. Spatial variation in basic chemistry of streams draining a volcanic landscape on Costa Rica's Caribbean slope. *Hydrobiologia* **206**:73–85.

Pringle, C. M., G. L. Rowe, F. J. Triska, J. F. Fernandez, and J. West. 1993. Landscape linkages between geothermal activity and solute composition and ecological response in surface waters draining the Atlantic slope of Costa Rica. *Limnology and Oceanography* **38**:753–774.

Rutherford, J. E., and H. B. N. Hynes. 1987. Dissolved organic carbon in streams and ground water. *Hydrobiologia* **154**:33–48.

Sigman, D. M., M. A. Altabet, R. Michener, D. C. McCorkle, B. Fry, and R. M. Holmes. 1997. Natural abundance-level measurement of the nitrogen isotopic composition of oceanic nitrate: an adaptation of the ammonia diffusion method. *Marine Chemistry* **57**:227–242.

Sklash, M. G., and R. N. Farvolden. 1979. The role of ground water in storm runoff. *Journal of Hydrology* **43**:45–65.

Smith, R. L., B. L. Howes, and J. H. Duff. 1991. Denitrification in nitrate-contaminated ground water: Occurrence in steep vertical geochemical gradients. *Geochimica et Cosmochimica Acta* **55**:1815–1825.

Smith, R. L., S. P. Garabedian, and M. H. Brooks. 1996. Comparison of denitrification activity measurements in ground water using cores and natural-gradient tracer tests. *Environmental Science and Technology* **30**:3448–3456.

Spaulding, R. F., and M. E. Exner. 1993. Occurrence of nitrate in groundwater—a review. *Journal of Environmental Quality* **22**:392–402.

Triska, F. J., V. C. Kennedy, R. J. Avanzino, G. W. Zellweger, and K. E. Bencala. 1989. Retention and transport of nutrients in a third-order stream in northwestern California: Hyporheic processes. *Ecology* **70**:1893–1905.

Valiela, I., J. Costa, K. Foreman, J. M. Teal, B. L. Howes, and D. Aubrey. 1990. Transport of ground water-borne nutrients from watersheds and their effects on coastal waters. *Biogeochemistry* **10**:177–197.

Valiela, I., G. Collins, J. Kremer, K. Lajtha, M. Geist, B. Seely, J. Brawley, and C. H. Shaw. 1997. Nitrogen loading from coastal watersheds to receiving estuaries: New method and application. *Ecological Applications* **7**:358–380.

Vervier, P., J. Gibert, P. Marmonier, and M.-J. Dole-Olivier. 1992. A perspective on the permeability of the surface freshwater-groundwater interface. *Journal of the North American Benthological Society* **11**:93–102.

Waddington, J. M., N. T. Roulet, and A. R. Hill. 1993. Runoff mechanism in a forested ground water discharge swamp. *Journal of Environmental Quality* **25**:743–755.

Wallis, P. M., H. B. N. Hynes, and S. A. Telang. 1981. The importance of ground water in the transportation of allochthonous dissolved organic matter to the streams draining a small mountain basin. *Hydrobiologia* **79**:77–90.

Ward, J. V. 1989. The four-dimensional nature of lotic ecosystems. *Journal of the North American Benthological Society* **8**:2–8.

Ward, J. V., and J. A. Stanford. 1982. Thermal responses in the evolutionary ecology of aquatic insects. *Annual Review of Entomology* **27**:97–117.

White, D. S., C. H. Elzinga, and S. P. Hendricks. 1987. Temperature patterns within the hyporheic zone of a northern Michigan river. *Journal of the North American Benthological Society* **6**:85–91.

Zeikus, J. G., and T. D. Brock. 1972. Effects of thermal additions from the Yellowstone geyser basins on the bacteriology of the Firehole River. *Ecology* **53**:283–290.

6

Surface–Subsurface Exchange and Nutrient Spiraling

Patrick J. Mulholland* and Donald L. DeAngelis[†]

*Environmental Sciences Division
Oak Ridge National Laboratory
Oak Ridge, Tennessee

†National Biological Service, South Florida Field Laboratory
Department of Biology
University of Miami
Coral Gables, Florida

I. Introduction   149
II. Empirical Studies   151
III. A Stream Nutrient Spiraling Model with Subsurface Transient
Storage   154
   A. Model Derivation   154
   B. Calculation of Nutrient Uptake Length Using
the Model   155
   C. Model Experiments   157
IV. Results of Model Experiments   158
   A. Nutrient Concentration   158
   B. Nutrient Uptake Length   158
V. Relevance of Model Experiments   161
VI. Future Research Needs   163
   References   164

I. INTRODUCTION

More than 20 years ago in a landmark paper, Hynes (1975) showed how stream ecosystems were highly influenced by the characteristics of their catchments, particularly in terms of inputs of water and materials. In more recent years, we have come to appreciate the special importance of the environments immediately adjacent to streams, such as riparian areas (Gregory *et al.*, 1991). In addition to a lateral perspective, stream ecologists have begun to look vertically and have identified the important role of subsurface environments as determinants of biogeochemical processes in stream ecosystems (Hendricks and White, 1991; Findlay, 1995; Jones and Holmes, 1996). In fact, the interactions and coupling between surface and subsurface processes are so strong that we now view stream ecosystems as including the zone within streambed sediments where there is penetration of surface water and mixing between ground and surface waters.

The development of a coupled experimental and modeling approach to the study of stream hydrodynamics by Bencala and co-workers (Bencala, 1983; Bencala and Walters, 1983) precipitated many of the recent advances in our understanding of the role of surface-subsurface interactions in stream ecosystems (see Harvey and Wagner, Chapter 1 in this book, and Packman and Bencala, Chapter 2 in this book). This approach involves experimental additions of conservative tracers to streams and subsequent application to the experimental data of an advection–dispersion transport model that includes terms for transient storage to derive important hydrodynamic characteristics of streams. Among the parameters quantified from application of the model to experimental data are exchange rate of surface water with a transient storage zone ($\propto$), cross-sectional area of the surface water zone (A), and cross-sectional area of a transient storage zone (A_s). The transient storage zone can include surface-water pools at the stream channel margins that have slow exchange of water with the primary mass of flowing water (Harvey *et al.*, 1996), or even interstitial water within dense algal mats (Mulholland *et al.*, 1994). In many streams, however, the storage zone is formed principally by subsurface water that exchanges with the surface water. The coupled field experiment-modeling approach provides a relatively straightforward methodology for evaluating the hydrodynamic properties involving surface–subsurface exchange in small to medium-size streams.

Significant hydrologic exchange between surface and subsurface water should have a strong effect on nutrient dynamics and nutrient spiraling in stream ecosystems (see Duff and Triska, Chapter 8 in this book, and Hendricks and White, Chapter 9 in this book). This expectation is based on two factors: (1) high ratios of surface area of sediments to volume of water within sediments should result in large effects of microbial processes on subsurface water, and (2) relatively slow advective flow of water within the subsurface zone retards the downstream movement of soluble materials compared with the surface environment. In this chapter, we review the empirical evidence that surface–subsurface exchange plays an important role in nutrient dynamics and nutrient spiraling in streams and present simulations of a stream nutrient model that explicitly incorporates this exchange. We demonstrate with this simple model that the size of the subsurface zone and the rates of water exchange between surface and subsurface zones have substantial effects on nutrient dynamics and uptake lengths in stream ecosystems.

II. EMPIRICAL STUDIES

Empirical evidence for the effect of surface–subsurface exchange on chemical dynamics in streams is both indirect and direct. Direct evidence primarily comes from experimental additions of conservative and reactive solutes to surface water or subsurface water and measurement of solute concentrations over time and space (Bencala, 1984; Triska *et al.*, 1989, 1990). These studies have generally emphasized the potential for chemical transformation and retention in the subsurface zone and the effect of subsurface hydrodynamic properties on transformation and retention. Indirect evidence includes observations of spatial and/or temporal patterns in physical and chemical characteristics in streams that suggest effects of subsurface processes and water exchange between the surface and subsurface zones (Valett *et al.*, 1990, 1996; Hendricks and White, 1991; Findlay *et al.*, 1993; Holmes *et al.*, 1994; Jones *et al.*, 1995; Findlay and Sobczak, 1996; Wondzell and Swanson, 1996). Studies of this type have generally shown the subsurface zone to be a source of nutrients to surface water and a sink for organic carbon from the surface zone.

Among the earliest observational studies were those of Rutherford and Hynes (1987) and Ford and Naiman (1989) that focused on areas of groundwater inflow to streams and showed that concentrations of dissolved organic carbon (DOC) and nitrogen in ground water were higher than in stream surface water. These and other observations suggested that the subsurface zone functions as a sink for DOC and nitrogen as ground water flows into streams (Fiebig and Lock, 1991).

More recent observational studies have focused on the role of the subsurface in controlling surface-water nutrient dynamics and generally have shown the subsurface zone as a source of inorganic nutrients to surface water. These studies have identified flowpaths from the surface zone into and through the subsurface zone, usually using conservative tracer additions, and have found changes in nutrient concentration along those flowpaths. For example, in a Michigan stream, inorganic nitrogen concentration increased along a subsurface flowpath indicating net mineralization (Hendricks and White, 1991). In a comparison between upwelling and downwelling areas in an Arizona stream with highly permeable sediments, the subsurface zone was a source of inorganic nitrogen to the surface zone as a result of mineralization and nitrification of organic nitrogen supplied via downwelling (Valett *et al.*, 1990; Jones *et al.*, 1995). Similarly, in an Oregon stream net mineralization of surface water dissolved organic nitrogen was observed in the sub-surface zone (Wondzell and Swanson, 1996). In a stream in New York, changes in DOC concentration along subsurface flowpaths derived from downwelling surface water indicated that the subsurface zone was a sink for surface water DOC.

Studies involving solute injections to streams have demonstrated the potential for nutrient uptake in the subsurface zone in streams. In an early study, Bencala (1984) demonstrated the usefulness of field experiments involving co-injection of conservative and reactive solutes to quantify and distinguish hydrologic and chemical processes that determine solute transport and transformation in the subsurface zone of streams. The study by Bencala provided an approach for future work involving biologically important solutes (e.g., nutrients) and the determination of hydrological versus biological controls on their transport. Using a co-injection of nitrate and chloride (conservative tracer) to the surface water of Little Lost Man Creek, California, the subsurface zone accounted for considerable uptake of added nitrate based on declines in nitrate concentration relative to chloride concentration in hyporheic zone wells (Triska *et al.*, 1989). Co-injections of ammonium and chloride to Little Lost Man Creek also indicated considerable nitrification within aerobic subsurface zones (Triska *et al.*, 1990). Co-injections of nitrate and bromide (conservative tracer) to three streams in New Mexico with contrasting subsurface zone sizes indicated that nitrate uptake length, measured as the decline in surface water nitrate concentrations over distance, was inversely related to relative storage zone size (A_s/A; Valett *et al.*, 1996). Moreover, based on samples collected from wells installed within the stream channels, considerable subsurface uptake of nitrate occurred in all three streams in New Mexico, with subsurface uptake estimated to account for about one-half of the total uptake of nitrate in these streams. Subsurface uptake of nitrate was inversely related to subsurface dissolved oxygen concentration, indicating that much of the apparent nitrate uptake was the result of dissimilatory reduction (denitrification). Although these studies have indicated considerable rates of nutrient immobilization in the subsurface zone of streams, they involved addition of nutrients at concentrations well above background levels and thus may not accurately reflect nutrient dynamics at ambient concentrations (Mulholland *et al.*, 1990; Stream Solute Workshop, 1990).

The use of radiotracers allows nutrient dynamics to be assessed at ambient concentrations. In one such radiotracer study, $^{33}PO_4$ and ^3H (conservative tracer) were injected into two small, forested streams with contrasting transient storage zones but similar sizes and nutrient characteristics (Mulholland *et al.*, 1997). Hugh White Creek, North Carolina, had a relatively large transient storage zone (A_s/A ratio of 1.5) presumably as a result of large sediment accumulations behind several debris dams, whereas Walker Branch, Tennessee, had a small storage zone (A_s/A ratio of 0.09) because the channel bottom consisted of bedrock or shallow cobble and thin sediment deposits over bedrock. The phosphate uptake rate was about 2.5 times higher, and the phosphate uptake length was 5 times shorter in Hugh White Creek than in Walker Branch. Examination of the surface water ^{33}P:^{3}H ratio profiles in each stream suggested that about 43% of the ^{33}PO$_4$ uptake in Hugh White Creek but very little of the ^{33}PO$_4$ uptake in Walker Branch occurred within the transient storage zone. The uptake rate in the transient storage zone estimated for Hugh White Creek (231 µg P m^{-2} h^{-1}) represented gross phosphorus uptake, however, and phosphorus mineralization rates could exceed gross uptake rates resulting in net release of phosphorus to water within the transient storage zone.

Recent conceptual models and reviews of the biogeochemistry of stream ecosystems emphasizing the influence of surface–subsurface interactions have highlighted the effect of hydrodynamic factors. Vervier *et al.* (1992) focused on the importance of the permeability of the subsurface zone arguing that permeability controls the flux of water and hence the distance over which nutrient transformations occur, as well as the spatial distribution of redox conditions in the subsurface. In highly permeable zones, exchanges between the surface and subsurface will be large, but nutrient transformations might occur over longer distances. In low permeable zones, exchanges between the surface and subsurface are lower, but transformations occur over a shorter distance. Bencala (1993) presented a conceptual model of a stream as an integral part of a larger catchment system consisting of multiple flowpaths of subsurface and surface transport. The stream subsurface zone (which has historically been referred to as the hyporheic zone) is viewed as being composed of several subsurface flowpaths (vertical, lateral, and downvalley) with bidirectional hydrologic linkages with surface water. Bencala argues that a better understanding of the hydrodynamic properties is needed to understand solute dynamics in the hyporheic zone. Findlay (1995) emphasized the role of water residence time in the subsurface zone as the primary determinant of dissolved oxygen concentration and, in turn, the types of metabolism and rates of biogeochemical processes within stream ecosystems. In reviewing the literature on the role of stream subsurface zones as a source or sink for surface water nitrate, Jones and Holmes (1996) emphasize the importance of hydraulic conductivity and water residence time as well as surface water nitrate concentrations and organic matter supply. When surface water nitrate concentration is high or hydraulic conductivity of the subsurface zone is low (leading to low dissolved oxygen levels), the subsurface zone tends to be a sink for nitrate. Alternatively, when nitrate concentration is low and hydraulic conductivity in the subsurface zone is high, it tends to be a source for nitrate, assuming a sufficient supply of organic nitrogen or ammonium.

III. A STREAM NUTRIENT SPIRALING MODEL WITH SUBSURFACE TRANSIENT STORAGE

A. Model Derivation

A simple nutrient spiraling model (nutrient cycling in conjunction with downstream transport) was developed to evaluate the effects of the relative size and water exchange rates of subsurface transient storage zones on nutrient uptake lengths in streams (Fig. 1). The model is similar to one developed previously by DeAngelis *et al.* (1995) for stream periphyton communities. Both models create a transient storage zone; however, in this model, the surface and transient storage zones each contain nutrient cycling between water and biomass. This model consists of a longitudinal series of cells each consisting of a flowing surface water zone with a soluble nutrient compartment (N_w) and biomass nutrient compartment (B_w) and a stationary (nonflowing) subsurface zone (transient storage) with a soluble nutrient compartment (N_s) and a biomass nutrient compartment (B_s). The biomass compartments include living and nonliving organic matter. Exchange of soluble nutrients between the surface and subsurface water compartments is represented as a first-order process. Nutrient uptake by biomass compart-

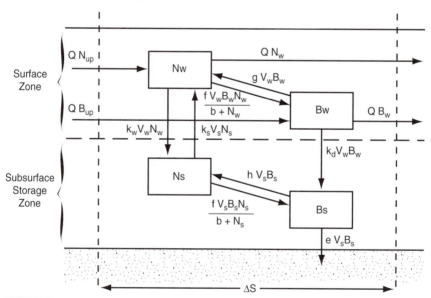

FIGURE 1 Schematic of each longitudinal cell of a stream nutrient cycling model with surface and subsurface storage zones. Soluble nutrient compartments are denoted by *N* and biomass nutrient compartments are denoted by *B*. The subscripts w and s refer to the surface and subsurface zones, respectively.

ments from soluble compartments follows Monod-type kinetics, and release of nutrients from biomass to soluble pools follows first-order, donor-controlled kinetics. There is a flux of biomass from the surface zone to the subsurface zone representing burial of detritus, and there is a loss of subsurface biomass from the system representing long-term storage or downstream export. Note that there is no explicit cycling or transformation of organic carbon; rather, nutrient regeneration from biomass back to available forms in water implicitly involves either excretion or respiratory losses of carbon from biomass.

Equations describing the rate of change in the state variables (N_w, B_w, N_s, B_s) for each longitudinal cell are as follows (see Fig. 1 for identification of each term):

$$\frac{d(V_w B_w)}{dt} = Q N_{up} - Q N_w - k_w V_w N_w + k_s V_s N_s - \frac{f V_w B_w N_w}{b + N_w} + g V_w B_w , \quad (1)$$

$$\frac{d(V_w B_w)}{dt} = Q B_{up} - Q B_w + \frac{f V_w B_w N_w}{b + N_w} - k_d V_w B_w - g V_w B_w , \quad (2)$$

$$\frac{d(V_s N_s)}{dt} = k_w V_w N_w - k_s V_s N_s + h V_s B_s - \frac{f V_s B_s N_s}{b + N_s} , \quad (3)$$

$$\frac{d(V_s N_s)}{dt} = k_d V_w N_w - e V_s B_s - h V_s B_s + \frac{f V_s B_s N_s}{b + N_s} , \quad (4)$$

where V_w and V_s are unit cell volumes for the free-flowing water and storage zone, respectively (Table I). The cross-sectional areas of the surface zone (A_w) and the subsurface zone (A_s) used in the standard transient storage zone model of Bencala (1983) are therefore simply $V_w/\Delta S$ and $V_s/\Delta S$, respectively, where ΔS is the unit cell length (0.1 m) used in the model. There is a similar set of equations for each cell, and the cells are linked by water flow (Q) transporting soluble nutrients (N_w) and biomass nutrients (B_w) in the surface zone only. The model was solved by numerically integrating from the upstream boundary conditions (Table I) downstream over thousands of linked cells. Values for the model exchange rate coefficients were based on data from Walker Branch, Tennessee (Table I).

B. Calculation of Nutrient Uptake Length Using the Model

As described by the nutrient spiraling concept (Newbold *et al.*, 1981, 1983), nutrient uptake length is defined as the distance traveled by a nutrient atom in water before it is taken up by stream biota. Thus, nutrient uptake length is an index of nutrient cycling in streams, describing the efficiency

TABLE 1   Parameters and Exchange Coefficients Used in the Stream Nutrient Cycling Model with Surface–Subsurface Exchange

Parameter	Description	Value
Dependent variables		
N_w	Nutrient concentration in surface zone water	
N_s	Nutrient concentration in subsurface zone water	
B_w	Biomass nutrients in surface zone	
B_s	Biomass nutrients in subsurface zone	
Boundary conditions (inputs)		
Q	Streamflow	1000 m³/d
N_{up}	Upstream nutrient concentration in water	0.05 g/m³
B_{up}	Upstream biomass nutrient concentration, surface zone	0.05 g/m³
Surface–subsurface volumes		
V_w	Volume of surface zone per cell ($A_w \, \Delta S$)	2.5 m³ ($A_w = 0.25$ m²)
Vs	Volume of subsurface zone per cell ($A_s \, \Delta S$)	0.25–12.5 ($A_s = 0.025$–1.25m²)
Hydrologic exchange coefficients		
k_w	Water exchange flux, surface to subsurface	5–100 d⁻¹
k_s	Water exchange flux, subsurface to surface	1–1000 d⁻¹
Nutrient cycling rate coefficients		
f	Maximum biotic nutrient uptake rate	10–25 d⁻¹
b	Half-saturation constant for biotic uptake	0.01 g/m³
g	Nutrient regeneration zone, surface zone	0.2 d⁻¹
h	Nutrient regeneration zone, subsurface zone	0.2 d⁻¹
Transport/loss rate coefficients		
k_d	Transfer of biomass from surface to subsurface	0.1 d⁻¹
e	Loss of biomass from subsurface, long term	0.01 d⁻¹

with which stream biota use nutrient inputs. Low values of nutrient uptake length indicate that a stream ecosystem is relatively efficient in nutrient uptake. Short nutrient uptake lengths can result from high nutrient uptake rates, low nutrient inputs, low water velocity, or a combination of all these factors.

Mean nutrient uptake length (S_w) was computed as the probability that a given molecule is taken up by biota (either in the surface or subsurface zone) in a cell at point x integrated over the entire stream length:

$$S_w = \int_0^\infty x' U(x') L(x') dx'. \qquad (5)$$

$U(x)$ can be thought of as the probability that a unit of soluble nutrient is not taken up by biota by the time it reaches location x (i.e., the fraction of nutrient remaining in solution at location x), and $L(x)$ is the probability that a unit of soluble nutrient entering the cell at location x will be taken up in that cell. $U(x)$ is found by numerically solving the equation:

$$\frac{dU(x)}{dx} = -L(x)U(x),\qquad(6)$$

where $U(0) = 1.0$. The probability of a unit of soluble nutrient being taken up at point x along the stream, $L(x)$, is for this model:

$$L(x) = P_{wupt} + (P_{ws}\, P_{supt}),\qquad(7)$$

where P_{wupt} is the probability of uptake by surface zone biota, P_{ws} is the probability of transfer from surface to subsurface, and P_{supt} is the probability of uptake by subsurface zone biota. Mathematically, $L(x)$ can be expressed as

$$L(x) = \frac{1}{Q}\left[\frac{A_w B_w f}{b + N_W} + k_w A_w \frac{fA_s\left(\dfrac{B_s}{b + N_s}\right)}{k_s A_s + \left(\dfrac{fB_s}{b + N_s}\right)}\right]\Delta S.\qquad(8)$$

C. Model Experiments

To evaluate the effects of subsurface storage zone size on nutrient uptake length, the surface zone size was held constant ($A_w = 0.25$ m^3, a typical value for small streams) and the model was solved for two contrasting cases: (1) minimal subsurface storage zone size ($A_s = 0.025$ m^3) relative to the surface zone ($A_s/A_w = 0.1$) and (2) large subsurface storage zone size ($A_s = 1.25$ m^3) relative to the surface zone ($A_s/A_w = 5$). Steady-state values of nutrients in water (N_w, N_s) and in biomass (B_w, B_s) were computed from the longitudinal trends in these parameters. Nutrient uptake length (S_w) was then computed for each case. To more closely examine the effect of water exchange rate (k_w, k_s), the model was solved, and S_w was calculated for a range of water exchange rates for each case. Because A_w, flow (Q), and nutrient input (N_{up}) were held constant, the effects of subsurface zone size and water exchange rates on S_w could be evaluated independently of the effects of water velocity and nutrient input. To evaluate the effects of the degree of nutrient demand, two values for the nutrient uptake coefficient ($f = 25$ d^{-1} and 10 d^{-1}, Fig. 1, Table I) were used in the model for each case (higher values of f denote stronger nutrient demand and hence stronger nutrient limitation).

IV. RESULTS OF MODEL EXPERIMENTS

A. Nutrient Concentration

Steady-state values of nutrient concentration in water (N_w, N_s) and in biomass (B_w, B_s) were obtained generally within about 500–1000 m downstream from the boundary conditions. Steady-state nutrient concentration was similar in the surface water and subsurface water (0.0002 g/m^3) and was independent of the relative size of the subsurface storage zone (A_s/A_w) and the water exchange rates (k_w, k_s). Steady-state concentration of nutrients in biomass was also independent of relative subsurface zone size, but because biomass is expressed on a volumetric basis, total biomass nutrient is considerably greater for the case with a large subsurface zone than for the case with a small subsurface cone. Although nutrient concentration of surface biomass (B_w, 0.05 g/m^3) was independent of water exchange rates, nutrient concentration of subsurface biomass (B_s) was inversely related to water exchange rates (Fig. 2).

Steady-state nutrient concentration in surface and subsurface waters was directly related to the degree of nutrient demand, with a 2.5-fold decline in the maximum nutrient uptake rate (f) resulting in a 2.5-fold increase in nutrient concentration in each zone. In contrast, biomass nutrient levels in the surface and subsurface zones were not affected by the degree of nutrient limitation.

B. Nutrient Uptake Length

When nutrient demand is strong ($f = 25$ d^{-1}), nutrient uptake length consistently declines with increasing surface water exchange rate, although the rate of decline in S_w with k_w is greater when subsurface storage zones are large than when they are small (Fig. 3a). At low values of k_w (< 20 d^{-1}), the effect of subsurface storage zone size on S_w is minimal, but at higher values of k_w, S_w is shorter for the larger storage zone cases. At k_w of 100 d^{-1}, S_w is about 25% shorter for a stream with $A_s/A_w = 1$ compared with a stream with $A_s/A_w = 0.1$. The effect of increasing storage zone size on S_w, however, appears to be greatest at relatively small storage zone sizes (A_s/A_w between 0.1 and 1.0), with little additional effect of increasing storage zone size above $A_s/A_w = 1.0$ (Fig. 3a).

Under weaker nutrient demand ($f = 10$ d^{-1}), S_w is considerably longer for all values of k_w compared with strong nutrient limitation (Fig. 3b). This result is as expected given that S_w is a measure of nutrient use efficiency and a greater fraction of the nutrient supply is taken up per unit time under more intense nutrient demand. In contrast to the case with strong nutrient demand, under weaker nutrient demand the effects of subsurface zone size on nutrient uptake length are not consistent between the two cases of differing

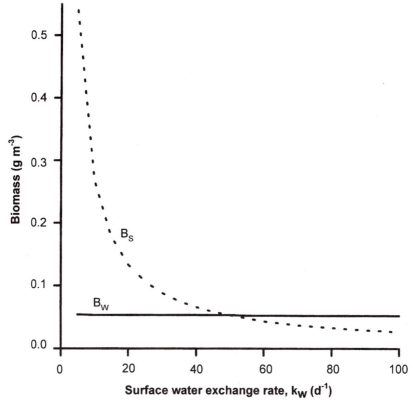

FIGURE 2   Relationship between biomass nutrients (surface zone, B_w, and subsurface storage zone, B_s) and water exchange rate between the surface and subsurface zones (k_w).

storage zone size. Although S_w declines with increasing k_w when subsurface storage zones are relatively large ($A_s/A_w \geq 1.0$), in the case of a small storage zone ($A_s/A_w = 0.1$), S_w reaches a minimum at a k_w of approximately 50 d^{-1} and then increases with increasing k_w. Therefore, although the effect of storage zone size on S_w is minimal at low k_w ($< 10\ d^{-1}$), it becomes quite large at high values of k_w. At a k_w of 100 d^{-1}, S_w is about 2.5 times shorter for the large storage zone cases ($A_s/A_w \geq 1.0$) than for the small storage zone case ($A_s/A_w = 0.1$). Again, the effect of increasing storage zone size on S_w is greatest for relatively small to moderate storage zone sizes ($A_s/A_w \leq 1.0$), with little effect of increases in storage zone size above $A_s/A_w = 1.0$.

The explanation for the shape of the S_w versus k_w curve for the stream with weaker nutrient demand and a small subsurface storage zone appears to be related to the distribution of nutrient uptake between the surface and subsurface zones. Under lower nutrient demand (weaker limitation), a greater fraction of total stream nutrient uptake occurs in the subsurface zone.

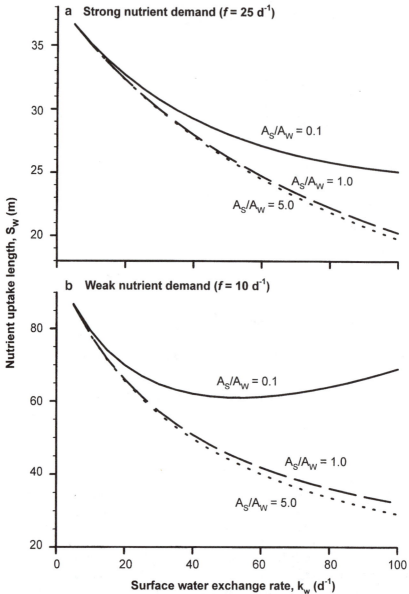

FIGURE 3 Relationship between nutrient uptake length (S_w) and water exchange rate between surface and subsurface zones (k_w) for three subsurface storage zone sizes ($A_s/A_w = 0.1$, 1.0, and 5.0). Relationships are given for (a) strong nutrient demand ($f = 25$ d^{-1}) and (b) weak nutrient demand ($f = 10$ d^{-1}).

Also, as k_w increases, k_s must also increase by the same factor such that the steady-state nutrient fluxes into and out of the subsurface zone are equal ($k_w V_w N_w = k_s V_s N_s$). Thus, in the case of $A_s/A_w = 0.1$, k_s increases from 50 to 1000 d^{-1} as k_w increases from 5 to 100 d^{-1}. As k_s increases, the residence time of water in the subsurface storage zone ($1/k_s$) declines, and a lower fraction of the nutrient entering the subsurface can be taken up with each pass. Thus, the effect of a shorter subsurface water residence time eventually compensates for increasing rates of water entry to the subsurface (k_w), and S_w increases with further increases in k_w and k_s. Under strong nutrient limitation, nutrient uptake is rapid relative to the water residence times in either zone, and the effect of a more rapid water exchange with the subsurface remains stronger than the effect of shorter subsurface water residence time as k_w and k_s increase. Eventually, at very large k_w (and k_s) values, S_w should begin to increase under all nutrient limitation and subsurface zone size scenarios.

In summary, our model results indicated that the relative size of subsurface storage zones and the rate of water exchange between surface and subsurface zones can have a significant effect on nutrient spiraling. Given similar water velocities and nutrient inputs, streams with larger subsurface storage zones and greater surface–subsurface water exchange rates generally had shorter nutrient uptake lengths.

V. RELEVANCE OF MODEL EXPERIMENTS

The values that we have used for the hydrodynamic properties in our model (e.g., A_s/A_w, k_w, k_s) are generally within the range of empirically derived values in streams. A_s/A_w in several streams at the Coweeta Hydrologic Laboratory, North Carolina ranged from < 0.1 to 2.0 (D'Angelo *et al.*, 1993) and A_s/A_w in three streams in New Mexico varied from 0.08 to 4.6 (Morrice *et al.*, 1997). The A_s/A_w values used in our model (0.1, 1.0, and 5.0) are within this range. Similarly, k_w ranges of 43–72 d^{-1} and 3–5 d^{-1} were reported for the Coweeta streams (D'Angelo *et al.*, 1993) and New Mexico streams (Morrice *et al.*, 1997), respectively. Our model simulated a range in k_w of 5–100 d^{-1}. Finally, in the New Mexico streams, k_s ranged from about 1 to 50 d^{-1}, whereas k_s was about 72 d^{-1} in Walker Branch, Tennessee (Mulholland *et al.*, 1997). The range of k_s values in our model simulations (1–1000 d^{-1}) was considerably greater than these real-world values, largely as a consequence of increasing k_w values for the case where $A_s/A_w = 0.1$. Normally, as k_w increases (greater rate of exchange of surface water with subsurface water), A_s/A_w would be expected to increase as well. Thus, our model results for the case of a relatively small subsurface storage zone ($A_s/A_w = 0.1$) may be unrealistic for high values of k_w and k_s (including the reversal in the S_w versus k_w curve for $A_s/A_w = 0.1$ and weak nutrient demand). Nonetheless, nutrient uptake length would be expected to decline with increasing k_w (and k_s) as shown in Fig. 3 because A_s/A_w would tend to increase with increases in k_w and k_s.

Modeling and field data support the hypothesis that increasing size of subsurface storage zones and/or water exchange rates between surface and subsurface zone water results in shorter nutrient uptake lengths (more efficient nutrient retention). Our model simulations show that nutrient uptake length decreases with an increase in the relative size of the subsurface storage zone (A_s/A_w) at a given water exchange rate (k_w). Our model also indicates that nutrient uptake length declines as k_w increases at a given A_s/A_w. The few empirical studies that have addressed this subject have generally supported this finding. In a comparison of small, forested streams, phosphorus uptake lengths were considerably shorter and phosphorus uptake rates were considerably higher in a stream with large values of A_s/A_w and k_w compared with a stream with lower values of A_s/A_w and k_w (Mulholland *et al.*, 1997). Similar findings for nitrate were found in a study of New Mexico streams (Valett *et al.*, 1996).

These results are not surprising. In our model, the subsurface storage zone is stationary. When the rate that water molecules and their dissolved nutrient load enter the subsurface storage zone is increased (i.e., k_w is increased) downstream transport is at least temporarily retarded, and nutrients can be taken up without further downstream movement. Although nutrient uptake rates by biota in the surface zone may be high relative to the subsurface zone because of higher light levels (stimulating autotrophs) or higher dissolved oxygen levels (stimulating heterotrophs), downstream transport in surface water is also high, and this tends to increase nutrient uptake lengths and reduce ecosystem nutrient use efficiency. Whereas nutrient uptake rates per unit of living biomass may be lower in the subsurface zone than in the surface zone, total nutrient uptake in the subsurface zone can be substantial due to high particle surface-to-volume ratios and longer water residence times. Although in reality there is often advective downstream transport within the subsurface zones of streams, as shown by the studies of Valett *et al.* (1990), Castro and Hornberger (1991), and Findlay *et al.* (1993) among others, subsurface water velocity is generally much lower than surface water velocity, and the net effect of surface–subsurface water exchange is to retard downstream transport and increase water residence times. Thus, our model is representative of the situation in which subsurface water velocity is negligible compared with the surface water velocity.

In our model, we have separated the effects of k_w and A_s/A_w on nutrient uptake length. The effect of A_s/A_w is greatest at smaller values of A_s/A_w (≤ 1.0) and larger values of k_w, and the effect tends to be greater under weaker nutrient demand (Fig. 3). The effect of k_w is greater for higher values of A_s/A_w, regardless of nutrient limitation status. The steady-state constraint that water fluxes into and out of the subsurface storage zone must be in balance, however, means that k_w and A_s/A_w are not independent (Fig. 1). Rather $\Delta Sk_w A_w$ must equal $\Delta Sk_s A_s$, and therefore A_s/A_w is equivalent to k_w/k_s. Thus, if A_s/A_w increases, k_w must also increase (i.e., increase in water exchange rate) or k_s must decline (i.e., increase in subsurface water residence time), both of which will tend to increase the importance of the subsurface storage zone as a site for nutrient uptake in streams (and reduce nutrient uptake lengths). Alternatively, if k_w increases, then either k_s or A_s/A_w must increase, again enhancing the effect of the subsurface storage zone on stream nutrient dynamics.

Our model indicates that the effect of k_w and k_s on nutrient uptake length could be substantially influenced by the degree of nutrient limitation of stream biota. In the case of strong nutrient demand (Fig. 3a), nutrient uptake length is short and declines consistently with increasing k_w and k_s, the rate of decline increasing at greater relative subsurface storage zone size (higher A_s/A_w). Under weaker nutrient demand (Fig. 3b), nutrient uptake length is longer at given k_w and A_s/A_w values. More importantly, however, under weak nutrient demand and for relatively small storage zone sizes ($A_s/A_w = 0.1$), nutrient uptake length is minimum at intermediate values of k_w and then increases at higher k_w, although high k_w with low A_s/A_w may not represent a realistic condition as described earlier. Thus, under weak nutrient demand, the importance of nutrient uptake in the subsurface zone becomes more strongly dependent on the residence time of water in this zone ($1/k_s$). As k_s becomes very large (at high k_w when $A_s/A_w = 0.1$), rates of nutrient uptake in the subsurface zone become small. For example, at a k_w of $80 \ d^{-1}$ for the case of $A_s/A_w = 0.1$, k_s is $800 \ d^{-1}$, and thus subsurface zone water residence time is only 1.8 min, allowing for little nutrient uptake before water is released back to the surface zone.

VI. FUTURE RESEARCH NEEDS

Additional empirical studies, comparative as well as experimental, are needed to test the model results we have presented here. Simultaneous measurements of surface–subsurface water exchange properties and nutrient uptake lengths can be made using conservative tracer and nutrient injections at low concentrations (Mulholland *et al.*, 1997). Measurements are needed in streams spanning a broad range in subsurface zone size, water exchange rates, and quantity and quality of subsurface organic matter storage. Experimental studies involving enhancement or reduction in the size of subsurface zones via manipulation of debris dams and other stream features that control sediment accumulation are also needed to allow a robust test of the inverse relationship between uptake length and subsurface zone size/exchange rate that we propose.

Additional modeling studies should attempt to account for differences in the nutrient cycling processes of heterotrophic organisms, which dominate in the subsurface zone, and autotrophic organisms, which can make up a substantial portion of the attached community in surface zones. Here we have treated nutrient uptake in both zones similarly using the same Monod-type uptake function and coefficients for both zones. We also have used similar biomass-specific nutrient regeneration rates for both zones. These assumptions seemed reasonable for this initial attempt at developing a nutrient spiraling model that incorporates surface–subsurface exchange; however, future models should explore the effects of using alternative formulations for nutrient uptake and regeneration processes in the subsurface.

It is clear from this modeling exercise and from the few empirical studies that have related surface–subsurface exchange to nutrient uptake length that the physical characteristics of subsurface zones are very important to nutrient spiraling in streams. The efficiency with which streams utilize nutrient inputs and retard the loss of nutrients downstream is a product not only of the biological communities in streams but also the physical characteristics of stream channels, and particularly the hydrodynamics of surface–subsurface interactions.

REFERENCES

Bencala, K.E. 1983. Simulation of solute transport in a mountain pool-and-riffle stream with a kinetic mass transfer model. *Water Resources Research* **19**:732–738.

Bencala, K. E. 1984. Interactions of solutes and streambed sediment. 2. A dynamic analysis of coupled hydrologic and chemical processes that determine solute transport. *Water Resources Research* **20**:1804–1814.

Bencala, K. E. 1993. A perspective on stream-catchment connections. *Journal of the North American Benthological Society* **12**:44–47.

Bencala, K. E., and R. A. Walters. 1983. Simulation of solute transport in a mountain pool-and-riffle stream: A transient storage model. *Water Resources Research* **19**:718–724.

Castro, N. M., and G. M. Hornberger. 1991. Surface-subsurface water interactions in an alluviated mountain stream channel. *Water Resources Research* **27**:1613–1621.

D'Angelo, D. J., J. R. Webster, S. V. Gregory, and J. L. Meyer. 1993. Transient storage in Appalachian and Cascade mountain streams as related to hydraulic characteristics. *Journal of the North American Benthological Society* **12**:223–235.

DeAngelis, D. L., M. Loreau, D. Neergaard, P. J. Mulholland, and E. R. Marzolf. 1995. Modelling nutrient-periphyton dynamics in streams: The importance of transient storage zones. *Ecological Modelling* **80**:149–160.

Fiebig, D. M., and M. A. Lock. 1991. Immobilization of dissolved organic matter from groundwater discharging through the stream bed. *Freshwater Biology* **26**:45–55.

Findlay, S. 1995. Importance of surface-subsurface exchange in stream ecosystems: The hyporheic zone. *Limnology and Oceanography* **40**:159–164.

Findlay, S., and W. V. Sobczak, 1996. Variability in removal of dissolved organic carbon in hyporheic sediments. *Journal of the North American Benthological Society* **15**:35–41.

Findlay, S., D. Strayer, C. Goumbala, and K. Gould. 1993. Metabolism of streamwater dissolved organic carbon in the shallow hyporheic zone. *Limnology and Oceanography* **38**:1493–1499.

Ford, T. E., and R. J. Naiman. 1989. Groundwater-surface water relationships in boreal forest watersheds: Dissolved organic carbon and inorganic nutrient dynamics. *Canadian Journal of Fisheries and Aquatic Sciences* **46**:41–49.

Gregory, S. V., F. J. Swanson, W. A. McKee, and K. W. Cummins. 1991. An ecosystem perspective of riparian zones. *BioScience* **41**:540–551.

Harvey, J. W., B. J. Wagner, and K. E. Bencala. 1996. Evaluating the reliability of the stream tracer approach to characterize stream-subsurface water exchange. *Water Resources Research* **32**:2441–2451.

Hendricks, S. P., and D. S. White. 1991. Physicochemical patterns within a hyporheic zone of a northern Michigan river, with comments on surface water patterns. *Canadian Journal of Fisheries and Aquatic Sciences* **48**:1645–1654.

Holmes, R. M., S. G. Fisher, and N. B. Grimm. 1994. Parafluvial nitrogen dynamics in a desert stream ecosystem. *Journal of the North American Benthological Society* **13**:468–478.

Hynes, H. B. N. 1975. The stream and its valley. *Verhandlungen Internationale Vereinigung für Theoretische und Angewandte Limnologie* **19**:1–15.

Jones, J. B., and R. M. Holmes. 1996. Surface-subsurface interactions in stream ecosystems. *Trends in Ecology and Evolution* **11**:239–242.

Jones, J. B., S. G. Fisher, and N. B. Grimm. 1995. Nitrification in the hyporheic zone of a desert stream ecosystem. *Journal of the North American Benthological Society* **14**:249–258.

Morrice, J. A., H. M. Valett, C. N. Dahm, and M. E. Campana. 1997. Alluvial characteristics, groundwater-surface water exchange and hydrological retention in headwater streams. *Hydrological Processes* **11**:253–267.

Mulholland, P. J., A. D. Steinman, and J. W. Elwood. 1990. Measurement of phosphorus uptake length in streams: Comparison of radiotracer and stable PO_4 releases. *Canadian Journal of Fisheries and Aquatic Sciences* **47**:2351–2357.

Mulholland, P. J., A. D. Steinman, E. R. Marzolf, D. R. Hart, and D. L. DeAngelis. 1994. Effect of periphyton biomass on hydraulic characteristics and nutrient cycling in streams. *Oecologia* **98**:40–47.

Mulholland, P. J., E. R. Marzolf, J. R. Webster, and D. R. Hart. 1997. Evidence that hyporheic zones increase heterotrophic metabolism and phosphorus uptake in forest streams. *Limnology and Oceanography* **42**:443–451.

Newbold, J. D., J. W. Elwood, R. V. O'Neill, and W. Van Winkle. 1981. Measuring nutrient spiralling in streams. *Canadian Journal of Fisheries and Aquatic Sciences* **38**:860–863.

Newbolt, J. D., J. W. Elwood, R. V. O'Neill, and A. L. Sheldon. 1983. Phosphorus dynamics in a woodland stream ecosystem: A study of nutrient spiralling. *Ecology* **64**:1249–1265.

Rutherford, J. E., and H. B. N. Hynes. 1987. Dissolved organic carbon in streams and groundwater. *Hydrobiologia* **154**:33–48.

Stream Solute Workshop. 1990. Concepts and methods for assessing solute dynamics in stream ecosystems. *Journal of the North American Benthological Society* **9**:95–119.

Triska, F. J., V. C. Kennedy, R. J. Avanzino, G. W. Zellweger, and K. E. Bencala. 1989. Retention and transport of nutrients in a third-order stream in northwestern California: Hyporheic processes. *Ecology* **70**:1893–1905.

Triska, F. J., V. C. Kennedy, R. J. Avanzino, G. W. Zellweger, and K. E. Bencala. 1990. In situ retention-transport response to nitrate loading and storm discharge in a third-order stream. *Journal of the North American Benthological Society* **9**:229–239.

Valett, H. M., S. G. Fisher, and E. H. Stanley. 1990. Physical and chemical characteristics of the hyporheic zone of a Sonoran Desert stream. *Journal of the North American Benthological Society* **9**:201–215.

Valett, H. M., J. A. Morrice, C. N. Dahm, and M. E. Campana. 1996. Parent lithology, groundwater-surface water exchange and nitrate retention in headwater streams. *Limnology and Oceanography* **41**:333–345.

Vervier, P., J. Gibert, P. Marmonier, and M. J. Dole-Olivier. 1992. A perspective on the permeability of the surface freshwater-groundwater ecotone. *Journal of the North American Benthological Society* **11**:93–102.

Wondzell, S. M., and F. J. Swanson. 1996. Seasonal and storm dynamics of the hyporheic zone of a 4th-order mountain stream. II. Nitrogen cycling. *Journal of the North American Benthological Society* **15**:20–34.

7

Emergent Biological Patterns and Surface– Subsurface Interactions at Landscape Scales

C. M. Pringle* and Frank J. Triska[†]

*Institute of Ecology
University of Georgia
Athens, Georgia

[†]U.S. Geological Survey
Water Resources Division
Menlo Park, California

I. Introduction   167
II. The Balance of Physical and Chemical Factors on the Geologic Template and Emergent Biological Patterns   169
III. Hydrothermal Systems as Models   171
IV. Human Impacts on Surface–Subsurface Interactions   175
    A. Nutrient and Toxicant Loading   177
    B. Land-Use Activities   178
    C. Hydrological Modifications   179
    D. Modifications to the Atmosphere—Atmospheric Deposition and Global Warming   184
V. Synthesis and Recommendations for Future Studies   185
    References   189

I. INTRODUCTION

The interaction between surface and subsurface water influences biological patterns in streams at landscape scales. Yet, our understanding of biological patterns in "natural" landscapes is poor, in part due to the inherent complexity of surface–subsurface systems and to the extent to which the earth has been transformed by human actions (Turner *et al.*, 1990). Too often, severe landscape disturbance by humans precedes our awareness of the nature and extent to which surface–subsurface interactions are important in maintaining the biological integrity of the landscape.

Streams represent a complex integration of the physics, chemistry, and biology of the landscape. Some aspects are expressed locally, such as light penetration through the riparian canopy which influences primary production. Others, such as dissolved organic carbon and nutrient composition, result from complex geochemical and biogeochemical processes within the catchment which control transport across the ground water–surface water interface and eventually down the stream channel. Many flowpath connections exist between the stream and the catchment, from small-scale upwelling in a single microhabitat to the large-scale discharge of alluvial aquifers (Brunke and Gonser, 1997). Moreover, streams can receive considerable nutrient input from these subsurface connections (Wallis *et al.*, 1981; Ford and Naiman, 1989; Fiebig and Lock, 1991; Pringle and Triska, 1991). Because of continual transport, which characterizes ground water–surface water systems, an effect originating in one part of the landscape may be expressed at a distant geographic location—often with a significant lag time. Clearly, a broad perspective is required to understand how surface–subsurface interactions influence biological patterns in streams at landscape scales. In this chapter, we use *landscape* to refer to scales that are greater than a river reach or low-order catchment. We also include regional and global patterns in our discussion.

The underlying geologic template and the location of surface and ground waters in the landscape largely determine surface–subsurface water interactions and the resulting biological pattern. Physical and chemical processes acting on the geological template determine biological patterns in lakes and streams. Catchment geology may represent a landscape constraint on the magnitude of surface–subsurface exchange (e.g., Freeze and Cherry, 1979; Valett *et al.*, 1997) and determine emergent biological patterns. Predictions are commonly made regarding nutrient properties of waters based on regional geology (e.g., Freeze and Cherry, 1979). Where phosphorus is low in underlying geological parent material, it is reflected by low concentrations in surface water. Correspondingly, predictions of phosphorus limitation of primary production are commonly made for surface waters draining these areas (e.g., Omernik, 1977). By contrast, in regions with abundant geologic sources of phosphorus, nitrogen typically limits primary production (e.g., Grimm *et al.*, 1991). Thus, the regional limitation of specific elements can determine biotic patterns in algal production and community composition, as well as biogeochemical element transformations at the surface–subsurface water interface where control is largely "bottom-up."

Human activities threaten the ecological integrity of biotic pattern in streams through nutrient, toxicant, and organic loading or by reducing connectivity (decoupling) which alters critical exchange processes (Brunke and Gonser, 1997). The extent and magnitude of anthropogenic modifications (e.g., urbanization, stream channelization, dams, interbasin transfers, atmospheric deposition) variably decouples riverine systems from the natural landscape, thereby altering surface–subsurface water interaction at landscape scales and affecting biological patterns.

In this chapter, we focus on emergent biological patterns in riverine ecosystems at landscape scales resulting from surface–subsurface water interaction. Our objectives are to examine (1) how the balance of physical and chemical factors on the "natural" geologic template affects biological patterns, (2) how natural hydrothermal systems can be used as a model for understanding surface–subsurface interactions and biological patterns in streams, and (3) how anthropogenic influences decouple the stream from the landscape by altering the nature of surface–subsurface water interactions and affecting biological patterns. We conclude with a synthesis and recommendations for future studies.

II. THE BALANCE OF PHYSICAL AND CHEMICAL FACTORS ON THE GEOLOGIC TEMPLATE AND EMERGENT BIOLOGICAL PATTERNS

Subsurface water that enters fluvial systems is ecologically important over a range of spatial scales. On the scale of a stream reach, the distribution of some aquatic macrophytes is influenced by the chemical and physical properties of subsurface water during interaction with surface water (Fortner and White, 1988; White *et al.*, 1992). Similarly, nutrient-rich groundwater upwellings in nutrient-poor streams can be important nutrient sources to benthic algal communities (Pringle *et al.*, 1986, Pringle and Triska, 1991). Hyporheic delivery of nutrients can increase algal standing crop and shorten recovery times of primary producers after spates (Grimm *et al.*, 1991; Valett *et al.*, 1992, 1994).

As discussed earlier, most studies have considered the biological significance of subsurface waters on relatively small spatial scales (e.g., stream reach). Relatively little attention has been focused on ecological patterns of subsurface water inputs into streams at landscape scales. However, the spatial distribution of small-scale surface–subsurface water interactions (e.g., Dent *et al.*, Chapter 16 in this book; Holmes, Chapter 5 in this book) may result in emergent physical and chemical patterns that are reflected by the distribution or productivity of aquatic biota at landscape and regional scales.

The capacity of the terrestrial environment to retain nutrients (or to lose them via hydrologic transport) and to induce or sustain biological patterns stems from landscape properties including soil lithology, infiltration capacity, and hydraulic conductivity, especially along the riverine corridor. Both hyporheic (Duff and Triska, Chapter 8 in this book) and riparian areas (Hill, Chapter 3 in this book) have demonstrated capacity for uptake and transformation of nutrients, especially nitrogen. Landscapes characterized by soils with rapid infiltration and high hydraulic conductivity readily transport nutrients, especially nitrate due to its high solubility and limited capacity for sorption. Thus, groundwater contamination—and eventually stream-water contamination—is most common beneath well-drained soils (Spalding and Exner, 1993). On the other hand, landscapes characterized by poorly drained soils especially in riparian areas can have a larger capacity for nitrate loss via denitrification (Gambrell *et al.*, 1975).

Jordan *et al.* (1997) contrasted transport-retention characteristics in soils of two different landscapes (piedmont and coastal plain) in the eastern United States for the delivery of nitrogen to streams draining into the Chesapeake Bay. In watersheds with similar proportions of cropland and nutrient application rates, there were systematic differences in nitrogen in stream discharge. In coastal plain soils, crops are typically located on well-drained uplands above poorly drained riparian forests (Gilliam and Skaggs, 1988; Correll, 1991). This type of landscape pattern can retain 70–90% of nitrogen inputs, especially nitrate, as either plant uptake or denitrification primarily in riparian zones which constitute the terrestrial–aquatic interface (Lowrance *et al.*, 1984; Peterjohn and Correll, 1984; Jacobs and Gilliam, 1985; Lowrance, 1992; Haycock and Pinay, 1993). However, in piedmont soils groundwater flows are deeper (up to 30 m) beneath riparian soils (Pavich *et al.*, 1989), and riparian zones are narrower (Jordan *et al.*, 1997), resulting in direct nitrate emergence into the stream and a limited opportunity for riparian or hyporheic transformation. Thus, the soil type in a landscape has a direct impact on surface–subsurface biogeochemistry and sensitivity to shifts in nutrient-related biological patterns, especially as human-dominated land-use intensifies.

Differences in temperature between upwelling ground- and surface-waters can dramatically affect the thermal regime of rivers and streams and habitat for biota. For example, upwelling of cold ground waters creates cold water refugia for striped bass (*Morone saxatilis*) and other stenotherms in a southeastern United States coastal river that drains into the Gulf of Mexico (Van Den Avyle and Evans, 1990). Cool water upwellings in warm climates are also essential for the survival of juvenile salmonids such as anadromous steelhead (*Oncorhynchus mykiss*) and coho salmon (*Oncorhynchus kisutch*) in northern California. During winter, upwelling ground water is often warmer than receiving surface waters and prevents freezing. Warm water upwellings in the Chilkat River of Alaska allow Chum Salmon (*Oncorhynchus keta*) to spawn during winter. As a result of this regional surface–subsurface water interaction, two biological patterns—salmon spawning and the migratory patterns of the largest bald eagle population in the world—have become linked (Keller and Kondolf, 1990).

III. HYDROTHERMAL SYSTEMS AS MODELS

An understanding of hydrothermal systems (e.g., Lowell, 1991) can yield important information on how groundwater–surface-water interactions influence emergent biological patterns on regional spatial scales (Pringle *et al.*, 1993). Convective hydrothermal systems require a heat source and a groundwater circulation system. The heat may be supplied by the regional groundwater or shallow magmatic intrusions. Solute levels and ratios in hydrothermal systems identify hydrologic flowpaths because of their distinctive chemical signature, despite dilution and transport over long distances. Analyses of hydrothermal systems can yield insight into flowpaths of anthropogenic contaminants. Often, pathways of such contaminants are little-known and are only identified after they have deleterious biological effects in surface waters or contaminate potable ground water.

Geothermal activity is the natural process of convective circulation of hot fluids through upper portions of the earth's crust. It occurs in tectonically active zones and can dramatically alter the biogeochemistry of aquatic systems. Surface waters are affected when they receive subsurface inputs of solute-rich, geothermally derived fluids. In continental systems, hydrothermally influenced fluids are typically deeply penetrating ground waters of meteoric origin. The composition of geothermal fluids is determined by many processes including addition of volatiles and metals from crystallizing and degassing magmas, the boiling of liquids as they rise to the surface, water–rock reactions, and mineral dissolution and precipitation reactions along surface and subsurface flowpaths (Ellis and Mahon, 1964, 1967).

The chemical composition of geothermally modified water is often very distinctive and produces a distinctive chemical–biological pattern in landscapes underlain by geothermal activity. For example, distinctive spatial patterns occur in the nutrient chemistry of streams draining the Central American volcanic arc. They result from geothermal activity associated with tectonic activity (the northeastward subduction of the Cocos Ridge under the Caribbean Plate along the Middle American Trench). Analysis of the solute chemistry of geothermal waters has allowed geochemists to make inferences regarding how geothermal systems operate and pathways by which flows of geothermally modified waters become distributed in the landscape (e.g., Henley and Ellis, 1983; Henley, 1985).

Streams draining the volcanic mountains of Costa Rica show a wide range of solute compositions and biological effects that partly reflect the contribution of different types of solute-rich geothermal waters (Pringle *et al.*, 1990, 1993; Pringle, 1991). Three major physical transport vectors affect flows of geothermally derived solutes and play an important role in determining emergent biogeochemical patterns in surface waters. By "physical transport vectors," we refer to those forces that drive flows of materials between elements of the landscape (Forman and Godron, 1986). These vectors impose directional fluxes of materials on the general background of resource concentration gradients and passive diffusion (Wiens *et al.*, 1985). They include (1) thermally driven convection of volcanic gases and geothermal fluids; (2) lateral and gravity-driven downward transport of geothermal fluids; and (3) wind dispersion of ash, gases, and acid rain (Table I).

Specific vector combinations interact to determine landscape patterns in stream solute chemistry and biota. Indicator taxa of algae and bacteria reflect factors such as high concentrations of various solutes, precipitation reactions, hydrologically transported acidity, and high temperatures. Geology, climate, and land use are additional variables that interact to affect the pathway of physical transport vectors, ultimately determining the location, distribution, and biogeochemistry of geothermally modified surface waters (Table I; Pringle *et al.*, 1993).

For example, on the Caribbean slope of Costa Rica's dormant Barva Volcano, some high-elevation streams receive subsurface inputs of solutes that are produced by the interaction of H_2S-rich steam with near-surface ground water. The resultant acid-sulfate streams have a biotic community characterized by acidophilic algae and sulfur-oxidizing bacteria. In contrast, low-elevation streams at the base of the volcano receive subsurface inputs of mixed geothermal waters of the $Na-Cl-HCO_3$ type (Fig. 1A). It is quite common for this type of water to be discharged at low elevations—at great distances from upflow areas where acid-sulfate waters occur (Henley, 1985). The $Na-Cl-HCO_3$ waters result from the lateral downward flow of $Na-Cl$ waters (which are generally produced by incorporation of magmatic fluids rich in chloride and alkali metals). The $Na-Cl$ waters ascend to the upper part of the volcano and then flow downward, incorporating carbon dioxide (CO_2) which results in the formation of carbonic acid, which in turn reacts with rock along the subsurface flowpath to produce cations and bi-

TABLE I Landscape and Regional Variables That Interact to Affect the Pathway of Physical Transport Vectors Associated with Geothermal Activity to Determine the Location, Distribution, and Biogeochemistry of Geothermally Modified Surface Waters

Landscape and regional variables	Physical transport vectors	Geothermally modified surface water
Geology	Thermally driven convection of volcanic gases and geothermal fluids	Location
Climate	Gravity-driven lateral and downward transport of geothermal fluids	Distribution
Topography	Wind dispersion of ash, gases, and acid rain	Biogeochemistry
Land use		

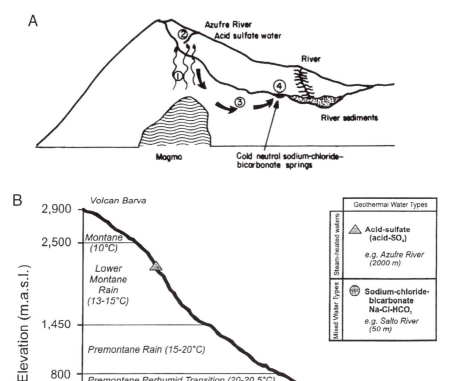

FIGURE 1 (A) Hydrothermal systems provide excellent models for understanding groundwater–surface-water interactions on landscape scales because of the distinctive chemical signature of hydrothermally altered waters. This schematic diagram of Volcan Barva in Costa Rica's Central Cordillerfa illustrates how (1) H_2S rich vapors from a deep, actively cooling magma body are absorbed into surficial ground waters. H_2S oxidation occurs when waters are exposed to the atmosphere, emerging as (2) acid–SO_4 waters at high elevations (2000 m.a.s.l.). (3) Extensive, subsurface lateral flow of acid geothermal waters occurs during which CO_2 is incorporated; and (4) neutral $Na–Cl–HCO_3$ springs emerge at the base of the volcano 35 m above seal level). Figure modified from Pringle *et al.* (1993). (B) Figure illustrating the location of geothermally modified ground waters with respect to major bioclimatic life zones (Holdrige *et al.*, 1971) along the Caribbean slope of Volcan Barva, Costa Rica. Acid–sulfate waters emerge at high elevations in lower montane rainforest and phosphorus-rich, sodium-chloride-bicarbonate springs emerge at low elevations in the "tropical wet forest" life zone which is characterized by extremely high biotic diversity. Figure modified from Pringle and Ramirez (1998).

carbonate and other weathering products. These subsurface sodium-chloride bicarbonate waters are biologically important because they are also rich in phosphorus (up to 400 μgP L^{-1}) and emerge at the base of the volcano in a very productive and biologically diverse life zone—lowland tropical wet forest (Fig. 1B).

The particular environmental conditions and trophic structure of stream ecosystems where solute-rich geothermal waters arise (Figs. 1B and 2) play a major role in determining biotic effects of introduced solutes. A major "gateway" by which solute-rich ground waters have an impact on the food web of lowland tropical streams at the base of Barva Volcano is through microbial communities at the surface–subsurface interface. Rates of algal growth (Pringle *et al.*, 1986; Pringle and Triska, 1991) and microbially mediated decomposition (A. D. Rosemond, University of Georgia, unpublished data) are presumably phosphorus-saturated in receiving streams. Relatively high temperatures in these warm (24–27°C) lowland streams further stimulate microbial activity. Where phosphorus-rich subsurface waters arise in stream systems draining pasture with high light penetration, algal growth may be stimulated. Alternatively, heterotrophic microbial processes may predominate if receiving streams drain densely shaded interior forest. Our current studies are examining the extent to which higher trophic levels (e.g., benthic insects, shrimps, and some fishes), that are dependent on algae or decomposing allochthonous inputs as food resources (Pringle and Hamazaki, 1997, 1998; Rosemond *et al.*, 1998) are affected by subsurface solute inputs (Fig. 2).

A general indication of the prevalence and pattern in the occurrence of many different types of distinctive geothermal water types in the Central American region can be found by examining maps of the volcanic mountain ranges of the region (Fig. 3). Certain river names that are indicative of geothermal water types consistently recur. Examples include Agrio (sour); Aufre (sulfur); Azul (blue); Caliente (hot); Salitral (salty); and Gata (which refers to the milky blue color of cat eyes—a color often seen in streams characterized by benthic precipitation of aluminum silicate resulting from changes in pH; Fig. 3).

In summary, hydrothermal systems provide excellent models to examine how subsurface waters influence biological patterns in streams on landscape scales for two major reasons. First, the scale of hydrothermal systems is regional: the movement of continental plates results in global patterns of geothermal activity with consequent effects on groundwater–surface-water interactions and emergent biological pattern. Second, chemical signatures are often distinct which make them relatively easy to trace in the landscape (see Genereux and Pringle, 1997). Although the effects of most landscapes on stream chemistry and biota may be more subtle than the distinctive chemistry resulting from hydrothermal systems, it is likely that biological assemblages

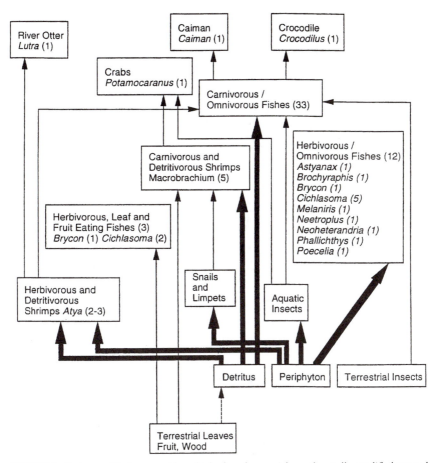

FIGURE 2 Pathways (dark arrows) by which phosphorus-rich geothermally modified ground water can affect the trophic structure of lowland streams draining tropical wet forest at the base of Volcan Barva, Costa Rica. Stimulation of bacterially mediated detrital decomposition and algal growth rates are two ways that phosphorus-rich ground water can affect higher tropic levels in receiving streams. For example, of approximately 45 taxa of fishes, 12 consume significant amounts of algae, and many fish taxa are omnivorous, consuming detritus.

in most regions of distinct geology reflect weathering processes, groundwater transport, and riparian and hyporheic interactions in a similar way.

IV. HUMAN IMPACTS ON SURFACE–SUBSURFACE INTERACTIONS

Anthropogenic influences often overwhelm or decouple riverine systems from the natural landscape (i.e., reduce connectivity), thereby affecting bio-

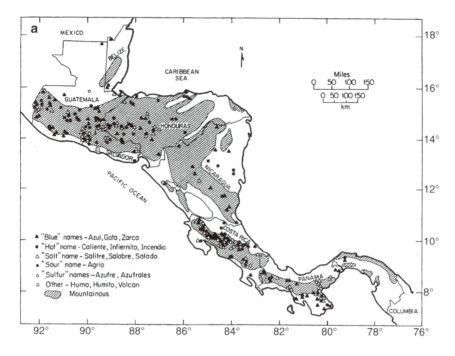

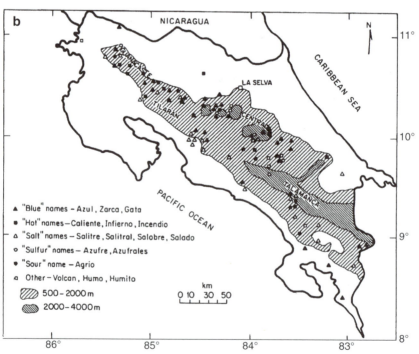

logical patterns. The magnitude and extent of human impacts (e.g., nutrient and toxicant loadings, land-use activities, hydrologic modifications, climate change) are altering the nature of subsurface interactions with stream surface water on landscape, regional, and global spatial scales (Pringle, 1999). Many of these impacts have complex and widespread cumulative effects that are interrelated.

A. Nutrient and Toxicant Loading

Nutrient application, in support of agricultural production, has increased dramatically over the past 50 years to the extent that agricultural nitrogen fixation currently exceeds natural global fixation (Vitousek *et al.*, 1997). In the Mississippi River, large, long-term increases in nitrogen and phosphorus concentrations at the mouth have occurred in response to 20- and 4-fold increases, respectively, in the application of fertilizers in the catchment (Turner and Rabelais, 1991). As a landscape property, nutrient discharge to streams is now directly related to the agricultural component of land use in a catchment for nitrogen (Hill, 1978; Neill, 1989; Mason *et al.*, 1990), phosphorus (Dillon and Kirchner, 1975), or both (Rekolainen, 1990; Correll *et al.*, 1992; Nearing *et al.*, 1993; Kronvang *et al.*, 1995).

Organic and toxic contamination can also be transferred from ground water into stream water and from the stream to the ground water in influent reaches, affecting biological patterns. For example, springs and streams fed by ground water in the Upper Rhine Valley, that originate from infiltrated polluted water from the channelized Rhine, have markedly altered aquatic vegetation. As a result of contamination of the aquifer, the macrophyte community is characterized by the accumulation of toxic metals such as mercury. Furthermore, the macrophyte community integrates discontinuous releases of stored eutrophicants and metals resulting from varying redox conditions in the aquifer (Carbiener and Trémolières, 1990; Trémolières *et al.*, 1993).

FIGURE 3 Maps of (a) Central America and (b) enlargement of Costa Rica showing mountainous areas and approximate locations of streams that have names suggesting that they may be affected by geothermal ground waters. Volcanic activity is common throughout Central America, as a result of the northeastward subduction of the Cocos ridge beneath the Caribbean plate. Thus, geothermal waters can be predicted to occur frequently. The distribution of these streams is clustered along the central axis of the mountains. Data for this map were gathered using the map collection in Olin library at Cornell University. Spanish names of streams that we felt were most likely to indicate geothermal impacts (see legend) were located and mapped. In the case of Nicaragua, we also included native languages in our search. In Figure 3a, streams were located for all of the countries using topographic maps (1:50,000) except for Nicaragua, El Salvador, and Belize. For these latter countries, Gazeteer maps (1:100,000) were used, which may explain why there appears to be a lower density of geothermally impacted streams in Nicaragua.

B. Land-Use Activities

Land-use activities can alter surface–subsurface water interactions in fluvial systems on regional scales. In a study of 23 sites along the Raisin River, which drains a 2800-km² basin in southeastern Michigan, Allan *et al.* (1997) found that the proportion of agricultural land use in a basin was the single best predictor of biotic integrity in streams. Biotic integrity was measured using Karr's (1991) "Index of Biotic Integrity" which is a summation of ten individual metrics, based on occurrence frequency of various taxonomic groups in the fish assemblage. Biotic integrity declined as the proportion of agriculture increased within the upper basin and was greatest in landscapes characterized by high proportions of wetlands and forest. Correlations of a second index, the habitat index (Michigan Department of Natural Resources, 1991) as well as the biotic integrity index, were strong with regional land use but weaker as spatial scales were reduced. Site conditions were not significant predictors of either local habitat index or biotic integrity (Roth *et al.*, 1996). However, in a second six-site study in these tributaries (Lammert, 1995), the relationship between stream integrity and land use was weaker; local conditions, particularly the extent of riparian vegetation, were better predictors of biotic integrity. Allan *et al.* (1997) conclude that the apparent contrasting results of Roth *et al.* (1996) and Lammert (1995) were actually complementary, pointing to the influence of local or regional landscape scale over site conditions in predicting habitat quality and indices of biotic integrity. In fact, landscape-based interactions such as nutrient supply, sediment delivery, hydrology, and channel characteristics (features most likely to interact at the surface-water–groundwater interface), override site factors such as vegetative cover in predicting biotic or habitat integrity.

An extreme example of human-induced changes in surface–subsurface interactions at the landscape scale involves the effects of the pulp and paper industry on the hydrology of riverine systems in Brazil. Vast monocultures of exotic species such as eucalyptus are being established in South America to satisfy the growing worldwide demand for paper. Brazil is the second largest area of industrial, fast-growing tree plantations in the world and is now the major world supplier of eucalyptus pulp. The high water consumption of eucalyptus trees (e.g., *Eucalyptus grandis*) has lowered groundwater tables on a regional scale accompanied by the disappearance of streams and rivers (Carrere, 1996). In the coastal state of Espiritu Santo, it is estimated that 156 streams have disappeared as a result of eucalyptus plantations, numerous wells have dried, and the San Domingos River has stopped flowing (Carrere, 1996). With increasing paper consumption worldwide, this trend of dropping groundwater tables and the desiccation of fluvial systems in paper–pulp producing regions of the world can only be expected to intensify.

Urbanization (and concomitant increases in impervious surface in the landscape) results in changing patterns of runoff, groundwater recharge, soil loss and hydrologic shifts in surface–subsurface flowpaths. This land-use activity is having increasingly large and adverse effects on natural biotic patterns along river corridors (Baer and Pringle, 1999).

C. Hydrological Modifications

Hydrological modifications (e.g., stream-water diversions, wetlands drainage, groundwater extraction, channelization, and damming) are decoupling spatial distributions of both surface and ground waters from their zones of interface within river basins. In many cases, the nature of abiotic and biotic gradients that occur at the groundwater–surface-water interface have been radically altered, with dramatic effects on stream-water quality and biota.

Effects of stream dewatering range from reduction of species numbers, population abundances, and changes in biotic distribution to the extirpation of entire faunal complexes within a river basin (as a result of complete dewatering of rivers). Water abstraction can have regional effects on temperature patterns within stream networks as a result of changes in surface–subsurface water interactions. For example, in a coastal river draining into the Gulf of Mexico, water temperature has increased (i.e., loss of cold extremes) and a greater homogeneity of temperatures has occurred resulting from extensive water withdrawal in the catchment (Van Den Avyle and Evans, 1990). These changes in water temperature regime have resulted in a loss of cold water spring refugia for striped bass (*Morone saxatilis*). Water withdrawals thus have the potential to affect regional patterns of biodiversity by threatening the survival of biota (e.g., striped bass) dependent on these cold water refugia.

Water level governs the interaction between ground and surface water. As a result of groundwater extraction from the Ogallala Aquifer (at a rate 10,000 times faster than recharge), the groundwater level is dropping up to a meter a year, with concomitant drying of some rivers in the central United States. In a different hemisphere, Bangkok's water table has dropped 25 m since the late 1950s with concomitant changes in surface–subsurface water interactions, including saltwater intrusion into potable groundwater supplies (Lean *et al.*, 1990). This pattern of decoupling the stream from its landscape by dropping groundwater tables is being repeated throughout the world (Postel, 1992), and it can have extremely deleterious effects on regional distribution patterns of riparian vegetation (Perkins *et al.*,1984; Stromberg *et al.*, 1996).

Irrigation development has caused regional changes in how ground water and surface water interact in many regions of the world. Even though regional patterns of decline and rise in groundwater levels are being increasingly documented and quantified (e.g., Fig. 4), to our knowledge there are no corresponding studies which document concomitant regional changes in

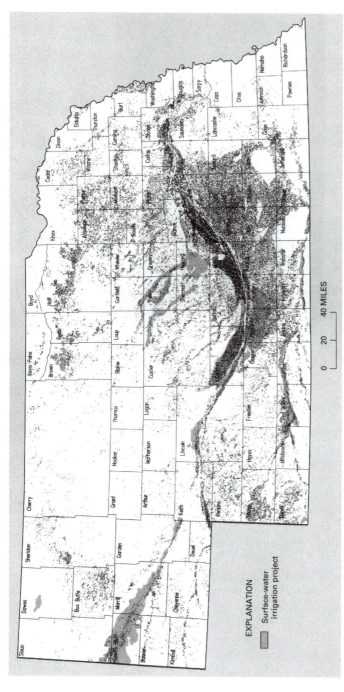

EXPLANATION

Surface-water
irrigation project

0   20   40 MILES

FIGURE 4 Irrigation water in the state of Nebraska is derived from both groundwater and surface-water sources. Surface-water irrigation projects are illustrated by grey and groundwater irrigation wells are illustrated by black dots which are so dense in many regions that irrigation wells appear as black shading (A). The use of both ground water and surface water for irrigation has resulted in significant alterations in groundwater–surface-water interactions in different parts of the state (B). For example, groundwater withdrawals by the large concentration of irrigation wells in Area E has caused declines in groundwater levels that could not be offset by recharge from precipitation and the presence of nearby flowing streams. In this area water withdrawals cause decreases in groundwater discharge to streams and/or induce flow from the streams to shallow groundwater. In contrast, effects of surface-water irrigation in Areas F and G in south-central Nebraska, probably caused the rises in groundwater levels observed here Map reprinted from Winter *et al.* (1998) and data provided by the University of Nebraska Conservation and Survey Division.

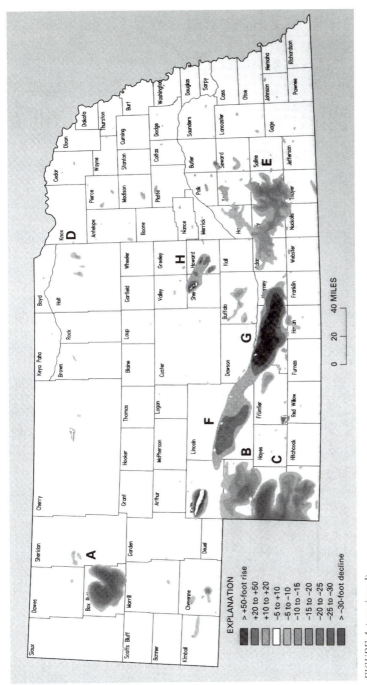

FIGURE 4 (continued)

aquatic biota resulting from changes in groundwater–surface-water interactions. For example, how have regional patterns in aquatic biota been altered within the state of Nebraska (Fig. 4), one of the most extensively irrigated states in the United States?

The location of riverine and groundwater systems within the landscape determines how hydrologic alterations influence surface–subsurface interactions. For example, channel straightening, river regulation, and draining wetlands in temperate zones drastically reduce contact zones between surface and subsurface waters (e.g., between stream channel and floodplain), whereas in arid areas, these alterations can result in an increase in contact zones (Zaletaev, 1997; Gibert *et al.*, 1997). Resultant alterations in biological patterns in streams at the landscape level in both temperate and arid areas can be very dramatic, as illustrated by the following examples.

The first example involves decoupling between surface water and riparian ground water in a channelized portion of the Rhine River, deprived of its floodplain. The loss of associated groundwater–surface-water interactions between the river and its floodplain has had an impact on the transfer of nutrients and micropollutants, decreasing both surface and groundwater quality and impacting riverine communities (Trémolières *et al.*, 1997). Riverside ground water is contaminated by phosphate, chloride, and mercury over a 1- to 2-km-wide fringe (Robach *et al.*, 1991; Roeck *et al.*, 1991). The increase in the contamination of riverside ground water is directly linked to the disconnection of the riparian flooded zone from the river and to the suppression of groundwater recharge during floods (Carbiener and Trémolières, 1990). The vegetation of groundwater-fed streams in the Rhine fringe is now characterized by eutrophic macrophyte communities (according to a bioindicator scale) as a result of the high levels of phosphate and ammonia–nitrogen in the ground water. The bryophyte, *Fontinalis*, had a level of mercury as high as in the Rhine channel. In contrast, a nearby river with intact groundwater–surface-water interaction provided large quantitites of good quality water to the groundwater table despite the highly contaminated river waters. Surface–subsurface exchange processes occurring within the intact floodplain improved water quality by stopping or slowing the transfer of contaminants. In contrast to the channelized Rhine, the reach with a functional floodplain was characterized by oligotrophic to mesotrohpic aquatic vegetation, and the mercury content of *Fontinalis* was also very low (Trémolières *et al.*, 1997).

In arid areas, spatial redistribution of ground waters in river basins through irrigation often leads to desertification (i.e., the degradation of natural ecosystems and landscapes with decreasing bioproduction). Seepage of water from irrigation canals creates surface-water–groundwater interactions, which influence patterns of both terrestrial and aquatic biota in the landscape. For example, about 20 million hectares of land are under irrigation in Central Asia, Kazakhstan and the southern part of the former Soviet

Union. Irrigation is achieved through extensive drainage canals which total up to 180,000 km in Uzbekistan alone. Only a very small proportion (12%) of these drainage canals have impermeable beds, and thus, water loss through seepage is about 45–55% (Zaletaev, 1997). This results in a rise in the groundwater table, with regional effects on the vegetation cover and biotic complexes in both surface and groundwater systems. Irrigation of desert lands in Central Asia, using waters derived from oases, resulted in the flooding of 300,000 hectares and 530,000 hectares became salty and marshy. Desert vegetation was replaced by halophytic shrubs and semishrubs when the groundwater level rose by 1.0–1.5 m. The change in vegetation type occurred in 5–6 years, during which desert Artemisia-shrub communities were replaced by meadow-bog and solonchak communities (Zaletaev, 1997).

Subsurface influences on the biological properties of surface waters in highly irrigated arid areas are often determined by mineralogical properties of soils. Intensive irrigation in arid regions can lead to leaching of minerals in the soil which become concentrated in such high amounts in ground water that they become toxic, resulting in emergent biological patterns in receiving surface waters. For example, the soils of the San Joaquin Valley in California, and other arid areas in the western United States, are naturally rich in selenium. Decades of irrigation from ground water and canals have leached from the soil and concentrated selenium in the rising saline groundwater table. In irrigation schemes, this contaminated ground water is often drained into surface water, leading to regional patterns in biological deformities of waterfowl. In the mid-1980s, 15,000 adult birds were dying each year at Kesterson Reservoir, California, as a result of selenium toxicity. This is not just a local problem specific to drainage schemes in California. Rising selenium levels have been found on irrigation projects in 17 states in the western United States. This is a compelling example of the important influence that surface–subsurface interactions exert at landscape scales—since high levels of selenium are poisonous not only to aquatic organisms and waterfowl but also to humans.

The role of agriculture as a landscape impact on hydrologic and biogeochemical patterns should continue to be a strong area of future research. Nutrient and pesticide applications continue to increase, as do hydrologic modifications via irrigation. The area of irrigated land in the world (currently at 222 million hectares) is expanding, with forecasts indicating that by the end of the twentieth century it might reach 400 million hectares. The impact of irrigated agriculture on biotic pattern as a result of raising water tables and accelerating runoff to streams will continue to increase.

Construction of dams and impounds has transformed many rivers into a series of artificial lakes (Arthington and Welcomme, 1995) connected by highly regulated flows, dramatically altering the spatial and temporal patterns of recharge and discharge between surface and subsurface waters within the landscape. The mixing relationships between surface water and ground water in the hyporheic zone dramatically alters transport-retention parameters both downstream of impoundments and within them. Bedload deficits caused by sediment retention by impoundments and increased transport capacity following channel straightening downstream of impoundments often incise river beds, lowering the adjacent groundwater level (e.g., Galay, 1983; Golz, 1994). Alteration of mixing relationships between surface water and ground water below dams can have severe impacts on the reproductive success of gravel-spawning fishes (Curry *et al.*, 1994). Dessication of the floodplains in highly regulated rivers below dams ultimately alters patterns of riverine and riparian biodiversity (e.g., Dister *et al.*, 1990; Keller and Kondolf, 1990).

In reservoirs, the abiotic and biotic gradients that previously characterized groundwater–surface-water interactions prior to impoundment are dramatically altered, often with negative implications for aquatic biota and human health. Surface–subsurface interactions in reservoirs shift from advective processes controlling nutrient flux (characteristic of lotic systems) to diffusional flux (characteristic of lentic systems). Imposition of lentic limnological properties, such as thermal stratification, often results in anoxic benthic conditions, slower exchange, and altered redox environment. The decay of organic matter can create severely anoxic conditions, altering the faunal composition. Low redox in conjunction with organic matter decay within reservoirs can also promote methylmercury production which can be a major problem in newly created reservoirs (Hecky *et al.*, 1991; Rosenberg *et al.*, 1997; McCully, 1996). Landscape patterns in levels of mercury in fishes have been documented both above and below dams in Canada (Rosenberg *et al.*, 1997).

The role of impoundments in determining biogeochemical patterns needs serious consideration due to global shifts in long-term storage, and thus the global transport of water, solutes, and sediments. Currently, 15% of the world's precipitation on land (after allowing for evaporation) is held in reservoirs of large dams (Stiassny, 1996) and five times the volume of all the rivers in the world are impounded in reservoirs. By the year 2000, more than 60% of the world's stream flow will be regulated.

D. Modifications to the Atmosphere—Atmospheric Deposition and Global Warming

Surface–subsurface water interactions and related biological responses are being influenced by anthropogenic activities through increasingly broader feedback loops including the alteration of climate. Predicted impacts of global climate change on water resources includes increases in global average precipitation and evaporation; changes in regional patterns of rainfall, snowfall, and snowmelt; changes in the intensity, severity, and timing of major storms; and rising sea levels and salt water intrusion into coastal aquifers

(Gleick, 1998). It is a safe prediction that these changes will have many complex and interrelated effects on surface–subsurface water interactions and patterns of biological response on landscape and regional scales.

For example, in boreal lakes and streams of Canada, climate warming and drought, accompanied by a lowered water table, are causing the oxidation of stored sulfur (Schindler, 1998). Storage of sulfur in a reduced form in soils and peatlands protects lakes and streams from acidification on landscapes where atmospheric deposition of sulfur oxides is high (Bayley *et al.*, 1986; Rochefort *et al.*, 1990). With increased exposure of peatlands and wet soils to atmospheric oxygen, sulfur is reoxidized, and pulses of sulfuric acid are released to streams and lakes. Acid pulses are especially severe during high streamflow following periods of drought (e.g., Lazerte, 1993; Devito, 1995). These pulses are more acidic in landscapes where anthropogenic sulfate deposition has been high for decades (e.g., eastern Ontario) than in areas with less sulfate deposition (e.g., northwestern Ontario; Schindler *et al.*, 1996). The potential consequence is a landscape scale effect on the distribution of aquatic biota sensitive to changes in pH.

The interaction of ground water and surface water is clearly important in determining the susceptibility of a given surface-water body to acid precipitation (e.g., Winter *et al.*, 1998). Emergent biological patterns have occurred in surface waters as a result of acid deposition, whereby those surface-water bodies with little inflow of alkaline ground water are highly vulnerable to acidification and become devoid of aquatic life that is intolerant of decreasing pH, whereas those surface-water bodies receiving significant inflows of ground water (which neutralize acidic waters) were buffered from changes in pH (Robson *et al.*, 1992).

V. SYNTHESIS AND RECOMMENDATIONS FOR FUTURE STUDIES

The critical importance of surface–subsurface interactions at the global scale is clear when one considers the role of the landscape on stream flow, temperature, and nutrient and ion chemistry, all of which exert a strong influence on biotic patterns in streams. Globally, streams and lakes comprise a very small fraction of the planet's freshwater; a mere 0.3%, with most of the remainder locked in ice, permanent snow, or ground water (Gleick, 1993). Despite the fact that 66 times more freshwater is located in ground water than in lakes, streams, soil moisture, and the atmosphere combined, we know relatively little about the physical, chemical, or biological properties of ground water or groundwater–surface-water interfaces. Two key factors that must be considered for protection of both surface- and groundwater systems are water exchanges and biological properties at the surface–subsurface interface (Gibert *et al.*, 1997).

The boundaries between river and groundwater ecological research are gradually breaking down, with both fields slowly merging toward a more comprehensive ecological understanding of the hydrological continuum (e.g., Brunke and Gonser, 1997; Gibert *et al.*, 1997; Winter *et al.*, 1998; and this volume). Progress at generalizing landscape-based biotic patterns in streams has been slow for at least three reasons:

1. The stream ecosystem may have an inherent capacity to impose pattern—regardless of the landscape. Our earliest and most enduring concept of biotic pattern along fluvial corridors, the River Continuum Concept (Vannote *et al.*, 1980), primarily conceptualizes community structure from source to mouth relative to site factors (e.g., shade cover which provides litterfall input and inhibits primary production). The major organizing principle of this concept is the input and processing of organic matter, and predicted patterns of community structure assume a fairly uniform bankside landscape. However, many questions arise when one considers the landscape. The River Continuum Hypothesis was developed from observations of biotic pattern in temperate streams: How do the longitudinal patterns predicted by the River Continuum Concept vary in different landscapes? For example, are biotic patterns in streams on geologically different templates similar or different? What is the dominant control of biotic pattern? Is it light, cover, and other local bankside properties or the geologic template? An important area of future research is the balance between landscape versus waterscape control on biotic pattern.

Future conceptual advances will also require integration of land use, geochemistry, local factors (e.g., allochthonous litterfall or light penetration to low-order streams), and channel geomorphic features which have temporal retention functions (e.g., beaver and debris dams; Fig. 5). Biotic patterns not only should include periphyton, invertebrates, and fishes but also should be extended to biogeochemical patterns and their alteration by landscape features such as riparian forests and wetlands.

2. Environmental problems in lotic environments are usually only identified after surface–subsurface interactions have been overwhelmed or the lotic system has been decoupled from the landscape. Even in the case of hydrothermal systems, which present natural shifts in the solute chemistry of receiving waters, streams receiving these subsurface inputs of ground water are often overwhelmed chemically and physically and decoupled from the immediate catchment. For example, hydrothermally altered, solute-rich waters that upwell in lowland streams in Costa Rica (Pringle *et al.*, 1993) represent natural inputs from outside of the catchment which are in excess of the riparian and hyporheic processing capabilities.

Most studies of agricultural nitrate and sediment loading indicate that water quality problems are highly correlated with the proportion of agricultural area in the catchment. The most significant predictor of nitrate and sediment loading in agricultural areas is thus a landscape-scale property. In

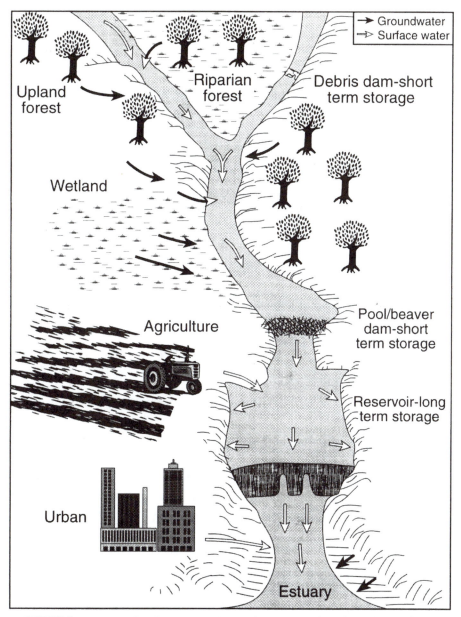

FIGURE 5 Pathways of surface-water and groundwater inputs from forested upland watersheds to larger anthropogenically modified streams and ultimately the estuary. In low-order reaches, biotic patterns remain intact and interruptions produce only local modification (e.g., beaver and debris dams). In higher-order streams, anthropogenic modification by dams, agriculture, channelization, urbanization, etc., decouple the landscape from the stream and may produce cumulative regional disruption to biotic pattern. Disruption of critical surface–subsurface water processes such as nitrogen biogeochemistry is increasingly extending to and impacting even the largest estuaries.

many cases, the capacity of hyporheic riparian interaction for nutrient transformation has been overwhelmed. In this regard, the study of Jordan *et al.* (1997) is extremely important because the role of a landscape-scale property (soil type) was critical in determining nitrate loads via subsurface pathways to streams under equal proportions of agriculture (e.g., in the coastal plain, terrestrial–aquatic interfaces were very effective in processing nitrate, whereas in piedmont soils they were not). In many agricultural areas water tables are high, and irrigation tiling of fields hydrologically decouples nutrient- and pesticide-rich agricultural runoff from potential surface–subsurface interaction. Future research should be directed toward determining the capacity (and predictors of that capacity) of various landscape types to prevent overwhelming or complete decoupling of the inherent biogeochemical capacity of riparian and hyporheic zones for controlling nutrient inputs to streams.

3. In large rivers, it is very difficult to separate anthropogenic effects from natural landscape controls along the river corridor. Relatively undisturbed streams are largely confined to low-order systems (Fig. 4). In higher-order streams, anthropogenic modification increasingly decouples the landscape from the channel (Fig. 4). At what point do anthropogenic effects on biotic pattern cease to be local and begin to impose a cumulative impact? For example, what is the overall impact of impoundments on patterns of biogeochemistry along a river corridor? The Serial Discontinuity Concept (Ward and Stanford, 1983), a conceptual expansion of the River Continuum Concept, considers the hydrologic effects of dams which retain sediments and organic matter. Alteration of flow patterns and deterioration of downstream habitat for fishes and invertebrates are important considerations of this concept. What about additional impacts on biogeochemical patterns associated with reservoir filling, raised water tables, and variations in water tables related to dam management? Further inclusion of such landscape-scale effects (including hydrologic regime, solute composition and levels, and channel modification which alter or destroy hydrological connectivity between surface and ground water) is now required. How can we separate the role of impoundments, the role of landscapes, and the inherent lotic capacity to determine biotic pattern? For example, how can the true theoretical framework envisioned by the Serial Discontinuity Concept be extended to determine the role of surface–subsurface interactions on biotic pattern along river corridors containing single or sequential impoundments? Does the relative importance of various geologic, hydrologic, and biogeochemical factors shift as impoundments undergo senescence-succession?

Contact zones between surface and subsurface waters will continue to shift spatially and change in biogeochemical functioning, given current trends and future projections of the magnitude and extent of human transformation of the landscape (e.g., Gleick, 1993, 1998). We are just beginning to elucidate some of the emergent biological patterns occurring in streams at landscape scales that result from these changes. Moreover, alterations of surface–subsurface water interactions have important social, political, and ecological implications for humanity which need to be addressed.

REFERENCES

Allan, J. D., D. L. Erickson, and J. Fay. 1997. The influence of catchment land-use on stream integrity across multiple spatial scales. *Freshwater Biology* 37:149–161.

Arthington, A. H., and R. L. Welcomme. 1995. The condition of large river systems of the world. *In* "Conditions of the World's Aquatic Habitats. Proceedings of the World Fisheries Congress. Theme 1" (N. B. Armantrout, ed.), pp. 411–423. Science Publishers, Lebanon.

Baer, K. E., and C. M. Pringle. 1999. Special problems of urban river conservation: The encroaching megalopolis. *In* "Global Perspectives on River Conservation: Science, Policy and Practice" (P. J. Boon, B. R. Davies, and G. E. Petts, eds.). Wiley, New York (in press).

Bayley, S. E., R. S. Behr, and C. A. Kelly. 1986. Retention and release of S from a freshwater wetland. *Water, Air, and Soil Pollution* 31:101–114.

Brunke, M., and T. Gonser. 1997. The ecological significance of exchange processes between rivers and groundwater. *Freshwater Biology* 37:1–33.

Carbiener, R., and M. Trémolières. 1990. The Rhine rift valley groundwater-river interactions: Evolution of their susceptibility to pollution. *Regulated Rivers* 5:375–389.

Carrere, R. 1996. Pulping the south: Brazil's pulp and paper plantations. *Ecologist* 26:206–214.

Correll, D. L. 1991. Human impact on the functioning of landscape boundaries. *In* "Ecotones: The Role of Landscape Boundaries in the Management and Restoration of Changing Environments" (M. M. Holland, P. J. Risser, and R. J. Naiman, eds.), pp. 90–109. Chapman & Hall, New York.

Correll, D. L., T. E. Jordan, and D. E. Weller. 1992. Nutrient flux in a landscape: Effects of coastal land-use and terrestrial community mosaic on nutrient transport to coastal waters. *Estuaries* 15:431–442.

Curry, R. A., J. Gehrels, D. L. Noakes, and R. Swainson. 1994. Effects of river fluctuations on groundwater discharge through brook trout, *Salvelinus fontinalis*, spawning and incubation habitats. *Hydrobiologia* 277:121–134.

Devito, K. J. 1995. Sulfate mass balances of Precambrian Shield wetlands: The influence of catchment hydrogeology. *Canadian Journal of Fisheries and Aquatic Sciences* 52:1750aendash}1760.

Dillon, P. J., and W. B. Kirchner. 1975. The effects of geology and land-use on the export of phosphorus from watersheds. *Water Research* 9:135–148.

Dister, E., D. Gomer, P. Obrdlik, P. Petermann, and E. Schneider. 1990. Water management and ecological perspectives of the Upper Rhine's floodplains. *Regulated Rivers* 5:1–15.

Ellis, A. J., and W. A. Mahon. 1964. Natural hydrothermal systems and experimental hot-water/rock interactions. *Geochimica et Cosmochimica Acta* 28:1323–1357.

Ellis, A. J., and W. A. Mahon. 1967. Natural hydrothermal systems and experimental hot water/rock interactions (Part II). *Geochimica et Cosmochimica Acta* 31:519–538.

Fiebig, D. M., and M. A. Lock. 1991. Immobilization of dissolved organic matter from groundwater discharging through the steam bed. *Freshwater Biology* 26:45–55.

Ford, T. E., and R. J. Naiman. 1989. Groundwater-surface water relationship in boreal forest watersheds: Organic carbon nutrient dynamics. *Canadian Journal of Fisheries and Aquatic Sciences* 46:41–49.

Forman, R. T., and M. Godron. 1986. "Landscape Ecology." Wiley, New York.

Fortner, S. L., and D. S. White. 1988. Interstitial water patterns: A factor influencing the distribution of some lotic aquatic vascular macrophytes. *Aquatic Botany* 31:1–12.

Freeze, R. A., and J. A. Cherry. 1979. "Groundwater." Prentice Hall, Englewood Cliffs, NJ.

Galay, V. J. 1983. Causes of river bed degradation. *Water Resources Research* 19:1057–1090.

Gambrell, R. P., J. W. Gilliam, and S. B. Weed. 1975. Nitrogen losses from soils of the North Carolina Coastal Plain. *Journal of Environmental Quality* 4:317–322.

Genereux, D., and C. Pringle. 1997. Chemical mixing model of streamflow generation at La Selva Biological Station, Costa Rica. *Journal of Hydrology* 199:319–330.

Gibert, J., F. Fournier, and J. Mathieu, eds. 1997. "Groundwater/Surface Water Ecotones: Biological and Hydrological Interactions and Management Options." Cambridge University Press, New York.

Gilliam, J. W., and R. W. Skaggs. 1988. Nutrient and sediment removal in wetland buffers. *In* "Proceedings of the National Wetland Symposium: Wetland Hydrology," pp. 174–177. Association of State Wetland Managers, Berme.

Gleick, P. H. 1993. "Water in Crisis: A Guide to the World's Freshwater Resources." Oxford University Press, New York.

Gleick, P. H. 1998. "The World's Water: The Biennial Report on Freshwater Resources." Island Press, Washington, DC.

Golz, E. 1994. Bed degradation—nature, causes, countermeasures. *Water Science and Technology* 29:325–333.

Grimm, N. B., H. M. Valett, E. H. Stanley, and S. G. Fisher. 1991. Contribution of the hyporheic zone to stability of an arid-land stream. *Verhandlungen Internationale Vereinigung für Theoretische ünd Angewandte Limnologie* 24:1595–1599.

Haycock, N. E., and G. Pinay. 1993. Nitrate retention in grass and polar vegetated riparian buffer strips during the winter. *Journal of Environmental Quality* 22:273–278.

Hecky, R. E., D. J. Ramsey, R. A. Bodaly, and N. E. Strange. 1991. Increased methylmercury contamination in fish in newly formed freshwater reservoirs. *In* "Advances in Mercury Toxicology" (T. Suzuki, N. Imura, and T. W. Clarkson, eds.), pp. 33–52. Plenum, New York.

Henley, R. W. 1985. The geothermal framework for epithermal deposits. *In* "Geology and Geochemistry of Epithermal Systems" (B. R. Berger and P. M. Bethke, eds.), pp. 1–24. Society of Economic Geologists, El Paso, TX.

Henley, R. W., and A. J. Ellis. 1983. Geothermal systems, ancient and modern. *Earth Science Review* 19:1–50.

Hill, A. R. 1978. Factors affecting the export of nitrate-nitrogen from drainage basins in southern Ontario. *Water Research* 12:1045–1057.

Holdridge, L. R., W. C. Grenke, W. H. Hatheway, T. Liang, and J. A. Tosi. 1971. "Forest Environments in Tropical Life Zones: A Pilot Study." Pergamon, Oxford.

Jacobs, T. C., and J. W. Gilliam. 1985. Riparian losses of nitrate from agricultural drainage waters. *Journal of Environmental Quality* 14:472–478.

Jordan, T. E., D. L. Correll, and D. E. Weller. 1997. Relating nutrient discharges from watersheds to land use and streamflow variability. *Water Resources Research* 33:2579–2590.

Karr, J. R. 1991. Biological integrity: A long-neglected aspect of water resource management. *Ecological Applications* 1:66–84.

Keller, E. A., and G. M. Kondolf. 1990. Groundwater and fluvial processes; selected observations. Groundwater geomorphology. The role of subsurface water in earth-surface processes and landforms. *Special Paper—Geological Society of America* 252:319–340.

Kronvang, B., R. Grant, S. E. Larsen, L. M. Svendsen, and P. Kristensen. 1995. Non-point-source nutrient losses to the aquatic environment in Denmark: Impact of agriculture. *Marine and Freshwater Research* 46:167–177.

Lammert, M. 1995. Assessing land-use and habitat effects on fish and macroinvertebrate assemblages: Stream biological integrity in an agricultural watershed. MS. Thesis, University of Michigan, Ann Arbor.

Lazerte, B. D. 1993. The impact of drought and acidification on the chemical exports from a minerotrophic conifer swamp. *Biogeochemistry* **18**:153–175.

Lean, G., D. Hinrichsen, and A. Marakham. 1990. "Atlas of the Environment." Prentice Hall, New York.

Lowell, R. P. 1991. Modeling continental and submarine hydrothermal systems. *Reviews of Geophysics* **29**:457–476.

Lowrance, R. 1992. Groundwater nitrate and denitrification in a coastal plain riparian forest. *Journal of Environmental Quality* **21**:401–405.

Lowrance, R., R. Todd, J. Fail, O. Hendrickson, R. Leonard, and L. Asmussen. 1984. Riparian forests as nutrient filters in agricultural watersheds. *BioScience* **34**:374–377.

Mason, J. W., G. D. Wegner, G. I. Quinn, and E. L. Lange. 1990. Nutrient loss via groundwater discharge from small watersheds in southwestern and south central Wisconsin. *Journal of Soil Water Conservation* **45**:327–331.

McCully, P. 1996. "Silenced Rivers: The Ecology and Politics of Large Dams." Zed Books, Atlantic Highlands, NJ.

Michigan Department of Natural Resources. 1991. "Qualitative Biological and Habitat Survey Protocols for Wadable Streams and Rivers," GLEAS Procedure No. 51. Michigan Department of Natural Resources, Surface Water Quality Division, Great Lakes Environmental Assessment Section.

Nearing, M. A., R. M. Risse, and L. F. Rogers. 1993. Estimating daily nutrient fluxes to a large Piedmont reservoir from limited tributary data. *Journal of Environmental Quality* **22**:666–671.

Neill, M. 1989. Nitrate concentrations in river waters in the southeast of Ireland and their relationship with agricultural practice. *Water Research* **23**:1339–1335.

Omernik, J. M. 1977. Nonpoint source stream nutrient level relationships: Nationwide survey. *U.S. Environmental Protection Agency, Office of Research and Development [Report] EPA* **EPA-600/3-77-105.**

Pavich, M. J., G. W. Leo, S. F. Obermeier, and J. R. Estabrook. 1989. Investigations of the characteristics, origin, and residence time of the upland residual mantle of the Piedmont of Fairfax Co., VA. *Geological Survey Professional Paper (U.S.)* **1352.**

Perkins, D. J., B. N. Carlson, M. Fredstone. 1984. The effects of groundwater pumping on natural spring communities in Owens Valley. *In* "California Riparian Systems: Ecology Conservation, and Productive Management" (R. E. Warner and K. M. Hendrix, eds.), pp. 515–527. University of California Press, Berkeley.

Peterjohn, W. T., and D. L. Correll. 1984. Nutrient dynamics in an agricultural watershed: Observations on the role of a riparian forest. *Ecology* **65**:1466–1475.

Postel, S. 1992. "Last Oasis: Facing Water Scarcity," The Worldwatch Environmental Alert Series. W. W. Norton, New York.

Pringle, C. M. 1991. Geothermally-modified waters surface at La Selva Biological Station, Costa Rica: Volcanic processes introduce chemical discontinuities into lowland tropical streams. *Biotropica* **23**:523–529.

Pringle, C. M. 1999. Riverine connectivity: Conservation and management: Implications for remnant natural areas in complex landscapes. *Verhandlungen Internationale Vereinigung für Theoretische und Angewandte Limnologie* (in press).

Pringle, C. M., and T. Hamazaki. 1997. Effects of fishes on algal response to storms in a tropical stream. *Ecology* **78**:2432–2442.

Pringle, C. M., and T. Hamazaki. 1998. The role of omnivory in structuring a neotropical stream: Separating diurnal versus nocturnal effects. *Ecology* **79**:269–280.

Pringle, C. M., and A. Ramirez. 1998. Use of both benthic and drift sampling techniques to assess tropical stream invertebrate communities along an altitudinal gradient, Costa Rica. *Freshwater Biology* **39**:101–115.

Pringle, C. M., and F. J. Triska. 1991. Effects of geothermal groundwater on nutrient dynamics of a lowland Costa Rican stream. *Ecology* **72**:951–965.

Pringle, C. M., P. Paaby-Hansen, P. D. Vaux, and C. R. Goldman. 1986. *In situ* nutrient assays of periphyton growth in a lowland Costa Rican stream. *Hydrobiologia* **134**:207–213.

Pringle, C. M., F. J. Triska, and G. Browder. 1990. Spatial variation in basic chemistry of streams draining a volcanic landscape on Costa Rica's Caribbean slope. *Hydrobiologia* **206**:73–86.

Pringle, C. M., G. L. Rowe, F. J. Triska, J. F. Fernandez, and J. West. 1993. Landscape linkages between geothermal activity, solute composition and ecological response in streams draining Costa Rica's Atlantic Slope. *Limnology and Oceanography* **38**:753–774.

Rekolainen, S. 1990. Phosphorus and nitrogen load from forest and agricultural areas in Finland. *Aqua Fennica* **19**:95–107.

Robach, F., E. Eglin, and R. Carbiener. 1991. L'hydrosystème rhenan: Evolution parallele de la végétation aquatique et de la qualité de l'eau (Rhinau). *Bulletin de l'Ecologie* **22**:227–241.

Robson, A., K. J. Beven, and C. Neal. 1992. Towards identifying sources of subsurface flow: A comparison of components identified by a physically based runoff model and those determined by chemical mixing techniques. *Hydrological Processes* **6**:199–214.

Rochefort, L., D. H. Vitt, and S. E. Bayley. 1990. Growth, production and decomposition dynamics of Sphagnum under natural and experimentally acidified conditions. *Ecology* **71**:1986–2000.

Roeck, U., M. Trémolières, A. Exinger, and R. Carbiener. 1991. Utilisation des mousses aquatiques dans une etude sur le transfert du mercure en tant que descripteur du fonctionnement hydrologique (échanges cours d'eau-nappe) en plaine d'Alsace. *Bulletin Hydroecologie* **3**:241–256.

Rosemond, A. D., C. M. Pringle, and A. Ramirez. 1998. Macroconsumer effects on insect detritivores and detrital processing in a tropical stream food web. *Freshwater Biology* **39**:515–524.

Rosenberg, D. M., F. Berkes, R. A. Bodaly, R. E. Hecky, C. A. Kelly, and J. W. M. Rudd. 1997. Large-scale impacts of hydroelectric development. *Environmental Review* **5**:27–54.

Roth, N. E., J. D. Allan, and D. E. Erickson. 1996. Landscape influences on stream biotic integrity assessed at multiple spatial scales. *Landscape Ecology* **11**:141–156.

Schindler, D. W. 1998. A dim future for boreal waters and landscapes. *BioScience* **48**:157–164.

Schindler, D. W., S. E. Bayley, B. R. Parker, K. G. Beaty, D. R. Cruikshank, E. J. Fee, E. U. Schindler, and M. P. Stainton. 1996. The effects of climatic warming n the properties of boreal lakes and streams at the Experimental Lakes Area, Northwestern Ontario. *Limnology and Oceanography* **41**:1004–1017.

Spalding, R. F., and M. E. Exner. 1993. Occurrence of nitrate in groundwater—a review. *Journal of Environmental Quality* **22**:392–402.

Stiassny, M. L. 1996. An overview of freshwater biodiversity with some lessons learned from African fishes. *Fisheries* **21**:7–13.

Stromberg, J. C., R. Tiller, and B. Richter. 1996. Effects of groundwater decline on riparian vegetation of semiarid regions: The San Pedro Arizona. *Ecological Applications* **6**:113–131.

Trémolières, M., I. Eglin, U. Roeck, and R. Carbiener. 1993. The exchange process between river and groundwater on the central Alsace floodplain (Eastern France). I. The case of the canalized river Rhine. *Hydrobiologia* **254**:133–148.

Trémolières, M., R. Carbinier, I. Eglin, F. Robach, U. Roeck, and J. M. Sanchez-Perez. 1997. Surface water/groundwater/forest alluvial ecosystems: functioning of interfaces. The case of the Rhine Floodplain in Alsace (France). *In* "Groundwater/Surface Water Ecotones: Biological and Hydrological Interaction and Management Options" (J. Gibert, J. Mathieu, and F. Fournier, eds.), pp. 91–101. Cambridge University Press, New York.

Turner, B. L., W. C. Clark, R. W. Kates, J. F. Richards, J. T. Mathews, and W. B. Meyer. 1990. "The Earth as Transformed by Human Action: Global and Regional Changes in the Biosphere Over the Past 300 Years." Cambridge University Press, New York.

Turner, R. E., and N. N. Rabelais. 1991. Changes in Mississsippi River water quality this century. *BioScience* **41**:140–147.

Valett, H. M., S. G. Fisher, N. B. Grimm, E. H. Stanley, and A. J. Boulton. 1992. Hyporheic-surface water exchange: implications for the structure and functioning of desert stream ecosystems. *In* "Proceedings of the First International Conference on Groundwater Ecology" (J. A. Stanford and J. J. Simons, eds.), pp. 395–405. American Water Resources Association, Bethesda, MD.

Valett, H. M., S. G. Fisher, N. B. Grimm, and P. Camill. 1994. Vertical hydrologic exchange and ecological stability of a desert stream ecosystem. *Ecology* 75:548–560.

Valett, H. M., C. N. Dahm, and M. E. Campana. 1997. Hydrologic influences on groundwater-surface water ecotones: Heterogeneity in nutrient composition and retention. *Journal of the North American Benthological Society* 16:239–247.

Van Den Avyle, M. J., and J. W. Evans. 1990. Temperature selection by striped bass in a Gulf of Mexico coastal river system. *North American Journal of Fisheries Management* 10:58–66.

Vannote, R. L., G. W. Minshall, K. W. Cummins, J. R. Sedell, and C. E. Cushing. 1980. The river continuum concept. *Canadian Journal of Fisheries and Aquatic Sciences* 37:130–137.

Vitousek, P. M, J. D. Aber, R. W. Howarth, G. E. Likens, P. A. Matson, D. W. Schindler, W. H. Schlesinger, and D. G. Tilman. 1997. Human alteration of the global nitrogen cycle: Causes and consequences. *Ecological Applications* 7:737–750.

Wallis, P. M., H. B. N. Hynes, and S. A. Telang. 1981. The importance of groundwater in the transportation of allochthonous dissolved organic water to the stream draining a small mountain basin. *Hydrobiologia* 79:77–90.

Ward, J. V., and J. A. Stanford. 1983. The serial discontinuity concept of lotic ecosystems. *In* "Dynamics of Lotic Ecosystems" (T. D. Fontaine and S. M. Bartell, eds.), pp. 29–42. Ann Arbor Science (Butterworth), Ann Arbor, MI.

White, D. S., S. P. Hendricks, and S. K. Fortner. 1992. Groundwater-surface water interactions and the distribution of aquatic macrophytes. *In* "Proceedings of the First International Conference on Groundwater Ecology" (J.A. Stanford and J.J. Simons, eds.), pp. 247–255. American Water Resources Association, Bethesda, MD.

Wiens, J. A., C. S. Crawford, and J. R. Gosz. 1985. Boundary dynamics: A conceptual framework for studying landscape ecosystems. *Oikos* 45:421–427.

Winter, T. C., J. W. Harvey, O. Lehn Franke, and W. M. Alley. 1998. Ground water and surface water: A single resource. *Geological Survey Circular (U.S.)* 1139.

Zaletaev, V. S. 1997. Ecotones and problems of their management in irrigation regions. *In* "Groundwater/Surface Water Ecotones: Biological and Hydrological Interaction and Management Options" (J. Gibert, J. Mathieu, and F. Fournier, eds.), pp. 185–193. Cambridge University Press, New York.

SECTION *TWO*

BIOGEOCHEMISTRY

Nutrients and Metabolism

8

Nitrogen Biogeochemistry and Surface–Subsurface Exchange in Streams

John H. Duff and Frank J. Triska

U.S. Geological Survey
Menlo Park, California

I. Introduction   197
II. Nitrogen Forms and Transformation Pathways in Fluvial
     Environments   199
III. Nitrogen Sources in Fluvial Environments   200
IV. Hydrologic Residence in Pristine Streams   201
V. The Redox Environment   203
VI. Ammonium Sorption to Hyporheic Sediments   204
VII. Linking Nitrogen Transformation to Hydrologic Exchange
     in Hyporheic Zone Research   205
     A. Little Lost Man Creek, California   205
     B. Sycamore Creek, Arizona   209
     C. Shingobee River, Minnesota   210
     D. Platte River, Colorado   212
VIII. What These Models Tell Us   214
IX. Future Directions for Research   216
     References   217

I. INTRODUCTION

Nitrogen, in conjunction with carbon, phosphorus, oxygen, and hydrogen, is one of the most important elements of living matter. Occurring primarily in proteins, nucleic acids, and other nitrogen-containing organic compounds, nitrogen constitutes about 10% dry mass in bacteria (Gottschalk, 1979), although it is less in higher plants. Green plants, algae, fungi, and bacteria assimilate nitrogen primarily as ammonium (NH_4^+) or nitrate (NO_3^-). Despite high demand, dissolved inorganic nitrogen (DIN) is usually scarce in pristine aquatic systems, often limiting production. Most nitrogen in aquatic systems is bound in organic matter and inaccessible for assimilation until it is mineralized to ammonium, which can be taken up directly or oxidized by bacteria to nitrate. Some microorganisms reduce nitrate, releasing nitrogen to the atmosphere. Others fix free nitrogen, harnessing it for passage to higher trophic levels. Virtually all nitrogen transformations are biotically mediated; thus, nitrogen cycling is a vital biologic component of aquatic systems.

Streams and rivers form a dynamic ecosystem, transporting water, solutes, and particulates derived from land to estuarine or marine environments. Physiochemical, biologic, and hydrologic attributes of riverine channels influence the composition, concentration, and supply of nitrogen reaching the oceans of the world. Even though biological nitrogen transformations are fundamentally the same from headwater streams through high-order rivers, the relative composition and concentration of nitrogen forms are likely to change during transport depending on the terrestrial source of nitrogen, rate of specific nitrogen transformations, biotic community structure, and transport-retention characteristics of the channel.

Hydrologic mixing of surface stream water with near-stream ground water beneath and adjacent to the channel can have a dramatic affect on the composition and concentration of nitrogen dissolved in surface water. This hydrologic exchange, which forms the hyporheic zone (Triska *et al.*, 1989b; Harvey *et al.*, 1996), influences many controlling variables of nitrogen cycling, especially the supply of dissolved oxygen (Triska *et al.*, 1993b; Findlay, 1995). Hydrologic exchange greatly increases the volume of sediment in contact with stream water (Harvey *et al.*, 1996), increases nitrogen contact with subsurface biota (D'Angelo *et al.*, 1993; Morrice *et al.*, 1997), enhances microbial activity rates (Mulholland *et al.*, 1997), reduces nitrogen cycling distances (Valett *et al.*, 1996), flushes DIN from intermittent channel sediments (McKnight *et al.*, 2000), and alters the chemical form of nitrogen (Triska *et al.*, 1989b; Duff and Triska, 1990; Bourg and Bertin, 1993; Holmes *et al.*, 1994a; Jones *et al.*, 1995a; Wondzell and Swanson, 1996b; Hill and Lymburner, 1998).

This chapter synthesizes studies in the last 10 years that link hydrologic exchange to nitrogen biogeochemistry in streams. We begin by reviewing the predominant nitrogen forms, transformation pathways, and terrestrial sources in fluvial environments. Next, we examine the variables related to hydrologic exchange that control the direction and rate of nitrogen transformations, including hydrologic residence time, the redox environment, and ammonium sorption to sediments. Then, using comprehensive studies from several physiographic settings, we synthesize models pertaining to the role of surface–subsurface interactions for nitrogen biogeochemistry. Finally, we summarize the main concepts and suggest relevant topics for future research.

II. NITROGEN FORMS AND TRANSFORMATION PATHWAYS IN FLUVIAL ENVIRONMENTS

Nitrogen occurs in fluvial environments in molecular form as dinitrogen (N_2), in reduced forms as ammonium and amine groups ($-NH_2$) in organic matter and as oxides including nitrate, nitrite (NO_2^-), and nitrous oxide (N_2O; Fig. 1). Between ammonium and nitrate, there is an eight-electron shift in valence state. As a consequence of this large range in redox potential, reactions that transform nitrogen from the reduced (valence of -3) to the oxidized state (valence of $+5$) produce energy used by chemolithotrophic bacteria. Reactions that transform nitrogen from the oxidized to the reduced state require energy from either sunlight or organic matter (Delwiche, 1970).

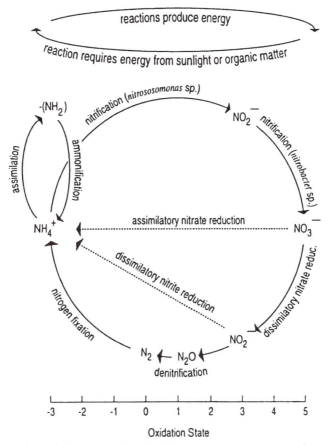

FIGURE 1   Nitrogen cycle, oxidation state, and energy relationship.

The most abundant form of nitrogen, elemental dinitrogen, directly enters biological pathways only through nitrogen-fixing bacteria (Fig. 1). Nitrogen-fixing bacteria form symbiotic associations with plant roots and are free living. Symbiotic associations are the most important in terms of the amount of nitrogen fixed (Postgate, 1982). Most of the fixed nitrogen is assimilated by plants and is later released to the stream as throughfall, as leachate from litterfall, or by plant decomposition (Wetzel, 1975). Heterotrophic bacteria liberate organically bound nitrogen as ammonia (NH_3) or ammonium (ammonification). In aquatic environments, ammonia is usually ionized to ammonium; the proportion of ionized to unionized ammonia is governed by pH. Chemolithotrophic nitrifying bacteria rapidly oxidize ammonium to nitrite (*Nitrosomonas* sp) and then to nitrate (*Nitrobactor* sp). Each stage of ammonium oxidation (nitrification) releases energy which is used to reduce carbon dioxide into organic matter. Unlike ammonium that can also bind to cation exchange sites on streambed sediments (Richey *et al.*, 1985; Triska *et al.*, 1994), nitrate is mobile and easily transported (Duff and Triska, 1990; Triska *et al.*, 1990; Bencala *et al.*, 1993; Holmes *et al.*, 1994a; Jones *et al.*, 1995a). Plants, algae, fungi, and bacteria reduce nitrate to ammonium prior to incorporation in cellular material (assimilatory nitrate reduction).

In addition to assimilatory nitrate reduction, obligate and facultative anaerobic bacteria dissimilatatively reduce nitrate to nitrite (dissimilatory nitrate reduction), nitrous oxide, and dinitrogen (denitrification) during the oxidation of organic matter in anoxic environments (nitrate, nitrite, and nitrous oxide serve as electron acceptors in electron transport phosphorylation). Unlike assimilation, nitrate is not used as a cellular nitrogen source, and products of dissimilatory reductions are excreted from cells. The energy produced from decomposition of organic matter is required for biosynthesis and other cellular functions. Whereas nitrogen fixation is a de novo source of reduced nitrogen in stream ecosystems, denitrification is a sink.

The reduction of nitrite to ammonium also occurs in fluvial environments. The reduction is assimilatory if the primary function is generation of physiological useful reduced nitrogen (Yordy and Ruoff, 1981). However, dissimilatory nitrite reduction may also occur with the primary function of detoxifying and dissipating excess electrons in the environment. Little information exists on the latter process in aquatic ecosystems, possibly because its role is minor compared to denitrification.

III. NITROGEN SOURCES IN FLUVIAL ENVIRONMENTS

Nitrogen cycling in headwater streams is closely linked to the bankside terrestrial environment (Triska *et al.*, 1984). Above ground terrestrial sources of nitrogen are typically particulate, including litterfall (Wetzel and

Manny, 1972; Meyer *et al.*, 1981) and lateral movement from the adjacent forest floor (Fisher and Likens, 1973; Triska *et al.*, 1984). Direct inputs into the stream of dissolved organic nitrogen (DON) in throughfall or DIN in precipitation are minor. Below ground sources of nitrogen include inputs of dissolved nitrogen in ground water (Triska *et al.*, 1984; Triska and Puckett, 1996). In small streams draining forests, a large proportion of nitrogen input to the stream is often organic (either particulate or dissolved), derived from production in the surrounding forest (Triska *et al.*, 1984, 1993a; Wondzell and Swanson, 1996b). Dissolved nitrogen in ground water constitutes the largest proportion of annual nitrogen input (Triska *et al.*, 1984).

In higher-order streams and rivers, linkages with terrestrial vegetation become more indirect as channel width increases and anthropogenic sources assume an increasingly greater role. Large, nonpoint-source nitrogen inputs in high-order drainages are often as nitrate introduced via agricultural practices (Lowrance *et al.*, 1983, 1985; Peterjohn and Correll, 1984, 1986; Jacobs and Gilliam, 1985; Pinay *et al.*, 1993, 1994; McMahon and Bohlke, 1996; Sjodin *et al.*, 1997), forestry practices (Swank and Caskey, 1982), or urban runoff (Bradley *et al.*, 1995; Sjodin *et al.*, 1997). Nitrate entering these ecosystems may be dissolved in ground water or enter directly as irrigation return flows or sewage outfalls. With massive loading, respiration rather than biosynthesis is the dominant nitrogen transformation at the surface–subsurface interface. Dissimilatory processes favoring nitrogen release become more significant than assimilatory processes favoring retention. Control of this biological balance between release and assimilation is closely linked to physical balance between hydrologic transport and retention.

IV. HYDROLOGIC RESIDENCE IN PRISTINE STREAMS

Streams develop and sustain a diverse and productive community through continual input and temporary retention of nitrogen. Retention of nitrogen by physical, chemical, hydrologic, and biologic mechanisms is well documented in the surface channel (Triska and Cromack, 1980; Vincent and Downes, 1980; Naiman, 1982; Richey *et al.*, 1985; Dahm *et al.*, 1987; Grimm, 1987; Triska *et al.*, 1989a, 1994; Duff *et al.*, 1984). Less well documented are hydrologic and biologic retention mechanisms in hyporheic flowpaths that constitute the surface–subsurface interface. In this section, we review and synthesize findings related to hydrologic exchange that influence the direction and rate of nitrogen transformations and microbial nitrogen metabolism in the hyporheic zone. The terms *hydrologic exchange, hyporheic exchange, surface-water exchange, surface–subsurface exchange,* and *stream-water exchange* are used interchangeably.

The ecology of the hyporheic zone, and its impact on benthic and water column nitrogen cycling, is largely influenced by hydrology. Hydrologic exchange ensures a continuous supply of nutrients to subsurface biota (Bencala *et al.*, 1993; Triska *et al.*, 1993a; Valett *et al.*, 1996, 1997; Wondzell and Swanson, 1996b; Morrice *et al.*, 1997; Mulholland *et al.*, 1997). Opportunities for nitrogen transformations are increased when a parcel of advected stream water has a long contact time with chemically or biotically active sediments (Triska *et al.*, 1989b; Holmes *et al.*, 1994a; Jones *et al.*, 1995a, Wondzell and Swanson, 1996b). Hydrologic retention occurs when surface water enters into slow flowing hyporheic flowpaths and the residence time of water in the stream channel is increased. Steep hydraulic gradients and high hydraulic conductivities promote surface–subsurface exchange that increases nitrogen contact with microbiota (Valett *et al.*, 1996; Morrice *et al.*, 1997).

In the most literal sense, nitrogen cycling begins and ends with the same chemical form in the same place. In streams, however, water velocity is high, and flow is unidirectional, so nitrogen is physically displaced downstream during transport. Because physical, chemical, hydrologic, and biologic processes interact with transported nitrogen, the downstream displacement of the nitrogen cycle has been viewed as a spiral (Newbold *et al.*, 1982; Elwood *et al.*, 1983). Spiraling length consists of uptake length, defined as the distance an atom travels before it is taken up by an organism, and turnover length, the distance an atom travels within the organic pool before returning to the dissolved inorganic form (Mulholland and DeAngelis, Chapter 6 in this book). Spiraling length is a useful index because it integrates retention processes occurring along a flowpath, including longitudinal and vertical connections to subsurface interstices (Munn and Meyer, 1988; Stream Solute Workshop, 1990; Valett *et al.*, 1996).

Nutrient uptake length is influenced by hydrologic retention associated with exchange between surface and interstitial environments and, presumably, biotic retention within the subsurface (Valett *et al.*, 1996; Mulholland *et al.*, 1997). In three watersheds in New Mexico, the shortest uptake length and highest uptake rate were associated with alluvium with the greatest hydraulic conductivity and surface–subsurface exchange (Valett *et al.*, 1996). In one watershed, Rio Calaveras, nitrate uptake lengths varied with time of day, with the smallest values occurring midday when whole stream respiration was greatest (C. S. Fellows, University of New Mexico, unpublished data). Nitrate uptake length varied with rate of stream metabolism in addition to changes in discharge. Further, hyporheic retention of ammonium and nitrate in Glen Major Stream and Duffin Creek, Ontario, was minor where the hyporheic zone was restricted by large groundwater pressure gradients (Hill and Lymburner, 1998). Thus, the extent to which a reach is gaining discharge also is an important factor for hydrologic exchange and residence time, and shifts temporally with discharge variation due to storm events (Harvey and Bencala, 1993).

Mulholland and DeAngelis (Chapter 6 in this book) present model simulations predicting that hyporheic exchange should reduce nutrient uptake length. Transient storage zone, or the cross-sectional area of the channel, can occasionally be as large or larger than the flowing surface water (Bencala *et al.*, 1984; Triska *et al.*, 1989a; D'Angelo *et al.*, 1993; Morrice *et al.*, 1997) and is a site of high metabolic activity (Findlay *et al.*, 1993; Hendricks, 1996; Mulholland *et al.*, 1997). An example of the effect of increased residence time on nitrogen cycling and composition was observed in McRae Creek, Oregon (Wondzell and Swanson, 1996a,b). Surface-water exchange and dissolved nitrogen concentration was measured in the stream, hyporheic zone, and ground water of a gravel bar to determine the effects of temporal changes in hydrologic exchange on nitrogen composition and concentration. Seasonally, exchange flow was not a significant nitrogen source to the surface channel in summer because the water table was hydrologically isolated from alder growing on the gravel bar. During summer, however, subsurface biota in gravel bars mineralized stream water-derived DON to nitrate in exchange flows, in a manner analogous to Sycamore Creek, Arizona (Jones *et al.*, 1995a). When subsurface flowpaths were reconnected with the alder gravel bar in autumn, nitrate and organic nitrogen derived from leaf fall and fine root turnover were transported to the channel.

V. THE REDOX ENVIRONMENT

Hydrologic exchange is important for the delivery of dissolved oxygen and other biotically labile solutes to hyporheic flowpaths (Leichtfried, 1988; Triska *et al.*, 1990; Duff and Triska, 1990; Valett *et al.*, 1990; Findlay *et al.*, 1993; Holmes *et al.*, 1994a; Findlay, 1995; Jones *et al.*, 1995a; Wondzell and Swanson, 1996b). Dissolved oxygen concentration strongly influences nitrogen transformations because of the wide range of available redox conditions for various reactions. Dissolved oxygen concentration in sediments is regulated by hydrologic residence time and the consumption of oxygen through respiration (Findlay, 1995) or chemolithotrophic processes (Triska *et al.*, 1990; Holmes *et al.*, 1994a; Jones *et al.*, 1995a). Where consumption of dissolved oxygen exceeds input by hydrologic exchange, anoxia develops forming an oxic–anoxic interface (Baker *et al.*, Chapter 11 in this book). Reduced and oxidized forms of nitrogen generally co-occur only under such conditions (Jones *et al.*, 1993a), and redox reactions stimulated at these interfaces can influence stream nutrient levels (Duff and Triska, 1990; Triska *et al.*, 1990; Duff *et al.*, 1996a,b).

As heterotrophic and chemolithotrophic metabolism exceed dissolved oxygen input to the hyporheic zone and as porewater becomes anoxic, organic matter mineralization proceeds through alternative electron acceptors beginning with nitrate (denitrification). Denitrifying bacteria are capable of using a variety of organic substrates as carbon and energy sources. The switch from oxic to anoxic conditions promotes synthesis of nitrate reductase enzymes by a consortium of denitrifying microorganisms (Payne, 1973). Because denitrifying enzymes are constitutive, the switch from dissolved oxygen to nitrate respiration involves a short lag time after anoxia becomes established. Although less energy is released with nitrate as the terminal electron acceptor than with dissolved oxygen, mineralization of organic matter can be substantial in anoxic sediments if nitrate is available, producing a significant nitrate sink in the hyporheic zones of low-order streams and high-order rivers (Duff and Triska, 1990; Pinay *et al.*, 1994; Bradley *et al.*, 1995; Duff *et al.*, 1996a; Holmes *et al.*, 1996; McMahon and Bohlke, 1996; Sjodin *et al.*, 1997).

Where inputs of nitrate from ground or surface water is low, mineralization of organic nitrogen may be the sole nitrate source for denitrifiers. Consequently, ammonification and nitrification become closely linked to denitrification in subsurface flowpaths (Duff and Triska, 1990; Bradley *et al.*, 1995; Duff *et al.*, 1996a; Holmes *et al.*, 1996; Triska and Duff, 1996). As mineralization proceeds, an oxic–anoxic interface may form creating spatial conditions favorable for a nitrification–denitrification coupling. Thus, in low-nitrate waters common to many headwater streams, mineralization and nitrification in the hyporheic zone are not only a nitrate source but also are redox controlling processes by consuming dissolved oxygen in porewater.

VI. AMMONIUM SORPTION TO HYPORHEIC SEDIMENTS

Sediments have the capacity to serve as a transient storage pool for ammonium by physical sorption. Many factors determine the capacity for sorption. These include the cation exchange capacity of sediment, sediment surface area in contact with stream water, and ammonium concentration of porewater (Triska *et al.*, 1994). Variations in these factors have produced conflicting results on the role of sorption in ammonium retention. For example, ammonium sorption to streambed sediments during a 90-day ammonium enrichment to Walker Branch, Tennessee, was insignificant (Newbold *et al.*, 1983). Ammonium disappearance could be accounted for by nitrification and biotic uptake associated with detritus. Similarly, ammonium decrease in hyporheic flowpaths during enrichment experiments to Glen Major Stream and Duffin Creek, Ontario, presumably resulted from biological transformations (Hill and Lymburner, 1998). In contrast, ammonium disappearance in excess of measured uptake or nitrification during an amendment study was due to sorption onto channel sediments at Bear Brook, New Hampshire (Richey *et al.*, 1985). The exchangeable ammonium was postulated to be nitrified to nitrate and subsequently transported from the system. Retention of ammonium by sorption may regulate availability and timing for biotic uptake and transformation. During a 9-day ammonium injection to an oxygenated hyporheic flowpath of Little Lost Man Creek, California, ammonium transport was retarded relative to a sulfate tracer, and porewater concentrations declined gradually after cutoff (Triska *et al.*, 1990). Presumably, ammonium was adsorbed to hyporheic sediments during the amendment pulse and desorbed slowly following cutoff. Although the amount of ammonium sorbed per unit sediment was small, the volume of sediment in contact with porewater was large (Bencala *et al.*, 1984), resulting in a large and temporally variable sorbed ammonium pool. In Little Lost Man Creek, the coastal mountain range consists of uplifted marine sediments rich in minerals with higher cation exchange capacity than those associated with igneous parent material. Triska *et al.* (1994) concluded that retention of ammonium by sorption transiently regulated the availability for biotic uptake and nitrification, thus altering the timing of DIN transport to the channel.

VII. LINKING NITROGEN TRANSFORMATIONS TO HYDROLOGIC EXCHANGE IN HYPORHEIC ZONE RESEARCH

Dissolved inorganic nitrogen concentration is typically low in headwater catchments; availability often depends on input pathways, hydrologic and biologic retention processes, and microbial cycling. Ground water may contribute sizable quantities, albeit variable, of organic and inorganic nitrogen. Autochthonous production is another source of reduced nitrogen in streams, with mineralization of both particulate and dissolved organic nitrogen occurring in subsurface flowpaths. Anthropogenic sources may contribute substantial nitrogen through both point and nonpoint discharges. Oxidation–reduction processes in near-stream ground water can reduce nitrogen inputs minimizing impacts on receiving waters. In this section, we present four conceptual models on the interaction of surface–subsurface exchange and hyporheic nitrogen biogeochemistry. The three low-order sites largely differ with respect to gradient, ground water–stream water interaction, and nitrogen transport–retention strategies (Table I). The fourth site is a high-order drainage with significant groundwater discharge, high river flow, and high inorganic nitrogen loading.

A. Little Lost Man Creek, California

Hyporheic studies in pristine catchments indicate that interaction between hydrologic exchange and microbial nitrogen metabolism can be vital for releasing DIN to the channel for uptake by benthic photoautotrophs. Triska *et al.* (1989b) presented the first conceptual model of inorganic nitrogen cycling in lateral hyporheic environments at Little Lost Man Creek (Fig. 2), demonstrating that reduced nitrogen in near-stream ground water

TABLE I Comparison of Four Research Sites Linking Nitrogen Transformations to Hydrologic Exchange in the Hyporheic Zone

Site location	Stream order	Discharge $(L\ s^{-1})$	Stream water exchange	Gradient $(m\ m^{-1})$	Sediment characteristics	Major discharge events
Little Lost Man Creek, California	1–3	25	Lateral Outwelling, 18 m; Downwelling, 100 cm	0.066	Poorly mixed sand, gravel, and cobbles	Winter rain
Sycamore Creek, Arizona	3–5	50	Lateral outwelling, 10 m; Downwelling, 60 cm	0.027	Coarse sand and gravel (1–5 mm)	Late summer-autumn and winter rain
Shingobee River, Minnesota	1–3	180	Lateral outwelling, insignificant; Downwelling, 1–10 cm	0.004	Medium to coarse sand (0.25–1 mm)	Spring snow melt and summer rain (buffered by wetlands and lakes)
Platte River, Colorado	4–6	330–6000	Lateral outwelling, 10–1000 m; Downwelling, 60–90 cm		Coarse sand and gravel floodplains	Hourly effluent releases

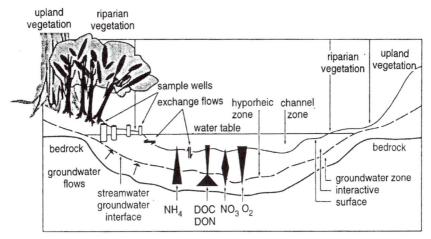

FIGURE 2 Conceptual model of hyporheic exchange and nutrient linkage in Little Lost Man Creek, California. Adapted from Triska *et al.* (1989b), *Ecology* 70:1893–1905, copyright by the American Geophysical Union.

can be oxidized to nitrate and subsequently transported to the channel for assimilation.

Initially, stream tracer studies at Little Lost Man Creek provided indirect evidence that nitrogen dissolved in stream water penetrating the hyporheic zone could undergo biotic transformation. Chloride (Cl^-), a conservative hydrologic tracer, and nitrate were co-injected into surface water for 20 days to examine solute retention and the fate of nitrate during surface and subsurface transport (Triska *et al.*, 1989a,b). Lateral outwelling of stream water, determined by chloride penetration to a series of shallow bankside wells, indicated hydrologic linkage at almost all wells. Chloride transport ranged from >90% stream water in wells <4 m from the wetted channel to 0% at a well 10 m from the channel. Co-injected nitrate was not conservative. Wells either were higher or lower in nitrate concentration than predicted by chloride. Triska *et al.* (1989b) proposed that wells with elevated nitrate (compared to that predicted by chloride) resulted from subsurface flow patterns where nitrification occurred during transport, whereas sites with lower than predicted nitrate resulted from denitrification during transport.

Hyporheic tracer studies at Little Lost Man Creek (Triska *et al.*, 1990) confirmed that ammonium dissolved in stream water added directly to a hyporheic flowpath could be nitrified. During a 9-day ammonium injection to an oxygenated hyporheic flowpath, nitrate concentration increased by 55 $\mu gN\ L^{-1}$ approximately 5 m down gradient and remained elevated for 6 days after cutoff, due to nitrification of ammonium slowly released from the sediments. Simultaneously, sediment slurry assays indicated significant potential for nitrification in sediments from the stream and hyporheic zone where dissolved oxygen would be supplied by surface-water advection. Nitrification potentials were absent in subsurface sediment without a dissolved oxygen supply, even though they were incubated aerobically (Triska *et al.,* 1990).

Hyporheic tracer studies at Little Lost Man Creek also confirmed that nitrate added to a hyporheic flowpath could be denitrified (Duff and Triska, 1990). During a 48-hour subsurface injection of nitrate and acetylene-amended water, nitrous oxide increased 13 m down gradient, demonstrating denitrification at background dissolved organic carbon (DOC) and dissolved oxygen levels during nitrate transport through the hyporheic zone. Denitrifying potential at *in situ* nitrate concentration was detectable in sediment slurry assays from all wells in the hyporheic zone at Little Lost Man Creek (Duff and Triska, 1990). Denitrifying potential increased with distance from the stream channel as did DOC and nitrate concentrations in porewater, in contrast to dissolved oxygen, which decreased. In wells with highest sediment denitrification potentials, dissolved oxygen, DOC, and nitrate all decreased over the summer. Maintenance of denitrifying capacity in the hyporheic zone was dependent on the flux of DOC in lateral groundwater inflows and the generation of nitrate from nitrification.

Hydrologic exchange under natural conditions resulted in high dissolved oxygen levels in wells with both ammonium and organic nitrogen present. Lateral mixing of dissolved oxygen-saturated water was extensive in spring but declined by autumn, presumably due to reduced stream discharge, microbial respiration, and nitrification (Triska *et al.,* 1990). Organic nitrogen dissolved in ground water was mineralized to ammonium along hyporheic flowpaths, with maximum concentrations of organic nitrogen and ammonium (both 30–40 μgN L^{-1}) occurring in August. By October, nitrate concentration increased from about 150 μgN L^{-1} to annual peak concentrations over 1 mgN L^{-1} in wells 4–12 m from the channel, presumably from nitrification. Subsequently, much of this nitrate was denitrified. Because ammonium was readily adsorbed to clay sediments at Little Lost Man Creek (Triska *et al.,* 1990, 1994), a large reservoir of sorbed ammonium in equilibrium with the porewater sustained nitrification throughout the dry, low flow period.

By studying trends in surface-water chemistry at Little Lost Man Creek, it was possible to determine if a reach was a source or sink for dissolved N. Channel nitrate concentration exhibited seasonal, diel, and upstream–downstream trends (Triska *et al.,* 1989a). Seasonally, nitrate in channel water increased over summer and had diel decreases midday of 25–40% of night concentration. The decreased concentration, attributed to uptake by photoautotrophs, constituted a significant transient sink for nitrate in transport. Along the same reach, however, instantaneous nitrate concentration was higher at successive downstream stations indicating a nitrate source, presumably hyporheic nitrification. Thus, shifts in nitrate loading along the reach were a net product of temporally and spatially variable channel and hyporheic biogeochemistry, resulting from groundwater and stream-water mixing.

B. Sycamore Creek, Arizona

Jones *et al.* (1995a) presented a conceptual model of hyporheic nitrogen cycling at Sycamore Creek, demonstrating that reduced nitrogen in organic matter can be oxidized to nitrate in subsurface flowpaths and transported to the channel for assimilation (Fig. 3). Jones *et al.* (1995a) demonstrated that dissolved oxygen-saturated water and labile organic matter derived from algal production were transported into hyporheic sediments at downwelling zones of the streambed, where heterotrophic microorganisms mineralized the organic nitrogen to ammonium. Nitrifying organisms subsequently oxidized the ammonium during subsurface transport, elevating nitrate concentration. Nitrification rates were significantly correlated with respiration rates in both downwelling and upwelling zones (Jones, 1995) but were higher in downwelling zones. These results indicated that organic matter advected into the hyporheic zone fueled nitrification, providing a critical nitrate source for channel production.

An analogous cycle was described for parafluvial zones at the same site (Holmes *et al.*, 1994a,b). Nitrate concentration increased along parafluvial flowpaths that were linked to the surface stream at their upstream and downstream ends. This increase was attributed to nitrification. Because ammonium concentration was low and declined along the flowpath, surface-derived material, either ammonium or dissolved organic matter, presumably fueled the process. Similar to the findings in the hyporheic zone (Valett *et al.*, 1990; Grimm *et al.*, 1991; Jones *et al.*, 1995a), parafluvial flowpaths were nitrate sources to Sycamore Creek.

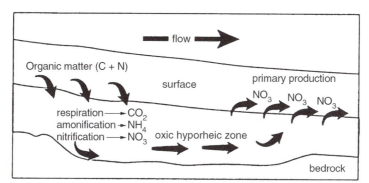

FIGURE 3   Conceptual model of surface stream and hyporheic linkage in Sycamore Creek, Arizona. Adapted from Jones *et al.* (1995a), with permission.

In Sycamore Creek, nitrate increased and dissolved oxygen declined during subsurface transport in hyporheic and parafluvial flowpaths, but dissolved oxygen concentrations rarely dropped below 5 mgO$_2$ L^{-1} (Grimm *et al.*, 1991; Holmes *et al.*, 1994a,b; Valett *et al.*, 1994b; Jones *et al.*, 1995a). Theoretically, denitrification should be restricted to active anoxic microsites. However, denitrification consumed an estimated 5–40% of nitrate produced by nitrification (Holmes *et al.*, 1996). Denitrification potential was generally highest at hyporheic and parafluvial downwelling zones. Downwelling zones also had high respiration and large labile organic pools (Jones *et al.*, 1995b), which may explain why downwelling zones would be suitable for denitrification if anaerobic microsites formed from intense local respiration and nitrification.

The concentration of surface-water nitrate in Sycamore Creek was positively correlated with vertical hydraulic gradient in upwelling areas, and hyporheic sources of nitrate alleviated nitrogen limitation at upwelling sites (Valett *et al.*, 1990, 1992, 1994). Consequently, algal productivity in Sycamore Creek was higher in upwelling than in downwelling zones (Valett *et al.*, 1990). Decreased nitrate concentration in surface water during channel transport reflected algal nitrate assimilation between upwelling and downwelling sites. Thus, nitrate-rich water flowed through sediments to upwelling zones where it was transported to the surface, stimulating algal growth (Valett *et al.*, 1990; Grimm *et al.*, 1991). In Sycamore Creek, shifts in nitrate loading along the reach were a product of longitudinally variable hyporheic and parafluvial biogeochemistry, resulting from surface-water transport through the bed.

C. Shingobee River, Minnesota

Hyporheic studies in catchments with high gaining reaches indicate that stream-water exchange is low because large groundwater pressure gradients limit surface-water exchange (Jackman *et al.*, 1996; Hill and Lymburner, 1998). In this section, we present a conceptual model from the Shingobee River suggesting that hyporheic processes can reduce nitrogen loads entering the stream in groundwater discharge.

Retention of groundwater DIN in the hyporheic zone of the Shingobee River was calculated from the difference between groundwater DIN flux across the oxic–anoxic interface of the hyporheic zone (estimated from the discharge rate and ammonium concentration of ground water), and the DIN flux across the sediment–stream water interface (estimated from the upstream–downstream change in DIN mass divided by the area of the streambed; Fig. 4). Groundwater DIN was mainly in the form of ammonium. The flow-weighted mass balance revealed that 60% of the ammonium flux in ground water was retained in the hyporheic zone (510 μmoles m^{-2} h^{-1}), suggesting that in summer the hyporheic zone was a sink for ground-

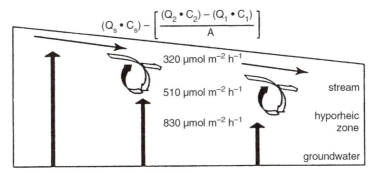

$$
(Q_s \cdot C_s) - \left[\frac{(Q_2 \cdot C_2) - (Q_1 \cdot C_1)}{A} \right]
$$

320 µmol m^{-2} h^{-1}

510 µmol m^{-2} h^{-1}

830 µmol m^{-2} h^{-1}

stream

hyporheic zone

groundwater

FIGURE 4   Dissolved inorganic nitrogen retention in the hyporheic zone of the Shingobee River, Minnesota.

water-derived DIN. Of the groundwater-derived ammonium flux across the oxic–anoxic interface, 40% was released to the channel (320 µmoles m^{-2} h^{-1}). Of the DIN released to the channel, 60% was released as nitrate (nitrified during transport), and 40% was released as ammonium. By calculating the mass balance from the DIN released to channel, benthic rates of mineralization and nitrification were calculated. The uncertain fate of ammonium retained in the hyporheic zone includes immobilization by microorganisms decomposing organic matter, sorption onto sediments, uptake by rooted aquatic macrophytes or by episamic and epiphytic diatoms, and coupled nitrification–denitrification.

Direct evidence of a surface–subsurface exchange zone in the Shingobee River was demonstrated by elevating chloride in the stream channel for 3 days to create an artificial chloride gradient in the streambed. Elevated chloride levels extended into the streambed about 10 cm (Duff *et al.*, 1998). During this and other chloride injections, porewater samples collected 10 cm deep every 20 m along the study reach indicated that reaches with chloride penetration were inversely related to high groundwater discharge areas based on the presence of bankside groundwater seeps (F. J. Triska, unpublished data). Chloride penetration, and the volume of the surface–subsurface exchange zone, was variable throughout the reach.

Porewater concentration profiles also demonstrated that the hyporheic zone was 10–20 cm thick in the Shingobee River (Duff *et al.*, 1998). The specific conductance profile delimited the mixing zone between ground water and stream water and the lower boundary of the hyporheic zone. Steep solute gradients for dissolved oxygen, DIN, and DOC fell within the mixing zone, indicating the presence of a relatively thin but biologically active hyporheic zone.

Close-interval profiles of dissolved oxygen, nitrate, and ammonium showed that an oxic–anoxic interface formed in the hyporheic zone, promoting conditions favorable for denitrification. Although nitrate was typically absent from porewater profiles, the profiles suggested a close connection between nitrification and denitrification. Denitrifying enzyme activity was present in the sediments, and potential rates were at least five times greater than either ammonification or nitrification potentials. Thus, we believe that ammonium lost during groundwater passage through the hyporheic zone was immediately denitrified after being oxidized to nitrate.

Linking hydrologic exchange to nitrogen retention implies both longitudinal and vertical connectivity of channel to subchannel sediments. This view is consistent with observations at Sycamore Creek, McRae Creek, and Little Lost Man Creek, where hyporheic zones constantly exchange water with the wetted channel and the shallow hyporheic zone consists primarily of advected surface water. In the Shingobee River, ammonium and nitrate were transported conservatively down a 600-m reach (Jackman *et al.*, 1996). Similar to the conclusions from Glen Major and Duffin Creek, Ontario (Hill and Lymburner, 1998), Jackman *et al.* (1996) concluded that hyporheic retention of stream-water ammonium and nitrate was minor because large groundwater pressure gradients limited exchange with subsurface sediments. In catchments with large groundwater pressure gradients, nitrogen retention and transformation are largely associated with nutrients in near-stream ground water.

D. Platte River, Colorado

The Platte River receives large anthropogenic DIN inputs in the form of ammonium from wastewater treatment plants and in the form of nitrate from agricultural runoff in ground water. Hyporheic studies in the Platte River reveal how hydrologic exchange and nitrogen biogeochemistry influence residence time of river water, redox potential in the sediments, nitrogen transformation processes, and DIN levels in river and porewater.

Nitrification was an important dissolved oxygen sink in hyporheic sediments of the Platte River (McMahon *et al.*, 1995). Dissolved oxygen uptake rate in the hyporheic zone, generated by advective movement of river water into the sediments, was fast enough to potentially account for riverine dissolved oxygen decreases. Hydrologic exchange between the river and the bottom sediments was promoted by hourly variations in wastewater treatment plant effluent discharge, which fluctuated river stage 4–8 cm. Dissolved oxygen depletion occurred 30 cm below the sediment–water interface presumably due to nitrification (Fig. 5A). Ammonium concentration in the Platte River, on the order of 8 mgN L^{-1} downstream from the wastewater treatment plant, decreased exponentially as water flowed downstream (Fig. 5B). Nitrate concentration simultaneously increased over the same distance. Hyporheic porewater concentrations of ammonium and nitrate were similar to river water and changed down gradient similar to DIN in river water (Fig. 5C). Presumably, ammonium in river water was nitrified upon advection into

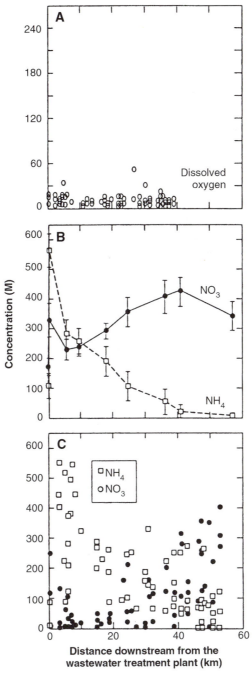

FIGURE 5 Average concentrations of dissolved oxygen, ammonium, and nitrate in the sediment porewater (A, C) and water column (B) of the Platte River, Colorado. Adapted from McMahon *et al.* (1995), with permission.

hyporheic sediments, producing nitrate and consuming dissolved oxygen. This system demonstrates how hydrologic exchange, dissolved oxygen distribution, and microbial nitrogen cycling interact to control ecosystem function.

Denitrification potentials in sediments collected 30 cm below the surface water–riverbed interface decreased with increasing distance downstream from the effluent discharge (Bradley *et al.*, 1995). Dissolved inorganic nitrogen in porewater and sediment organic carbon also declined downstream. The decline in DIN was due to ammonium depletion because nitrate increased slightly downstream. Inability to account for ammonium depletion as nitrate accumulated suggested the presence of a sink for nitrate, due either to assimilation or denitrification. The coupling of denitrification in the anoxic zone of bed sediments (30 cm deep) to nitrification in the oxidized zone was concluded to be a significant mechanism for removing anthropogenic nitrogen downstream of the effluent discharge (Bradley *et al.*, 1995).

Surface–subsurface exchange was shown to physically dilute high anthropogenic nitrate applied to the land in an agricultural setting of the Platte River and transported to it in ground water (McMahon and Bohlke, 1996). Using chloride and silicate concentrations as groundwater dilution tracers from wells in terrace, floodplain deposits, and riverbed sediments of the Platte River, surface–subsurface exchange was deduced to substantially reduced nitrate concentration between groundwater recharge areas and discharge areas in the floodplain and riverbed (McMahon and Bohlke, 1996). Although dilution itself does not reduce the total nitrate load, dilution during hydrologic exchange was more than twice as effective in lowering groundwater nitrate concentrations than denitrification (McMahon and Bohlke, 1996; Pfenning and McMahon, 1996).

VIII. WHAT THESE MODELS TELL US

Streams and rivers transport nitrogen derived from terrestrial environments to the oceans of the world. During the journey, nitrogen turns over numerous times as biotic communities interact with dissolved forms, or it is lost to the atmosphere through denitrification. Nitrogen turnover and loss is largely associated with benthic and hyporheic processes in both low- and high-order fluvial environments. Opportunities for nitrogen transformation increase when a parcel of surface water has a long contact time with chemically or biotically active sediments. All geomorphic features that retain water or promote hydrologic mixing with near-stream ground water increase the residence time, the volume of sediment in contact with dissolved nitrogen, and, consequently, dissolved nitrogen interaction with subsurface biota. As a result of these interactions, nitrogen uptake length decreases, nitrogen regeneration rate increases, and nitrogen turnover length shortens in fluvial systems with large exchange zones compared to systems with small exchange zones or brief residence time within the channel.

The hyporheic zone is the subsurface interface between the terrestrial environment and the stream. A large percentage of terrestrial nitrogen and carbon enters the hyporheic zone at the surface water–ground water interface. Surface water–ground water mixing creates redox environments that stimulate aerobic, anaerobic, and chemolithotrophic subsurface microbial processes. These processes can impact the composition and concentration of groundwater nutrients reaching the channel, particularly inorganic nitrogen and low molecular weight organic acids. Cycling and transformation of organic pools can be rapid at this interface as long as they are labile. However, there may be little processing of recalcitrant compounds when the residence time is relatively short.

Stream-water advection through subsurface flowpaths enhances microbial transformation of labile organic nitrogen into inorganic forms that can be assimilated by aquatic photoautotrophs. Coupling of stream-water advection and nitrogen mineralization is particularly important in pristine drainages where inorganic nitrogen is scarce. Stream-water advection through subsurface flowpaths also can influence nitrogen levels as ammonium associated with hyporheic sediments is released to porewater. Both processes, along with nitrogen loss associated with denitrification of nitrate transported in subsurface flowpaths, affect nitrogen subsidies to the reach through spatial and temporal controls on nitrogen cycling and release.

In pristine montane catchments like Little Lost Man Creek, Sycamore Creek, and McRae Creek, hyporheic zones constantly exchange water with the wetted channel, and the shallow hyporheic zone consists primarily of advected surface water. Generally, low nutrient concentrations in headwater streams, relatively high hydraulic conductivities of hyporheic sediments, and low DOC concentration prevents widespread anaerobiosis. Aerobic respiration is likely much greater than anaerobic respiration, and most DIN in surface and hyporheic waters is oxidized (nitrate). In low gradient catchments like Glen Major Creek, Duffin Creek, and Shingobee River, with large groundwater pressure gradients, the size of the hyporheic zone is restricted so that nitrogen retention and transformation mechanisms are largely associated with transformation of nutrients in ground water. Where inputs of organic matter and DIN are large, elevated levels of respiration in the shallow sediments can deplete dissolved oxygen supplies and promote conditions favorable for denitrification and fermentation. Under such conditions, ammonium dominates porewater DIN. Balance between hydrologic processes that supply dissolved oxygen and organic carbon and microbial processes that consume dissolved oxygen and organic carbon ultimately affects the nitrogen cycle by either producing redox conditions favorable for aerobic metabolic processes and nitrate accumulation via nitrification or redox conditions promoting ammonium accumulation in porewater and denitrification. This balance may vary with shifts in geology, basic geomorphologic structure of the channel, and seasonal variation in temperature and organic matter input.

In summary, hydrologic exchange enables microbial communities to cycle, retain, and utilize nitrogen at the surface–subsurface interface that might otherwise be exported from the system. The extent of hydrologic exchange has a significant effect on the system's ability to retain nitrogen by increasing sediment surface in contact with stream water and increasing the amount of nitrogen in contact with biotically and chemically active sediment surfaces. In pristine streams, community composition and productivity is dependent on nitrogen retention mechanisms because of the low dissolved nitrogen concentration in stream water, short residence time of channel water, and limited contact between nitrogen dissolved in stream water and sediment surfaces. In streams and rivers with high anthropogenic or natural nitrogen loading, hydrologic exchange has a significant impact on nitrogen loads as microbial reactions at the stream water–ground water boundary shift from aerobic to anaerobic metabolism, attenuating high nitrogen inputs.

IX. FUTURE DIRECTIONS FOR RESEARCH

The relative composition of stream, hyporheic, and groundwater nitrogen is well known for inorganic species, but relatively little information is available regarding the composition of total dissolved N. Additional resources and method development involving measurement of the concentration and quality of DON is sorely needed. Differences between DON released from terrestrial-derived organic matter compared to DON released from autochthonously derived organic matter need to be assessed for their ability to enter nitrogen cycling pathways. Denitrification is an easy process to measure by the C_2H_2 inhibition technique but is difficult to relate to areal estimates and other nitrogen cycling processes. Areal determinations are especially hampered by high site variation. Estimating closely coupled processes such as nitrification and denitrification over small depth intervals is technically difficult. Use of inhibitors and enzyme analogues are problematic because many of the same enzymes compete or are disrupted by the same inhibitors making it difficult to isolate denitrification from interrelated processes. At least five methods exist for identifying and quantifying nitrogen cycling processes in hyporheic sediments, including sediment enzyme assays, injection experiments, natural abundance measurements, concentration gradients, and mass balance calculations. These methods, in conjunction with recent advances in the use of nitrogen isotopes, may soon allow routine determination of nitrogen transformation rates at *in situ* concentrations. Isotopic analysis may also identify isotopic signatures in biotic tissues, chemical intermediates, or products (e.g., ^{15}N, ^{18}O in nitrate). Carefully integrating existing techniques and developing and perfecting new technologies to quantify and compare process rates between systems remains a major challenge. Such technologies related to nitrogen sources will be critical in scaling up to address critical large-scale nitrogen cycling issues, such as hypoxia in the Gulf of Mexico.

REFERENCES

Bencala, K. E., V. C. Kennedy, G. W. Zellweger, A. P. Jackman, and R. J. Avanzino. 1984. Interactions of solutes and a streambed sediments. 1. An experimental analysis of cation and anion transport in a mountain stream. *Water Resources Research* **20**:1797–1803.

Bencala, K. E., J. H. Duff, J. W. Harvey, A. P. Jackman, and F. J. Triska. 1993. Modeling within the stream-catchment continuum. *In* "Modeling Change in Environmental Systems" (A. J. Jakeman, M. B. Beck, and M. J. McAleer, eds.), pp. 163–187. Wiley, New York.

Bourg, A. C. M., and C. Bertin. 1993. Biogeochemical processes during the infiltration of river water into an alluvial aquifer. *Environmental Science and Technology* **27**:661–666.

Bradley, P. M., P. B. McMahon, and F. H. Chapelle. 1995. Effects of carbon and NO_3^- on denitrification in bottom sediments of an effluent-dominated river. *Water Resources Research* **31**:1063–1068.

Dahm, C. N., E. H. Trotter, and J. R. Sedell. 1987. Role of anaerobic zones and processes in stream ecosystem productivity. *In* "Chemical Quality of Water and the Hydrologic Cycle" (R. C. Averett and D. M. McKnight, eds.), pp. 157–178. Lewis Publishers, Chelsea, MI.

D'Angelo, D. J., J. R. Webster, S. V. Gregory, and J. L. Meyer. 1993. Transient storage in Appalachian and Cascade mountain streams as related to hydraulic characteristics. *Journal of the North American Benthological Society* **12**:223–235.

Delwiche, C. C. 1970. The nitrogen cycle. *Scientific American* **233**:136–146.

Duff, J. H., and F. J. Triska. 1990. Denitrification in sediments from the hyporheic zone adjacent to a small forested stream. *Canadian Journal of Fisheries and Aquatic Sciences* **47**:1140–1147.

Duff, J. H., F. J. Triska, and R. S. Oremland. 1984. Denitrification associated with stream periphyton: Chamber estimates from undisrupted communities. *Journal of Environmental Quality* **13**:514–518.

Duff, J. H., C. M. Pringle, and F. J. Triska. 1996a. Nitrate reduction in sediments of lowland tropical streams draining swamp forest in Costa Rica: An ecosystem perspective. *Biogeochemistry* **33**:179–196.

Duff, J. H, F. J. Triska, A. P. Jackman, and J. W. LaBaugh. 1996b. The influence of streambed sediments on the solute chemistry of ground-water discharge in the upper Shingobee River. *Water-Resources Investigations (U.S. Geological Survey), Report* **96–4215**, 143–148.

Duff, J. H., F. Murphy, C. C. Fuller, F. J. Triska, J. W. Harvey, and A. P. Jackman. 1998. A mini drivepoint sampler for measuring porewater solute concentrations in the hyporheic zone of sand-bottom streams. *Limnology and Oceanography* **43**:1378–1383.

Elwood, J. W., J. D. Newbold, R. V. O'Neill, and W. Van Winkle. 1983. Resource spiraling: An operational paradigm for analyzing lotic ecosystems. *In* "Dynamics of Lotic Ecosystems" (T. D. Fontaine and S. M. Bartell, eds.), pp. 3–27. Ann Arbor Science (Butterworth), Ann Arbor, MI.

Findlay, S. 1995. Importance of surface-subsurface exchange in stream ecosystems: The hyporheic zone. *Limnology and Oceanography* **40**:159–164.

Findlay, S., D. Strayer, and C. Goumbala. 1993. Metabolism of stream water dissolved organic carbon in the shallow hyporheic zone. *Limnology and Oceanography* **38**:1493–1499.

Fisher, S. G., and G. E. Likens. 1973. Energy flow in Bear Brook, New Hampshire: An integrative approach to ecosystem metabolism. *Ecological Monographs* **43**:421–439.

Gottschalk, G. 1979. "Bacterial Metabolism." Springer-Verlag, New York.

Grimm, N. B. 1987. Nitrogen dynamics during succession in a desert stream. *Ecology* 68:1157–1170.

Grimm, N. B., H. M. Valett, E. H. Stanley, and S. G. Fisher. 1991. Contribution of the hyporheic zone to stability of an arid-land stream. *Verhandlungen Internationale Vereinigung für Theoretische ünd Angewandte Limnologie* 24:1595–1599.

Harvey, J. W., and K. E. Bencala. 1993. The effect of streambed topography on surface-subsurface water exchange in mountain catchments. *Water Resources Research* 29:89–98.

Harvey, J. W., B. J. Wagner, and K. E. Bencala. 1996. Evaluating the reliability of the stream tracer approach to characterize stream-subsurface water exchange. *Water Resources Research* 32:2441–2451.

Hendricks, S. P. 1996. Bacterial biomass activity and production within the hyporheic zone of a north-temperate stream. *Archiv für Hydrobiologie* 136:467–487.

Hill, A. R., and D. J. Lymburner. 1998. Hyporheic zone chemistry and stream-subsurface exchange in two groundwater-fed streams. *Canadian Journal of Fisheries and Aquatic Sciences* 55:495–506.

Holmes, R. M., S. G. Fisher, and N. B. Grimm. 1994a. Parafluvial nitrogen dynamics in a desert stream ecosystem. *Journal of the North American Benthological Society* 13:468–478.

Holmes, R. M., S. G. Fisher, and N. B. Grimm. 1994b. Nitrogen dynamics along parafluvial flowpaths: Importance to the stream ecosystem. *In* "Proceedings of the Second International Conference on Ground Water Ecology" (J. A. Stanford and H. M. Valett, eds.), pp. 47–56. American Water Resources Association, Herndon, VA.

Holmes, R. M., J. B. Jones, S. G. Fisher, and N. B. Grimm. 1996. Denitrification in a nitrogen-limited stream ecosystem. *Biogeochemistry* 33:125–146.

Jackman, A. P., F. J. Triska, and J. H. Duff. 1996. Hydrologic examination of ground-water discharge in the upper Shingobee River. *Water-Resources Investigations (U.S. Geological Survey), Report* 96–4215, 137–142.

Jacobs, T. C., and J. W. Gilliam. 1985. Riparian losses of NO_3^- from agricultural drainage waters. *Journal of Environmental Quality* 14:472–478.

Jones, J. B. 1995. Factors controlling hyporheic respiration in a desert stream. *Freshwater Biology* 34:91–99.

Jones, J. B., S. G. Fisher, and N. B. Grimm. 1995a. Nitrification in the hyporheic zone of a desert stream ecosystem. *Journal of the North American Benthological Society* 14:249–258.

Jones, J. B., S. G. Fisher, and N. B. Grimm. 1995b. Vertical hydrologic exchange and ecosystem metabolism in a Sonoran Desert stream. *Ecology* 76:942–952.

Leichtfried, M. 1988. Bacterial substrates in gravel beds of a second-order alpine stream (Project Ritrodat-Lunz, Austria). *Verhandlungen Internationale Vereinigung für Theoretische ünd Angewandte Limnologie* 23:1325–1332.

Lowrance, R. R., R. L. Todd, and L. E. Asmussen. 1983. Waterborne nutrient budgets for the riparian zone of an agricultural watershed. *Agriculture, Ecosystems and Environment* 10:371–384.

Lowrance, R. R., R. A. Leonard, and L. E. Asmussen. 1985. Nutrient budgets for agricultural water-sheds in the southeastern coastal plain. *Ecology* 66:287–296.

McKnight D. M., R. L. Runkel, J. H. Duff, C. M. Tate, and D. Moorhead. 2000. Inorganic nitrogen and phosphorus dynamics of glacial meltwater streams as controlled by hyporheic exchange and benthic autotrophic communities. *Journal of the North American Benthological Society* (in press).

McMahon, P. B., and J. K. Bohlke. 1996. Denitrification and mixing in a stream-aquifer system: effects on NO_3^- loading to surface water. *Journal of Hydrology* 186:105–128.

McMahon, P. B., J. A. Tindall, J. A. Collins, and K. J. Lull. 1995. Hydrologic and geochemical effects on oxygen uptake in bottom sediments of an effluent-dominated river. *Water Resources Research* 31:2561–2569.

Meyer, J. L., G. E. Likens, and P. J. Sloane. 1981. Phosphorus, nitrogen, and organic carbon flux in a headwater stream. *Archiv für Hydrobiologie* 91:28–44.

Morrice, J. A., H. M. Valett, C. N. Dahm, and N. E. Campana. 1997. Alluvial properties, groundwater-surface water exchange and hydrologic retention in headwater streams. *Hydrologic Processes* **11**:253–267.

Mulholland, P. J., E. R. Marzolf, J. R. Webster, D. R. Hart, and S. P. Hendricks. 1997. Evidence that hyporheic zones increase heterotrophic metabolism and phosphorus uptake in forest streams. *Limnology and Oceanography* **42**:443–451.

Munn, N. L., and J. L. Meyer. 1988 Rapid flow through sediments of a headwater stream in the southern Appalachians. *Freshwater Biology* **20**:235–240.

Naiman, R. J. 1982. Characteristics of sediment and organic export for pristine boreal forest watersheds. *Canadian Journal of Fisheries and Aquatic Sciences* **39**:1699–1718.

Newbold, J. D., R. V. O'Neill, J. W. Elwood, and W. Van Winkle. 1982. Nutrient spiraling in streams: Implications for nutrient limitations and invertebrate activity. *American Naturalist* **120**:628–652.

Newbold, J. D., J. W. Elwood, M. S. Schulze, R. W. Stark, and J. C. Barmeier. 1983. Continuous NH_4^+ enrichment of a woodland stream: Uptake kinetics, leaf decomposition, and nitrification. *Freshwater Biology* **13**:193–204.

Payne, W. J. 1973. Reduction of nitrogenous oxides by microorganisms. *Bacteriological Review* **37**:409–452.

Peterjohn, W. T., and D. L. Correll. 1984. Nutrient dynamics in an agricultural watershed: Observations on the role of a riparian forest. *Ecology* **65**:1466–1475.

Peterjohn, W. T., and D. L. Correll. 1986. The effect of riparian forest on the volume and chemical composition of base-flow in an agricultural watershed. *In* "Watershed Research Perspectives" (D. L. Correll, ed.), pp. 244–262. Smithsonian Press, Washington, DC.

Pfenning K. S., and P. B. McMahon. 1996. Effect of nitrate, organic carbon, and temperature on potential denitrification rates in nitrate-rich riverbed sediments. *Journal of Hydrology* **187**:283–295.

Pinay, G., L. Roques, and A. Fabre. 1993. Spatial and temporal patterns of denitrification in a riparian forest. *Journal of Applied Ecology* **30**:581–591.

Pinay, G., N. E. Haycock, C. Ruffinoni, and R. M. Holmes. 1994. The role of denitrification in nitrogen retention in river corridors. *In* "Global Wetlands: Old World and New" (W. J. Mitsch, ed.), pp. 107–116. Elsevier, Amsterdam.

Postgate, J. R. 1982. "The Fundamentals of Nitrogen Fixation." Cambridge University Press, New York.

Richey, Y. J. S., W. H. McDowell, and G. E. Likens. 1985. Nitrogen transformations in a small mountain stream. *Hydrobiologia* **124**:129–139.

Sjodin, A. L., W. M. Lewis, and J. F. Saunders. 1997. Denitrification as a component of the nitrogen budget for a large plains river. *Biogeochemistry* **39**:327–342.

Stream Solute Workshop. 1990. Concepts and methods for assessing solute dynamics in stream ecosystems. *Journal of the North American Benthological Society* **9**:95–119.

Swank, W. T., and W. H. Caskey. 1982. Nitrate depletion in a second-order mountain stream. *Journal of Environmental Quality* **11**:581–584.

Triska, F. J., and K. Cromack. 1980. The role of wood debris in forests and streams. *In* "Forests: Fresh Perspectives from Ecosystem Analysis" (R. H. Waring, ed.), pp. 171–190. Oregon State University Press, Corvallis.

Triska, F. J., and J. H. Duff. 1996. Sediment-associated nitrification and denitrification potentials at the interface between a bankside ground-water seep and the channel of the Shingobee River. *Water-Resources Investigations (U.S. Geological Survey), Report* **96–4215**, 155–160.

Triska, F. J., and L. J. Puckett. 1996. United States Geological Survey Nitrogen-Cycling Workshop. *Geological Survey Open-File Report (U.S.)* **96–477**.

Triska, F. J., J. R. Sedell, K. Cromack, S. V. Gregory, and F. M. McCorison. 1984. Nitrogen budget for a small coniferous forest stream. *Ecological Monographs* **54**:119–140.

Triska, F. J., V. C. Kennedy, R. J. Avanzino, G. W. Zellweger, and K. E. Bencala. 1989a. Retention and transportation of nutrients in a third-order stream in northwestern California: Channel processes. *Ecology* **70**:1877–1892.

Triska, F. J., V. C. Kennedy, R. J. Avanzino, G. W. Zellweger, and K. E. Bencala. 1989b. Retention and transportation of nutrients in a third-order stream in northwestern California: Hyporheic processes. *Ecology* **70**:1893–1905.

Triska, F. J., J. H. Duff, and R. J. Avanzino. 1990. Influence of exchange flow between the channel and hyporheic zone on NO_3^- production in a small mountain stream. *Canadian Journal of Fisheries and Aquatic Sciences* **11**:2099–2111.

Triska, F. J., J. H. Duff, and R. J. Avanzino. 1993a. Patterns of hydrological exchange and nutrient transformation in the hyporheic zone of a gravel-bottom stream: Examining terrestrial-aquatic linkages. *Freshwater Biology* **29**:259–274.

Triska, F. J., J. H. Duff, and R. J. Avanzino. 1993b. The role of water exchange between a stream channel and its hyporheic zone in nitrogen cycling at the terrestrial-aquatic interface. *Hydrobiologia* **251**:167–184.

Triska, F. J., A. P. Jackman, J. H. Duff, and R. J. Avanzino. 1994. Ammonium sorption to channel and riparian sediments: A transient storage pool for dissolved inorganic nitrogen. *Biogeochemistry* **26**:67–83.

Valett, H. M., S. G. Fisher, and E. H. Stanley. 1990. Physical and chemical characteristics of the hyporheic zone of a Sonoran Desert stream. *Journal of the North American Benthological Society* **9**:201–215.

Valett, H. M., S. G. Fisher, N. B. Grimm, E. H. Stanley, and A. J. Boulton. 1992. Hyporheic-surface water exchange: Implications for the structure and functioning of desert stream ecosystems. *In* "Proceedings of the First International Conference on Ground Water Ecology" (J. A. Stanford and J. J. Simons, eds.), pp. 395–405. American Water Resources Associations, Bethesda, MD.

Valett, H. M., S. G. Fisher, and N. B. Grimm. 1994. Vertical hydrologic exchange and ecological stability of a desert stream ecosystem. *Ecology* **75**:548–560.

Valett, H. M., J. A. Morrice, and C. N. Dahm. 1996. Parent lithology, groundwater-surface water exchange and NO{NE}$_3^-$ retention in headwater streams. *Limnology and Oceanography* **41**:333–345.

Valett, H. M., C. N. Dahm, M. E. Campana, J. A. Morrice, M. A. Baker, and C. S. Fellows. 1997. Hydrologic influence on groundwater-surface water ecotones: Heterogeneity in nutrient composition and retention. *Journal of the North American Benthological Association* **16**:239–247.

Vincent, W. F., and M. T. Downes. 1980. Variation in nutrient removal from a stream by watercress (*Nasturtium officinale* R.Br.). *Aquatic Botany* **9**:221–235.

Wetzel, R. G. 1975. "Limnology." Saunders College Publishing, Philadelphia.

Wetzel, R. G., and B. A. Manny. 1972. Decomposition of dissolved organic carbon and nitrogen compounds from leaves in an experimental hard-water stream. *Limnology and Oceanography* **17**:927–931.

Wondzell, S. M., and F. J. Swanson. 1996a. Seasonal and storm dynamics of the hyporheic zone of a 4th-order mountain stream. I: Hydrologic processes. *Journal of the North American Benthological Society* **15**:3–19.

Wondzell, S. M., and F. J. Swanson. 1996b. Seasonal and storm dynamics of the hyporheic zone of a 4th-order mountain stream. II: Nitrogen cycling. *Journal of the North American Benthological Society* **15**:20–34.

Yordy, D. M., and K. L. Ruoff. 1981. Dissimilatory nitrate reduction to ammonia. *In* "Denitrification, Nitrification, and Atmospheric Nitrous Oxide" (C. C. Delwiche, ed.), pp. 171–190. Wiley, New York.

9

Stream and Groundwater Influences on Phosphorus Biogeochemistry

Susan P. Hendricks and David S. White

Hancock Biological Station and Center for Reservoir Research
Murray State University
Murray, Kentucky

I. Introduction   221
II. Sources and Forms of Phosphorus   222
III. Abiotic Phosphorus Retention by Bed Sediments   223
IV. Biotic Phosphorus Retention and Release within Bed
      Sediments   224
V. Fluvial Dynamics and Physical Retention Mechanisms   225
VI. Phosphorus and Surface–Subsurface Exchange: A Conceptual
      Model   226
VII. Summary and Research Needs   232
      References   233

I. INTRODUCTION

Along with carbon, oxygen, and nitrogen, there is an extensive literature on phosphorus in aquatic ecosystems (e.g., reviews in Stumm and Morgan, 1981; Wetzel, 1983; Brezonick, 1994). Whereas much of our knowledge of phosphorus biogeochemistry comes from marine, lentic, and wetland environments, there is also a growing literature on cycling within streams, particularly surface waters (Stream Solute Workshop, 1990; Allan, 1995). Perhaps least examined at this time are the patterns and processes occurring within streambed sediments, the hyporheic zone, an ecotone between ground water and surface water. In this chapter, we provide brief reviews of the forms and sources of phosphorus to streams and the potential role of abiotic–biotic interactions within streambed sediments on phosphorus retention. We focus principally on the biogeochemical dynamics of phosphorus within stream hyporheic zones that contribute to retention (sink) or release (source) and then conclude by presenting seasonal data from four study sites and proposing a conceptual model of surface–subsurface phosphorus patterns and processes.

II. SOURCES AND FORMS OF PHOSPHORUS

Landscape features such as geology, topography, soil, and vegetation types along with human influences control the amounts and forms of phosphorus reaching streams. Most soil phosphorus is present in insoluble particulate forms complexed with iron, aluminum or manganese hydroxides, clays, colloidal suspensions, or suspended solids in solution (Fox, 1993). Watershed land-use practices that lead to greater erosion of fine-grained surface soils or sediments contribute to phosphorus loading (Dunne and Leopold, 1978; Lowrance *et al.*, 1984; Klotz, 1985; Campbell *et al.*, 1995; Gächter *et al.*, 1998b). Inputs of phosphorus via precipitation are negligible in most cases (Dunne and Leopold, 1978). The decomposition of litter from terrestrial vegetation further contributes both dissolved and particulate inorganic phosphorus via runoff or groundwater discharge. Finally, a major source of phosphorus to lotic ecosystems is the weathering and dissolution of parent geologies, such as limestone and volcanic rock, and the transport of phosphate and mineral–phosphate complexes to streams via ground water (Munn and Meyer, 1990; Triska *et al.*, 1993; Martí and Sabater, 1996).

Phosphorus often is the limiting nutrient in lotic and lentic surface waters (Wetzel, 1983; Allan, 1995). Exceptions include some nitrogen-limited lakes and streams where there are abundant geologic sources of phosphorus or where human input leads to eutrophication (Grimm *et al.*, 1981; Wetzel, 1983; Munn and Meyer, 1988; Triska *et al.*, 1993). Phosphorus is present in both dissolved and particulate forms in aquatic environments. Dissolved forms consist of inorganic phosphate and organic compounds such as polyphosphates, colloidal organic phosphorus, and phosphate esters. Particulate phosphorus consists of (1) complexes with inorganic substances such as clays, iron hydroxides, carbonates, and detrital matter, and (2) cellular components such as nucleic acids, phosphoproteins, enzymes, and vitamins. Once phosphorus reaches a stream, concentrations in surface and interstitial waters are dependent on complex interactions among discharge, erosional and sedimentation patterns, streambed geomorphology, and sediment properties such as adsorption–desorption capacities, the interstitial redox environment, and phosphorus–biota interactions.

III. ABIOTIC PHOSPHORUS RETENTION BY BED SEDIMENTS

Phosphorus behavior within aquatic sediments is controlled by abiotic adsorption of phosphorus to oxides and biotic microbial metabolism (Enell and Lofgren, 1988; Davelaar, 1993). We discuss abiotic and biotic controls as separate topics only for organizational convenience. Sediment structural properties such as adsorption–desorption capacity, particle-size distribution, mineral composition, and precipitation–dissolution processes determine abiotic immobilization of phosphorus. Sorption of phosphorus onto inorganic particle surfaces occurs under oxic conditions and high redox potentials via chemical bonding, steric effects, or electrostatic forces (Stumm and Morgan, 1981). Ferric and aluminum oxyhydroxides, calcium compounds (calcite and arganite), and clay minerals found within heterogeneous sediment matrices have strong affinities for phosphorus (Enell and Lofgren, 1988). The interaction of phosphate (PO_4) with iron is particularly well known in aquatic ecosystems (Wetzel, 1983). Complex interactions occur between interstitial pH and redox, and iron and phosphate concentrations occur within hydrologically retentive hyporheic environments. Sediments exhibiting large differences in phosphorus concentration between the oxic–anoxic interface zones generally have high adsorption capacities due to iron content (Enell and Lofgren, 1988). Sorption of phosphate by iron will, in part, determine the amount of both free phosphate and iron in interstitial water. Stoichiometry and pH are the primary influences on the adsorption capacity of sediments. As pH decreases, the surface charge of solids changes, and the adsorption of phosphate to iron, aluminum, and some clays increases (Enell and Lofgren, 1988). Iron hydroxides can adsorb phosphate in stoichiometric ratios ($Fe:PO_4$) ranging from 2 to 40 (Jensen *et al.*, 1992; Griffioen, 1994) depending on pH and the redox environment. For example, at higher pH and within oxidizing environments where $Fe:PO_4$ ratios are ≥ 15, iron adsorbs most free phosphate; within reducing environments and where ratios are ≤ 10, iron hydroxide–phosphate complexes dissolve releasing phosphate into solution (Jensen *et al.*, 1992). Phosphate not adsorbed to iron hydroxides may remain in solution or may bind with humic compounds (Enell and Lofgren, 1988; Griffioen, 1994) depending on the amount of metal ion (iron or aluminum) associated with the humic molecule (Enell and Lofgren, 1988; Pettersson *et al.*, 1988). The onset of anoxia results in lower redox potentials causing phosphorus to desorb from dissolved solids and to return to solution. Where sulfate (SO_4) concentration is high, particularly in anoxic lake, estuarine, and wetland sediments, sulfate–iron interactions may facilitate the mobilization of phosphate into interstitial solution (Caraco *et al.*, 1993; Roden and Edmonds, 1997). The importance of sulfate in hyporheic environments, however, is not well understood because interstitial concentrations are rarely measured.

IV. BIOTIC PHOSPHORUS RETENTION AND RELEASE WITHIN BED SEDIMENTS

Sediment–bacterial interactions play a significant role in uptake (sink) and release (source) of phosphorus in lake sediments (Gächter *et al.*, 1988a). Interactions among sediment properties, hydrology, and microbial activity and respiration maintain a mosaic of oxic–anoxic conditions often observed within hyporheic zones (Dahm *et al.*, 1987, 1991), and no doubt they play an important role in subsurface stream phosphorus dynamics as well.

The functional role of aerobic microorganisms in freshwater sediments with regard to phosphorus includes (1) mineralization of organic matter whereby organic phosphorus is transformed from large organic compounds to soluble, bioavailable forms (e.g., phosphate), (2) biosynthesis whereby inorganic phosphate is assimilated into living organic matter, and (3) extracellular and intracellular phosphorus cycling activities (Davelaar, 1993). Biofilm microorganisms play a central role in modifying the interstitial redox environment by rapidly responding to changes in soluble organic matter supply, temperature, and availability of electron acceptors (Enell and Lofgren, 1988; Davelaar, 1993). During aerobic microbial respiration, the electron transport system involves a sequence of reduction reactions which occurs along a thermodynamic gradient (Curtis, 1983). For example, terminal electron acceptors are reduced in the following order:

$$O_2 > NO_3^- > Mn^{4+} > Fe^{3+} > SO_4^{2+} > CO_2$$

A succession of microbial functional groups parallels this redox gradient from strict aerobes to obligate anaerobes such as denitrifiers, manganese and iron reducers, sulfate reducers, fermentation bacteria, and methanogens (Davelaar, 1993). Thus, a biotic coupling between bacterial metabolism, organic matter mineralization, and manganese, iron, and sulfate redox reactions and phosphorus uptake and release occurs along an aerobic–anaerobic gradient within hyporheic sediments.

In general, organic phosphorus in sediments includes a wide range of complex structures such as phytic acid, nucleic acids, adenosine phosphates (DeGroot, 1993), and humic–phosphorus complexes (Pettersson *et al.*, 1988). The array of exoenzymes found in sediments to hydrolyze these compounds includes acid and alkaline phosphatases, phosphodiesterase, RNase, DNase, phytase, and phospholypase. Exoenzymes such as phosphatases are located on bacterial cell surfaces, dissolved in interstitial water, or associated with inorganic or organic surfaces (Frauillade and Dorioz, 1992; Marxsen and Schmidt, 1993; Scholz and Marxsen, 1996). High phosphatase activity in subsurface sediments strongly suggests a demand for phosphorus by the sediment-associated microflora, indicating fairly rapid cycling of phosphorus by the attached microbial-biofilm community (Marxsen and Schmidt, 1993; Scholz and Marxsen, 1996). Microorganisms normally do not require large amounts of phosphorus; under phosphorus-limiting conditions, however, phosphatases which cleave phosphate groups from organically complexed phosphorus compounds for subsequent biotic uptake are produced. Phosphorus limitation is not necessarily a prerequisite for all phosphatase activity as bacterial production of internal phosphatases facilitates recycling of phosphorus within the bacterial cell as well (Sayler *et al.*, 1979; Klotz, 1985, 1988). Some chemoheterotrophic sediment bacteria are able to flourish under alternating oxia and anoxia causing rapid uptake and release of phosphorus in highly organic enriched sediments (see Davelaar, 1993, for an extensive review of the polyphosphate bacteria).

V. FLUVIAL DYNAMICS AND PHYSICAL RETENTION MECHANISMS

Phosphorus dynamics are directly linked to stream capacity to physically retain solutes and particulates within both the water column and sediments (D'Angelo *et al.*, 1991). Surface water retention is dependent on discharge, velocity, and the number of physical retention devices within the channel. Geomorphological features along drainage basin continuums determine, in part, physical retention within a stream (Vannote *et al.*, 1980; Dahm *et al.*, 1987; Triska *et al.*, 1989). Pools and backwaters, gravel and sand bars, large debris dams, leaf and wood accumulations, and macrophyte beds reduce water velocity, thereby promoting longer contact time between surface water and the physical–biological components of a stream reach (Allan, 1995). Therefore, phosphorus retention and cycling in streams are functions of sources, transport patterns, water residence times, and abiotic and biotic processing rates (Valett *et al.*, 1996).

Less is understood about retention within streambed sediments. Because they are covered in other chapters, it is not our intent to thoroughly discuss hydrologic and geomorphic features of streams that establish surface–subsurface exchange patterns. There are, however, factors that need some discussion as they can establish physicochemical conditions important to phosphorus dynamics.

Subsurface flow and transient storage of water in hyporheic zones can be substantial (Bencala and Walters, 1983; Stream Solute Workshop, 1990). Tracer injection experiments in conjunction with advection-dispersion-transient storage zone models have been used to estimate the ratio of storage zone cross-sectional area to stream channel cross-sectional area. These ratios vary from 0.03 to 4.6 in small streams (D'Angelo *et al.*, 1993; Morrice *et al.*, 1997; Mulholland *et al.*, 1997). Although retention dynamics are expected to be different in first-order streams compared with fifth- or tenth-order streams, the relative magnitudes of the ratios appear to be independent of stream order and demonstrate that, depending on flow conditions and sediment characteristics, even some first-order streams may be capable of retaining considerable amounts of solutes in their storage zones. The most important conclusion, however, is that most streams demonstrate some degree of subsurface transient storage and, therefore, the potential for phosphorus–sediment (and other nutrients) interaction.

In highly porous streambeds (e.g., gravels), interstitial water is commonly expected to be oxic, and biological and geochemical immobilization of phosphorus dominate. Fine sediments in backwaters and ponded or sluggish pools may be largely anoxic due to organic decomposition and very slow surface–subsurface exchange. Under these conditions, primarily anaerobic and geochemical processes dominate phosphorus cycling. Where bed features alter surface flow and the bed is permeable (but not extremely porous), there is the potential for establishing slowly moving subsurface interstitial flow (underflow paths). Such underflow paths commonly occur beneath riffles and bars and between stream meanders (Vaux, 1968; White *et al.*, 1987; White, 1990, 1993; Harvey and Bencala, 1993; Jones and Holmes, 1996) resulting in gradients of physicochemical conditions dependent on the length of the path and extent of organic decomposition occurring along the path. Underflow paths may begin as downwelling of aerobic surface water containing dissolved organic carbon, phosphorus and other nutrients, and end as upwelling of hyporheic water (possibly mixed with ground water), cumulative of all biological and physicochemical processes occurring along the path (Fig. 1). Thus, underflow patterns provide a range of conditions and a context in which phosphorus dynamics may best be discussed, and we have taken this perspective in creating the following conceptual model.

VI. PHOSPHORUS AND SURFACE–SUBSURFACE EXCHANGE: A CONCEPTUAL MODEL

In this final section, we present the results of some hyporheic phosphorus studies and propose a conceptual model of phosphorus dynamics at the small reach scale along underflow paths beneath an idealized riffle-pool hyporheic zone. Although riffle-pool sequences represent an important geomorphic retention feature in streams, our discussion is not necessarily limited to them as similar dynamics would be expected to occur beneath other geomorphic features wherever stream water enters the hyporheic storage zone, travels for some distance, and exits back to the stream surface. As riffles are prominent features of second- and third-order streams, we have concentrated on factors in this reach size both for convenience and because more is known about them.

Increased contact time between solutes and sediments along underflow paths can result in steep thermal, dissolved oxygen, and biogeochemical gradients (White *et al.*, 1987; Hendricks and White, 1991, 1995). In general, studies have shown that phosphorus concentration is usually higher in in-

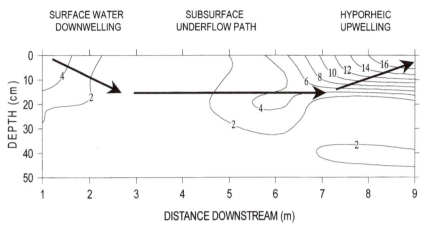

FIGURE 1 Isopleth contours of hyporheic soluble reactive phosphorus (SRP; μgP L^{-1}) patterns along a longitudinal transect beneath a riffle pool on July 16, 1989, in a northern Michigan stream. Isopleths are based on 135 determinations for SRP; $n = 3$ at each depth (5) at each meter (9) along the transect. Surface water SRP concentration was <2 μgP L^{-1}. Depth is magnified relative to distance in order to illustrate detail of the phosphorus contours. Depth 0 (*y*-axis) is the streambed surface. Meter 1–2 (*x*-axis) represent a downwelling zone. Meters 3–7 represent a subsurface flowpath. Meters 7–9 represent an upwelling zone. (From Hendricks and White, 1995, with permission.)

terstitial water than in surface water (Grimm and Fisher, 1984; Ford and Naiman, 1989; Valett *et al.*, 1990; Dahm *et al.*, 1991; Hendricks and White, 1991, 1995; Jones *et al.*, 1995). Table I summarizes examples of surface/hyporheic soluble reactive phosphorus (SRP) concentrations in four eastern streams during four seasons. Hyporheic phosphorus concentration, in general, is less variable from one season to the next than surface-water concentration. Higher seasonal variation in surface waters are most likely related to periphyton development.

In hydraulically retentive stream reaches where sediments and organic matter accumulate and decompose, anaerobic zones develop (Dahm *et al.*, 1987), often resulting in inverse correlations between interstitial dissolved oxygen, phosphorus, and many other dissolved substances (Valett *et al.*, 1990; Dahm *et al.*, 1991; Hendricks and White, 1991, 1995). High phosphate concentration commonly has been found in subsurface regions where dissolved oxygen is low (Hendricks and White, 1995), and although this trend is stronger in some streams than in others, similar distribution patterns have been observed in both the East and West Forks of Walker Branch, Tennessee (Fig. 2). As dissolved oxygen decreases to approximately ≤2 mgO$_2$ L^{-1}, increases in total phosphorus and SRP have been observed during both summer and winter (S. P. Hendricks, unpublished data).

Spatial–temporal differences in SRP concentrations have been observed

TABLE 1 Seasonal Comparison of Soluble Reactive Phoshorus Concentrations (μgP L^{-1}) in Stream Water and Hyporheic Water among Four Streams

Stream	Spring	Summer	Autum	Winter
Maple River, Michigan (third order)				
Surface water	2.0 ± 0.0 (3)	BDL	2.3 ± 1.2 (3)	6.3 ± 0.6 (3)
Hyporheic water	3.8 ± 1.6 (73)	2.0 ± 2.1 (83)	3.2 ± 1.9 (83)	9.4 ± 4.5 (83)
Walker Branch, Tennessee (first order)				
Surface water	3.2 ± 1.6 (10)	4.1 ± 1.1 (6)	3.4 ± 2 (12)	2.1 ± 0.7 (6)
Hyporheic water	6.9 ± 5.8 (55)	2.7 ± 4.8 (47)	4.6 ± 6.0 (92)	4.4 ± 4.9 (42)
Ledbetter Creek, Kentucky (second to third order)				
Surface water	7.5 ± 5.2 (6)	2.3 ± 0.5 (4)	3.5 ± 1.3 (4)	4.8 ± 2.2 (4)
Hyporheic water	2.3 ± 19.7 (81)	6.3 ± 2.4 (52)	8.1 ± 11.2 (54)	9.4 ± 4.8 (54)
Panther Creek, Tennessee (third order)				
Surface water	33.3 ± 1.2 (3)	16.6 ± 4.3 (6)	19.5 ± 2.4 (4)	33.8 ± 4.6 (4)
Hyporheic water	27.2 ± 11.3 (69)	33.8 ± 106.2 (104)	28.3 ± 18.8 (81)	36.2 ± 49.5 (77)

Data for the Maple River, MI, is from Hendricks and White (1995). Other streams are from S. P. Hendricks and D. S. White, unpublished data. BDL = detection limit (2 μgP L^{-1}). Numbers in parentheses are number of observations.

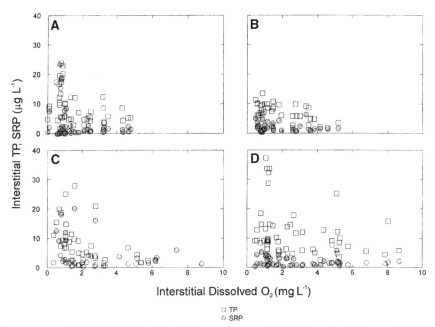

FIGURE 2 Hyporheic total phosphorus (TP) and soluble reactive phosphorus (SRP) concentrations (μgP L^{-1}) at 10–20 cm depth intervals within the sediments at Walker Branch, Tennessee. (A, C) West Fork, June and December 1993, respectively. (B, D) East Fork, July 1993 and January 1994, respectively (S. P. Hendricks, unpublished data).

as well (Hendricks and White, 1995; Jones *et al.*, 1995). For example, in a sandy northern Michigan stream, spatial pattern of increasing SRP concentration occurred along a 9-m longitudinal subsurface flowpath during summer and winter when strong hyporheic thermal gradients existed (e.g., Fig. 1; Hendricks and White, 1995). The SRP concentration in surface water and in the downwelling zone were near or below detection limits (2–4 μgP L^{-1}) during both summer and winter, whereas concentrations in the upwelling zone were 15 and 30 μgP L^{-1} during summer and winter, respectively. The longitudinal pattern was attributed to decomposition processes occurring along the subsurface flowpath and the vertical pattern to a combination of decomposition and thermal stratification that occurred during summer and winter in north temperate streams (Hendricks and White, 1995). Similarly, in Sycamore Creek, Arizona, SRP concentration was significantly greater in downwelling (57 μgP L^{-1}) than upwelling (73 μgP L^{-1}) hyporheic zones which was attributed to phosphorus enrichment via decomposition of organic matter (Jones *et al.*, 1995). Further, concentration differences between downwelling and upwelling zones were attributed to differences in water residence time (0.1 and 2.5 hours, respectively) as estimated from the relative magnitudes of vertical hydraulic gradients and respiration rates.

Where streambed contour is convex, such as at the heads of riffles, surface water infiltrates into the bed (downwelling; Fig. 3A). Where streambed contour is concave, such as at the ends of riffles and beginnings of pools, subsurface water (hyporheic return flow which may include a groundwater component) is deflected upward and exits the bed (upwelling; White *et al.*, 1987). Interstitial water in the downwelling zone tends to reflect surface water quality, whereas upwelling zone water reflects a combination of subsurface biogeochemical transformations and groundwater inputs occurring along the underflow path. The downwelling zone is generally characterized by higher dissolved oxygen concentration and redox potential that define it as largely oxic, although some anoxic microsites may exist at particle surfaces. Inorganic phosphate concentration in surface water and interstitial water are expected to be similar (e.g., Fig. 1; Hendricks and White, 1991, 1995). The downwelling redox environment favors the presence of oxidized forms of iron and manganese complexed with phosphorus depending on pH. Thus, phosphorus retention would be controlled primarily by chemical sorption.

By contrast, slower interstitial water velocities and greater potential for interaction of interstitial solutes with subsurface and surface sediment-associated microflora characterize the upwelling zone. The upwelling zone is characterized by lower dissolved oxygen concentration and redox potential that may be low enough to define it as a largely anoxic reducing environment. In this zone intensive microbial respiration and rapid recycling of phosphorus result in steep interstitial oxygen and SRP (and other solute) gradients if sufficient organic matter is present (Hendricks and White, 1995, 1995; Hendricks, 1993; Jones *et al.*, 1995). Lower redox potential favors the dissolution of iron–phosphorus and possibly other mineral–phosphorus complexes, such as aluminum–phosphorus and manganese–phosphorus, releasing reduced iron (or other metals) and phosphate into the interstitial water. Advective hyporheic flow results in the transport of phosphate to the sediment–water interface (oxidized microzone, Fig. 3B). Upon reaching this interface, dissolved phosphorus would be either (1) re-adsorbed to oxidized forms of iron (or manganese), (2) taken up rapidly and fixed into biomass by benthic microflora, or (3) diffuse/advect across the interface into stream water. Hyporheic conditions (1) and (2) could be considered phosphorus sinks, and (3) could be considered a phosphorus source. Diel patterns of phosphorus uptake and release at the sediment–water interface related to algal activity have been observed in lakes (Carlton and Wetzel, 1988), which potentially could occur in streams depending on redox conditions within the benthic microflora–biofilm complex.

In general, a longer residence time of water (and dissolved nutrients) within the hyporheic zone contributes to longer turnover time of water in the stream and spatially differential accumulations of phosphate where organic matter degradation and subsequent oxygen depletion occur. Phosphate accumulates in the interstices either from physicochemical dissolution of phosphate, from particles and dissolved solids, or from organic matter

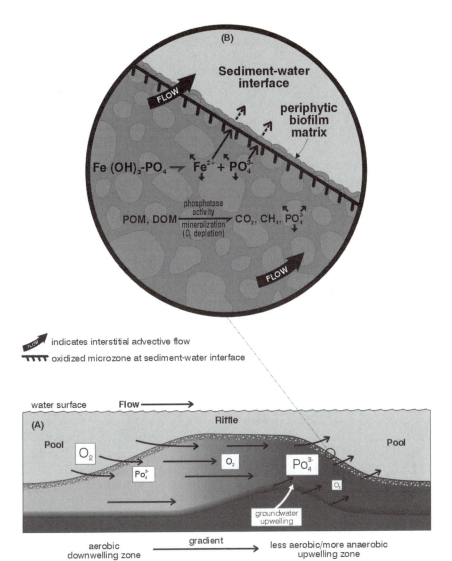

FIGURE 3 (A) Conceptual model of hyporheic flowpath through an idealized riffle-pool sequence. The upstream end of the riffle constitutes the downwelling zone where stream water infiltrates (see also Fig. 1). The downstream end of the riffle (and beginning of the pool) constitutes the upwelling zone where return flow exits the bed. The sizes of the boxes indicate the relative magnitudes in dissolved oxygen and phosphate along the flowpath. (B) magnification of the sediment–water interface and interstitial hyporheic environment immediately beneath the interface. As the interstitial environment becomes more anoxic, immobilized phosphorus is released from dissolved solids and/or dissolved or particulate organic matter. Phosphorus is released into the interstitial environment where advective forces (arrows) may carry it to the sediment–water interface for (1) subsequent uptake by the periphytic biofilm matrix, (2) recomplexation with iron (or other metal) oxides, or (3) release into surface water.

degradation or both and may be transported to overlying water across the sediment–water interface by diffusion, gas ebullition, bioturbation, and advective forces (Fig. 3B; Enell and Lofgren, 1988; Jensen *et al.*, 1992).

VII. SUMMARY AND RESEARCH NEEDS

Clearly, phosphorus behavior in stream hyporheic zones results from a complex interaction of several retention and release mechanisms functioning simultaneously within a mosaic of hydrological, geomorphological, and biological conditions. Retentiveness is temporally (diel and season) and spatially (upwelling versus downwelling) dependent on discharge, sediment permeability, and advection patterns into and within the sediments. Relationships between subsurface flow and the relative rates and processes involved in regulating overall phosphorus assimilation by the biota are important topics for future research (Jones and Holmes, 1996; Valett *et al.*, 1996).

Flux measurements of phosphorus in streambed sediments remain underrepresented in stream studies. Little is known about the effects of subsurface advective flow on phosphorus flux across the sediment–water interface. Coupling conservative tracer and reactive phosphorus release experiments (sensu Mulholland *et al.*, 1997) with mesocosm-scale laboratory or measurements using *in situ* sediment cores or chambers designed to measure microbially mediated phosphorus transformation and flux (e.g., exoenzyme activity) may provide a promising approach for gaining understanding of the linkage between subsurface phosphorus flux and surface–subsurface exchange phenomena (Marxsen and Fiebig, 1993; Jones and Holmes, 1996; Morrice *et al.*, 1997; Mulholland *et al.*, 1997; White and Hendricks, Chapter 15).

High bacterial biomass and productivity occurring in stream sediments, along with the high level of phosphatase activity, suggests a rapid flux of phosphorus (Marxsen and Schmidt, 1993). Does phosphorus-"deficient" water entering the hyporheic zone stimulate microbes to produce exoenzymes (e.g., phosphatases) that liberate phosphates from organic substrates for subsequent microbial uptake along a subsurface flowpath? Regulatory mechanisms (process kinetics, pH, and temperature optima), competitive strategies and interactions such as phosphatase–substrate diversity (Marxsen and Schmidt, 1993), biofilm development, and microbial processing rates in response to hydrological fluctuations and physicochemical gradient development are areas for further research in the hyporheic zone. Further, little is known about the role of sulfates in the dissolution of iron–phosphorus complexes in stream sediments.

Benthic organisms are known to play an important role in altering redox conditions and active transport of sediment particles and water across the sediment–water interface in lakes (Gallepp, 1979; Fukuhara and Sakamoto, 1987). Although the distributions of some surface-dwelling organisms have been linked to interstitial and sediment phosphorus concentrations (Meyer, 1979; White, 1990; White and Hendricks, Chapter 15), the role of benthic organisms (periphytic biofilms, macrophytes, chironomids, tubificids) in phosphorus flux from sediments to surface waters is not well documented in streams. Finally, cross-system comparisons of hyporheic processes under spatially and temporally variable flow regimes, among different watershed-scale land-use practices, and among differing parent geologies and ecoregions are essential to understanding phosphorus dynamics in streams.

REFERENCES

Allan, J. D. 1995. "Stream Ecology." Chapman & Hall, London.

Bencala, K. E., and R. A. Walters. 1983. Simulation of solute transport in a mountain pool-and-riffle stream: A transient storage model. *Water Resources Research* **19**:718–724.

Brezonick, P. L. 1994. "Chemical Kinetics and Process Dynamics in Aquatic Systems." Lewis Publishers, Boca Raton, FL.

Campbell, K. L., J. C. Capece, and T. K. Tremwel. 1995. Surface/subsurface hydrology and phosphorus transport in the Kissimmee river Basin, Florida. *Ecological Engineering* **5**:301–330.

Caraco, N. F., J. J. Cole, and G. E. Likens. 1993. Sulfate contol of phosphorous availability in lakes. *Hydrobiologia* **253**:275–280.

Carlton, R. G., and R. G. Wetzel. 1988. Phosphorus flux from lake sediments: Effect of epipelic algal oxygen production. *Limnology and Oceanography* **33**:562–570.

Curtis, C. D. 1983. Microorganisms and the diagenesis of sediments. *In* "Microbial Geochemistry" (W. E. Krumbein, ed.), pp. 263–286. Blackwell, Oxford.

Dahm, C. N., E. H. Trotter, and J. R. Sedell. 1987. The role of anaerobic zones and processes in stream ecosystem productivity. *In* "Chemical Quality of Water and the Hydrologic Cycle" (R. A. Averett and D. M. McKnight, eds.), pp. 157–178. Lewis Publishers, Chelsea, MI.

Dahm, C. N., D. L. Carr, and R. L. Coleman. 1991. Anaerobic carbon cycling in stream ecosystems. *Verhandlungen Internationale Vereinigung für Theoretische und Angewandte Limnologie* **24**:1600–1604.

D'Angelo, D. J., J. R. Webster, and E. F. Benfield. 1991. Mechanisms of stream phosphorus retention: An experimental study. *Journal of the North American Benthological Society* **10**:225–237.

D'Angelo, D. J., J. R. Webster, S. V. Gregory, and J. L. Meyer. 1993. Transient storage in Appalachian and Cascade mountain streams as related to hydraulic characteristics. *Journal of the North American Benthological Society* **12**:223–235.

Davelaar, D. 1993. Ecological significance of bacterial polyphosphate metabolism in sediments. *Hydrobiologia* **253**:179–192.

DeGroot, C. 1993. Some remarks on the presence of organic phosphates in sediments. *Hydrobiologia* **207**:303–309.

Dunne, T., and L. B. Leopold. 1978. "Water in Environmental Planning." Freeman, New York.

Enell, M., and S. Lofgren. 1988. Phosphorus in interstitial water: Methods and dynamics. *Hydrobiologia* **170**:103–132.

Ford, T. E., and R. J. Naiman. 1989. Groundwater–surface water relationships in boreal forest watersheds: Dissolved organic carbon and inorganic nutrient dynamics. *Canadian Journal of Fisheries and Aquatic Sciences* **46**:41–49.

Fox, L. E. 1993. The chemistry of aquatic phosphate: Inorganic processes in rivers. *Hydrobiologia* 253:1–16.

Frauillade, M., and J. M. Dorioz. 1992. Enzymatic release of phosphate in sediments of various origins. *Water Research* 26:1195–1201.

Fukuhara, H., and M. Sakamoto. 1987. Enhancement of inorganic nitrogen and phosphate release from lake sediment by tuvificid worms and chironomid larvae. *Oikos* 48:312–320.

Gächter, R., J. S. Meyer, and A. Mares. 1988a. Contribution of bacteria to release and fixation of phosphorus in lake sediments. *Limnology and Oceanography* 33:1542–1558.

Gächter, R., J. M. Ngatiah, and C. Stamm. 1998b. Transport of phosphate from soil to surface waters by preferential flow. *Environmental Science and Technology* 32:1865–1868.

Gallepp, G. W. 1979. Chironomid influence on phosphorus release in sediment-water microcosms. *Ecology* 60:547–556.

Griffioen, J. 1994. Uptake of phosphate by iron hydroxides during seepage in relation to development of groundwater composition in coastal areas. *Environmental Science and Technology* 28:675–681.

Grimm, N. B., and S. G. Fisher. 1984. Exchange between interstitial and surface water: Implications for stream metabolism and nutrient cycling. *Hydrobiologia* 111:219–228.

Grimm, N. B., S. G. Fisher, and W. L. Minckley. 1981. Nitrogen and phosphorus dynamics in a hot desert stream in southwestern U.S.A. *Hydrobiologia* 83:303–312.

Harvey, J. W., and K. E. Bencala. 1993. The effect of streambed topography on surface-subsurface water exchange in mountain catchments. *Water Resources Research* 29:89–98.

Hendricks, S. P. 1993. Microbial ecology of the hyporheic zone: A perspective integrating hydrology and biology. *Journal of the North American Benthological Society* 12:70–78.

Hendricks, S. P., and D. S. White. 1991. Physicochemical patterns within a hyporheic zone of a northern Michigan river, with comments on surface water patterns. *Canadian Journal of Fisheries and Aquatic Sciences* 48:1645–1654.

Hendricks, S. P., and D. S. White. 1995. Seasonal biogeochemical patterns in surface, water, subsurface hyporheic, and riparian groundwater in a temperate stream ecosystem. *Archiv für Hydrobiologie* 134:459–490.

Jensen, H. S., P. Kristensen, E. Jeppesen, and A. Skytthe. 1992. Iron:phosphorus ratio in surface sediment as an indicator of phosphate release from aerobic sediments in shallow lakes. *Hydrobiologia* 235/236:731–743.

Jones, J. B., and R. M. Holmes. 1996. Surface-subsurface interactions in stream ecosystems. *Trends in Ecology and Evolution* 11:239–242.

Jones, J. B., S. G. Fisher, and N. B. Grimm. 1995. Vertical hydrologic exchange and ecosystem metabolism in a Sonoran Desert stream. *Ecology* 76:942–952.

Klotz, R. I. 1985. Factors controlling phosphorus limitation in stream sediments. *Limnology and Oceanography* 30:543–553.

Klotz, R. L. 1988. Sediment control of soluble reactive phosphorus in Hoxie Gorge Creek, New York. *Canadian Journal of Fisheries and Aquatic Sciences* 45:2026–2034.

Lowrance, R. R., R. L. Todd, and L. E. Asmussen. 1984. Nutrient cycling in an agricultural watershed: II. Streamflow and artificial drainage. *Journal of Environmental Quality* 13:27–32.

Martí, E., and F. Sabater. 1996. High variability in temporal and spatial nutrient retention in Mediterranean streams. *Ecology* 77:854–869.

Marxsen, J., and D. M. Fiebig. 1993. Use of perfused cores for evaluating extracellular enzyme activities in stream-bed sediments. *FEMS Microbial Ecology* 13:1–12.

Marxsen, J., and H. H. Schmidt. 1993. Extracellular phosphatase activity in sediments of the Breitenbach, a central European mountain stream. *Hydrobiologia* 253:207–216.

Meyer, J. L. 1979. The role of sediments and bryophytes in phosphorus dynamics in a headwater stream ecosystem. *Limnology and Oceanography* 24:365–376.

Morrice, J. A., H. M. Valett, C. N. Dahm, and M. E. Campana. 1997. Alluvial characteristics, groundwater-surface water exchange and hydrologic retention in headwater streams. *Hydrological Processes* 11:253–267.

Mulholland, P. J., E. R. Marzolf, J. R. Webster, D. R. Hart, and S. P. Hendricks. 1997. Evidence of hyporheic retention of phosphorus in Walker Branch. *Limnology and Oceanography* **42**:443–451.

Munn, N. L., and J. L. Meyer. 1988. Rapid flow through the sediments of a headwater stream in the southern Appalachians. *Freshwater Biology* **20**:235–240.

Munn, N. L., and J. L. Meyer. 1990. Habitat-specific solute retention in two small streams: An intersite comparison. *Ecology* **71**:2069–2082.

Pettersson, K., B. Bostrom, and O. Jacabsen. 1988. Phosphorus in sediments—speciation and analysis. *Hydrobiologia* **170**:91–101.

Roden, E. E., and J. W. Edmonds. 1997. Phosphate mobilization in iron-rich anaerobic sediments: Microbial Fe(III) oxide reduction versus iron-sulfide formation. *Archiv für Hydrobiologie* **139**:347–378.

Sayler, G., M. Puziss, and M. Silver. 1979. Alkaline phosphatase assay for freshwater sediments: Application to perturbed sediment systems. *Applied and Environmental Microbiology* **38**:922–927.

Scholz, O., and J. Marxsen. 1996. Sediment phosphatases of the Breitenbach, a first-order Central European stream. *Archiv für Hydrobiologie* **135**:433–450.

Stream Solute Workshop. 1990. Concepts and methods for assessing solute dynamics in stream ecosystems. *Journal of the North American Benthological Society* **9**:95–119.

Stumm, W., and J. J. Morgan. 1981. "Aquatic Chemistry," 2nd edition. Wiley-Interscience, New York.

Triska, F. J., V. C. Kennedy, R. J. Avanzino, G. W. Zellweger, and K. E. Bencala. 1989. Retention and transport of nutrients in a third-order stream in northwestern California: Hyporheic processes. *Ecology* **70**:1893–1905.

Triska, F. J., C. M. Pringle, G. W. Zellweger, and J. H. Duff. 1993. Dissolved inorganic nitrogen composition, transformation, retention, and transport in naturally phosphate-rich and phosphate-poor tropical streams. *Canadian Journal of Fisheries and Aquatic Sciences* **50**:665–675.

Valett, H. M., S. G. Fisher, and E. H. Stanley. 1990. Physical and chemical characteristics of the hyporheic zone of a Sonoran Desert stream. *Journal of the North American Benthological Society* **9**:201–215.

Valett, H. M., J. A. Morrice, C. N. Dahm, and M. E. Campana. 1996. Parent lithology, surface-groundwater exchange, and nitrate retention in headwater streams. *Limnology and Oceanography* **41**:333–345.

Vannote, R. L., G. W. Minshall, K. W. Cummins, J. R. Sedell, and C. E. Cushing. 1980. The river continuum concept. *Canadian Journal of Fisheries and Aquatic Sciences* **37**:130–137.

Vaux, W. G. 1968. Intergravel flow and interchange of water in a streambed. *Fisheries Bulletin* **66**:479–489.

Wetzel, R. G. 1982. "Limnology," 2nd ed. Saunders, Philadelphia.

White, D. S. 1990. Biological relationships to convective flow patterns within streambeds. *Hydrobiologia* **196**:148–159.

White, D. S. 1993. Perspectives on defining and delineating hyporheic zones. *Journal of the North American Benthological Society* **12**:61–69.

White, D. S., C. H. Elzinga, and S. P. Hendricks. 1987. Temperature patterns within the hyporheic zone of a northern Michigan river. *Journal of the North American Benthological Society* **6**:85–91.

10

Surface and Subsurface Dissolved Organic Carbon

Louis A. Kaplan and J. Denis Newbold

Stroud Water Research Center
Avondale, Pennsylvania

I. Introduction   237
II. DOC Concentrations   238
    A. Stream-Water DOC Concentrations   238
    B. Ground Water DOC Concentrations   240
    C. Hyporheic Zone DOC Concentrations   241
III. Processes within the Hyporheic Zone   242
    A. Sources and Sinks   242
    B. Hydrologic Exchanges   243
    C. Interactions of DOC with Surfaces   244
    D. Profiles through the Hyporheic Zone   244
    E. Impact on Energy Flow and Carbon Cycling   250
References   253

I. INTRODUCTION

Dissolved organic carbon (DOC) is a potential source of carbon and energy for heterotrophic organisms and contributes significantly to stream ecosystem metabolism. Subsurface waters provide a vector for the movement of DOC from terrestrial environments to stream ecosystems (Fisher and Likens, 1973), and it has been suggested that subsurface DOC contributes significantly to stream ecosystem metabolism (Hynes, 1983). Stream ecologists have come to appreciate the hydrologically dynamic nature of stream channels (White, 1983) and the potential for hyporheic zone metabolism to equal or exceed metabolism at the streambed surface (Grimm and Fisher, 1984; Fuss and Smock, 1996; Mulholland *et al.*, 1997; Naegeli and Uehlinger, 1997). Downwelling surface waters enriched with DOC fuel hyporheic metabolism (Jones *et al.*, 1995) and upwelling ground waters enriched with DOC are subject to geochemical immobilization (Fiebig *et al.*, 1990) and subsequent metabolism (Freeman and Lock, 1995). Thus a reasonable expectation is a general linkage between surface and subsurface processes involving DOC metabolism. The generality of these interactions has not been explored, nor is it understood whether subsurface processes that either produce or consume DOC influence streambed surface processes and communities. In this chapter, we review what is known about (1) DOC concentration in stream waters, ground waters, and hyporheic zone waters; (2) processes involving DOC in hyporheic zones; and (3) the impact of hyporheic zone processes involving DOC on energy flow and carbon cycling in streams.

II. DOC CONCENTRATIONS

A. Stream-Water DOC Concentrations

The concentration of DOC in streams and rivers under baseflow conditions ranges from <0.5 mgC L^{-1} for streams in alpine (Moeller *et al.*, 1979) and evergreen tropical (Newbold *et al.*, 1995) forests to >30 mgC L^{-1} for streams and rivers draining wetland swamps (Mulholland, 1981) or extensive floodplain forests (Meyer, 1986). A classification of DOC concentration in streams and rivers from different biomes has been presented by Meybeck (1982) and has general applicability even though DOC concentration in lotic systems is not a simple function of rates of terrestrial primary productivity of the contributing watersheds (Kaplan and Newbold, 1993). Watershed vegetation, climate, and decomposer activity influence the resulting concentration of DOC in streams, as well as two additional factors that are equally, if not more, influential—watershed soils and hydrology (Aiken and Cotsaris, 1995).

A soil attribute of particular importance to DOC concentration is clay content. In two Australian watersheds with similar climate, vegetation, and land use but different soils, DOC concentration differed by 2.5-fold (Nelson *et al.*, 1990). Adsorption studies with the soils demonstrated that DOC adsorption capacity was related to specific surface area, which, in turn, was related to clay content (Nelson *et al.*, 1990). On a broader scale, 44% of the variation in the mean DOC concentration of 18 streams within catchments from the Mt. Lofty Ranges, Australia, was accounted for by the clay content of the soil A horizon (Nelson *et al.*, 1990). Studies in two adjacent catchments within the Otway Ranges in Australia showed that DOC concentration of the streams, which differed by 8.4-fold, also was a function of soil clay content and specific surface area (Nelson *et al.*, 1993). DOC complexation with aluminum and iron oxides facilitates adsorption to clays (podzolization) in mineral soils (Reve and Fergus, 1981; McDowell and Wood,

1984) and is the mechanism believed to explain decreases in DOC concentration with soil depth in forested soils of the Adirondack region (Cronan and Aiken, 1985). In fact, soil clay content probably explains the widespread pattern of low colored, low DOC streams throughout eastern North America including North-Central Ontario (Wassenaar *et al.*, 1991), the Piedmont Physiographic Province (Kirk *et al.*, 1978), parts of New England (McDowell and Likens, 1988; David *et al.*, 1992), and the Ridge and Valley Province (Mulholland and Hill, 1997), versus high DOC blackwater streams throughout the sandy Atlantic Coastal Plain including the New Jersey Pinelands (Lord *et al.*, 1990), Virginia (Metzler and Smock, 1990), South Carolina (Dosskey and Bertsch, 1994), and Georgia (Meyer, 1986).

Hydrology influences DOC concentration of stream water in two ways. One is the influence of wetland soils and the other is the pathways that water flows to streams. Wetland soils have high DOC concentration due to the extensive leaching of organic matter combined with slow rates of decomposition under saturated, anoxic conditions. Studies in streams draining the Precambrian Shield in Canada (Dillon and Molot, 1997; Hinton *et al.*, 1998) and Sweden (Ivarsson and Jansson, 1994), peaty soils overlying glacial gravels in New Zealand (Moore and Jackson, 1989), and glaciated catchments characterized by relatively shallow soils in Massachusetts (Hemond, 1990) have shown a direct relationship between DOC concentration and the amount of wetlands in the watershed. A curious aspect of this relationship is that wetlands have high DOC concentration because there is insufficient flow through them to oxygenate the water, yet flow is required to transport DOC to streams. In each of the studies cited, the wetlands are physically connected to the stream as stream channel wetlands (Hemond, 1990), or riparian peatlands (Ivarsson and Jansson, 1994), or through outflowing streams from beaver ponds and conifer swamps (Schiff *et al.*, 1997).

Obviously any allochthonous organic carbon source, not just wetland carbon, needs to be transported to streams or rivers to contribute to the organic matter budget. In addition to wetland sources, a high concentration of DOC is present in throughfall, the porewaters of unsaturated organic soils, especially in the upper litter layers (Cronan and Aiken, 1985; Cronan, 1990), and some ground waters (Cronan, 1990; Wallis *et al.*, 1981). In the absence of precipitation, throughfall or overland flow do not occur, and the quantitative importance of lateral unsaturated flow to stream flow is not known (Hemond, 1990). However, the pathways that water flows to streams, especially in the near-stream or riparian zone, are critical to understanding the terrestrial sources of DOC (Trumbore *et al.*, 1992). For example, ground water that moves to a stream through a wetland or through the upper soil horizon will have a significantly greater DOC concentration than ground water entering the stream through lower soil horizons (Hinton *et al.*, 1998) or directly from a fractured bedrock zone.

Storms and associated increases in discharge lead to an increase in DOC concentration by two- to as much as five- to tenfold in nearly all streams and rivers, including ecosystems in the tropics (Lewis and Saunders, 1989; Newbold *et al.*, 1995), temperate zone (Baker *et al.*, 1974; McDowell and Fisher, 1976; Foster and Grieve, 1982; Grieve, 1984; McDowell and Likens, 1988; Mulholland and Hill, 1997; Wehr *et al.*, 1997), and boreal forest (Lush and Hynes, 1978; Hinton *et al.*, 1997). An exception to this phenomenon may be arid land streams such as Sycamore Creek, Arizona. In one study, floods in Sycamore Creek were found to produce a DOC concentration similar to low flow levels (Jones *et al.*, 1996), although another study found significantly elevated DOC associated with flooding (Holmes *et al.*, 1998). In ecosystems where storms do increase stream-water DOC concentration, an alteration of flowpaths often results in the transport of relatively unprocessed DOC from the vadose zone to the stream (McDowell and Likens, 1988; Mulholland, 1993; Gremm and Kaplan, 1998). Surface transport from large floodplain forests also is an important process (Meyer, 1986; Lewis and Saunders, 1989). Additionally, experimental studies in a sluice have shown that some of the DOC increases at elevated flows result from within-stream disturbances (Casey and Farr, 1982). Lastly, in alpine (Hornberger *et al.*, 1994; Boyer *et al.*, 1995) and tundra (Michaelson *et al.*, 1998) streams, spring snowmelt flushes high levels of DOC from organic matter that has accumulated in the soils or tundra during the growing season.

B. Ground Water DOC Concentrations

All subsurface waters are ground water, but it is useful for purposes of description and modeling to distinguish between different subsurface water zones. These include water in unsaturated soils (vadose zone water), water in the portion of the phreatic zone that is above the regolith (shallow ground water or saturated soil water), and deep ground water that is in the geologic aquifer (bedrock zone water). These are, in fact, the subsurface waters identified in a three-component model of hydrologic flowpaths for Walker Branch watershed in the Tennessee Ridge and Valley Province (Mulholland, 1993), and they have general applicability in most watersheds. Wetlands are a special category because they can be at least intermittently unsaturated but have a high water content and saturated conditions much of the time (Hemond, 1990).

The DOC concentration of bedrock zone water is typically quite low, as evidenced by a median value of 0.7 mgC L^{-1} and an 85% quantile of less than 2 mgC L^{-1} in a survey of 100 wells and springs in 27 states within the United States (Leenheer *et al.*, 1974). There were exceptions in each of the five aquifer types sampled, however, with maximum concentrations of 3.2 mgC L^{-1} for sandstone, 5.0 mgC L^{-1} for limestone, 3.8 mgC L^{-1} for crystalline rock, 15.0 mgC L^{-1} for shallow sand and gravel, and 6.9 mgC L^{-1} for deep sand and gravel (Leenheer *et al.*, 1974). Flow from deep ground water is often equated with baseflow discharge, and chemical end-member mixing model analyses indicate that this deep source can contribute as much as 80% of the water under nonstorm conditions (Easthouse *et al.*, 1992; Mulholland, 1993).

Studies of watershed DOC sources often do not involve drilling deep wells into the bedrock but rather involve sampling shallow ground water from the saturated soils in the phreatic zone. Water from the shallow phreatic zone can be low in DOC concentration, including samples from the Amazon (Williams *et al.*, 1997), Atlantic Coastal Plain in South Carolina (Dosskey and Bertsch, 1994) and North Carolina (Mulholland, 1981), Ridge and Valley Province in Tennessee (Mulholland and Hill, 1997) and North Carolina (Wallace *et al.*, 1982; Meyer and Tate, 1983), Pennsylvania Piedmont (Kaplan *et al.*, 1980), Adirondacks (Cronan and Aiken, 1985), mountainous regions of Massachusetts (McDowell and Fisher, 1976) and New Hampshire (McDowell and Likens, 1988), northern Michigan (Hendricks and White, 1991), and Precambrian Shield (Trumbore *et al.*, 1992). Alternatively, other studies in ecosystems located in a Canadian subalpine forest (Wallis *et al.*, 1981), northwestern California (Duff and Triska, 1990), New Jersey Atlantic Coastal Plain (Lord *et al.*, 1990), mid-Wales (Fiebig *et al.*, 1990), and central Germany (Fiebig, 1995) all reported high DOC concentration in shallow phreatic zone waters. A pattern in these data is difficult to see, especially when, for example peaty soils can yield DOC concentration in the phreatic zone that are either low (Trumbore *et al.*, 1992) or high (Fiebig *et al.*, 1990). What can be said with some certainty is that low phreatic zone DOC levels are the result of extensive abiotic and biotic processing of terrestrial DOC sources within the vadose zone, whereas high phreatic zone DOC implies the opposite.

C. Hyporheic Zone DOC Concentrations

If hyporheic zone DOC concentration is simply the result of conservative mixing of ground water and stream water, the concentration of DOC within the hyporheic zone, by definition, would fall somewhere between those end members. In fact, that is not always the case (Hendricks and White, 1991), and a consideration of such deviations, discussed later, provides insight into the DOC dynamics within the hyporheic zone. Data for DOC concentration within hyporheic zones are limited and variable, including instances of higher, lower, and equivalent concentrations compared to surface waters and ground waters. In Sycamore Creek, a Sonoran Desert stream, the high hydraulic conductivity of streambed sediments supports rapid interstitial flow rates, and the resulting frequent hydrologic exchanges generate hyporheic zone DOC concentrations (range 1.5–8.7 mgC L^{-1}) that are similar to surface-water DOC concentrations (Jones *et al.*, 1996). Hyporheic zone DOC concentrations that exceeded surface-water concentrations 1.2-

to 2.0-fold were measured in three first-order desert streams where DOC ranged from 5.5 to 11.5 mgC L^{-1} in oxic and anoxic interstitial waters (Dahm *et al.*, 1991). In White Clay Creek, a southeastern Pennsylvania piedmont stream, average interstitial DOC concentration measured at five transects ranged from 0.3 to 1.9 mgC L^{-1} and was consistently lower than concentration in the water column; in many cases, concentration continued to diminish with increasing depth into the streambed (T. J. Battin, L. A. Kaplan, and J. D. Newbold, unpublished data).

There are other references providing data on hyporheic zone DOC concentration, which as a group contain values that range from 0.5 mgC L^{-1} (Crocker and Meyer, 1987) to 82.6 mgC L^{-1} (Dahm *et al.*, 1987), but no consistent pattern across sites exists. In a North Carolina mountain stream, interstitial DOC concentration exceeded stream water DOC (Crocker and Meyer, 1987). In a northern Michigan hyporheic zone, DOC was consistently greater than either stream-water or groundwater DOC (Hendricks and White, 1991). In three agriculturally impacted streams in southern Ontario, DOC concentration patterns were extremely variable temporally and spatially when measured over as many as seven depths (Rutherford and Hynes, 1987). Strikingly higher DOC concentration was observed in anoxic (maximum of 82.6 mgC L^{-1}) versus oxic (maximum of 20.6 mgC L^{-1}) interstitial waters in a second-order stream in the Coast Range of western Oregon, and both were more than an order of magnitude higher than the stream water DOC concentration (Dahm *et al.*, 1987). In larger streams and rivers containing gravel bars, hyporheic zone DOC concentration at the upstream end of the gravel bars was equal to or slightly higher than the water column concentration but declined with increasing distance downstream to values close to 1 mgC L^{-1} in the Stillaguamish River (Vervier and Naiman, 1992) and Wappinger Creek (Findlay *et al.*, 1993). In the polluted Garrone River, gravel-bar hyporheic zone DOC concentration never was much below 2 mgC L^{-1} (Vervier *et al.*, 1993).

III. PROCESSES WITHIN THE HYPORHEIC ZONE

A. Sources and Sinks

If a surface water is enriched with DOC relative to the hyporheic zone water, downwelling surface waters can fuel hyporheic zone metabolism (Jones *et al.*, 1995), and the hyporheic zone will function as a DOC sink. Similarly, when surface water downwells into the hyporheic zone and flows downstream, supporting metabolism before resurfacing, or moves from the stream into the riparian zone, the hyporheic zone functions as a DOC sink. Alternatively, if ground waters enriched in DOC relative to surface waters upwell into the hyporheic zone, the upwelling ground water is a DOC source, stimulating hyporheic zone metabolism (Fiebig, 1995), and unless all of the DOC is immobilized or metabolized, the hyporheic zone becomes a DOC source for the surface water. Likewise, if particulate organic matter trapped in the hyporheic zone releases DOC, and that DOC enters the surface water, the hyporheic zone functions as a DOC source. It is certainly conceivable that the hyporheic zone in a given stream functions simultaneously as both as source and a sink for DOC. To better appreciate the role of the hyporheic zone in DOC biogeochemistry, an understanding of the processes that influence DOC dynamics in surface and subsurface waters is important.

B. Hydrologic Exchanges

Dynamic hydrological, physical-chemical, and biological interactions of the water within the riparian, water column, hyporheic, and bedrock groundwater zones influence the concentration of DOC in stream water. The two dominant exchange processes involving the hyporheic zone are the upwelling of ground water from shallow and deep sources through the hyporheic zone and streambed into the surface water, and the downwelling of surface water through the streambed and into the hyporheic zone (Fig. 1). Additionally, transport involving the horizontal movement of hyporheic zone water downstream either beneath the main channel or outside the wetted perimeter is an important hydrological process, and in arid regions, water from the hyporheic zone can be lost from the stream channel through flow into the ri-

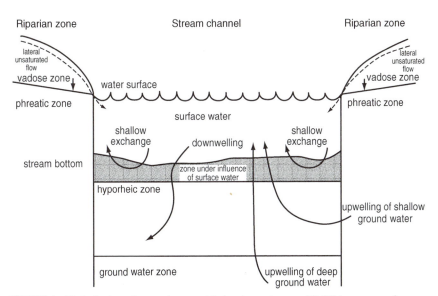

FIGURE 1 Hydrologic pathways that provide for the exchange of DOC between surface waters, the hyporheic zone, ground waters, and the riparian zone in streams.

parian zone. Most streams have a zone at the top of the streambed where the sediments are almost completely under the influence of surface water. This region is composed of >90% surface water, acts as a transition from the streambed surface to the hyporheic zone, and is characterized by shallow, rapid exchanges at the water–sediment interface, high rates of metabolic activity, and high bacterial biomass (Fig. 1). Below this zone, the physical dimensions and hydrologic character of hyporheic zones vary widely, as some streams have large volumes that participate in rapid mixing (Castro and Hornberger, 1991), whereas others are much more constrained (Wroblicky *et al.*, 1998). These characteristics partly control the importance of hyporheic zone processes to stream ecosystem metabolism by placing boundaries on the flux of materials and energy exchanging between the water column and the hyporheic zone.

C. Interactions of DOC with Surfaces

The dominant physical-chemical interactions controlling DOC concentration within the hyporheic zone are adsorption, desorption, and diffusion, whereas the dominant biological interactions are enzymatic hydrolysis and metabolism (Fig. 2). Free-living bacterial cells may be important for heterotrophic metabolism in the hyporheic zone, but any formation of biofilm communities will involve cell attachment, colonization, and microcolony development. As a water source enriched with DOC enters the hyporheic zone, either through upwelling or downwelling, organic molecules can be removed from solution by abiotic adsorption or biotic uptake. These are selective processes with adsorption favoring the most hydrophobic, unionized molecules and biological uptake favoring the most biologically labile constituents. Experiments with streambed sediments and leaf leachate have shown that microbial uptake of DOC from solution is kinetically slower, but quantitatively more important than abiotic adsorption (Dahm, 1981). Similar phenomena should occur within the hyporheic zone, although that view is not universally held (Fiebig, 1995). We will return to an examination of DOC-surface interactions when we evaluate conceptual models of DOC transformations within the hyporheic zone.

D. Profiles through the Hyporheic Zone

Despite the heterogeneity and complexity of the hyporheic zone in streams, selective reactions of DOC can lead to the formation of measurable biological and chemical gradients. Some gradients have been measured and reported; others can be deduced from descriptions of hyporheic zone processes. In this section, we identify four different profiles of DOC, oxygen, microbial biomass, and microbial activity that are likely to develop under different upwelling or downwelling scenarios, and match a particular pro-

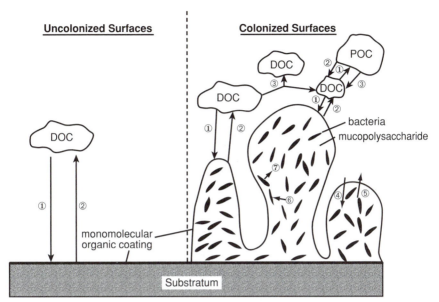

FIGURE 2 Dominant physical, chemical, and biological interactions of DOC and surfaces including (1) adsorption, (2) desorption, (3) enzymatic hydrolysis, (4) diffusion of adsorbed DOC into biofilms, (5) diffusion of excreted DOC out of biofilms, (6) biological uptake, and (7) excretion.

file to literature descriptions. For profiles that have not been described, we review what influences DOC processing, the available data on sediment organic matter, and the hydrology of upwelling ground waters to suggest whether a profile is plausible. In the process, we argue that DOC processing within the hyporheic zone is primarily microbial, and that the ability of hyporheic zone sediments to abiotically immobilize DOC from upwelling ground waters is regulated by biological activity.

The selective removal of DOC from water perfusing through a column of hyporheic zone sediments by biotic and abiotic mechanisms should generate gradients of declining DOC concentration, biological lability, and hydrophobic character with distance from the DOC source; these gradients should be steepest closest to the source. Microbial growth yield and, to some extent, growth rate correlate with the biological lability of DOC, so gradients of declining concentration will be accompanied by gradients of diminishing metabolic activity and heterotrophic bacterial biomass. Observations that support these predictions include steep gradients corresponding to amino acid uptake in perfusion experiments with hyporheic zone sediments (Fiebig, 1997), a reduction in hyporheic zone sediment respiration over time during a batch incubation of surface waters enriched with algal exudates (Jones, 1995), and gradients of DOC quantity and quality in biofilm plug-flow reactors colonized and nourished by stream water (Kaplan and New-bold, 1995; L. A. Kaplan and T J. Gremm, unpublished data).

Additionally, as the quality of DOC changes, there should be a gradient of bacterial species, with those most competitive for the labile constituents dominating the community closest to the source, and perhaps slower growing species that are able to metabolize refractory compounds dominating the community at some greater distance from the source. The strength of these gradients will depend upon the nature of the DOC, the consistency of the supply, and the persistence of a unidirectional hydrologic flow regime.

In streams where the surface water is enriched in DOC relative to the ground water, areas of downwelling will carry carbon and energy into the hyporheic zone. Under these circumstances, oxygen, DOC, and microbial biomass and activity should be highest at the top of the hyporheic zone and diminish with depth (Fig. 3A). Data suggesting this pattern have been reported for the hyporheic zone of a sandy-bed, north-temperate stream in northern Michigan, including bacterial biomass and production, especially during the spring season (Hendricks, 1996).

A special case of downwelling surface waters enriched with DOC is the gravel bar, mentioned earlier. In the hyporheic zone of a gravel bar, the primary flow vector is the horizontal component, generally parallel to the flow within the main channel. Horizontal water velocities within the hyporheic zone can be as low as 1 cm d^{-1}, as reported for Wappinger Creek in New York (Findlay *et al.*, 1993), 43 m d^{-1} for the Garrone River in France (Vervier *et al.*, 1993), and three orders of magnitude less than water column velocities in the Steina, a Black Forest mountain stream (Pusch and Schwoerbel, 1994). These low velocities extend the residence time of water and DOC within the hyporheic zone, providing opportunity for microbial uptake, and result in gradients of declining DOC and oxygen concentrations, and in some cases bacterial biomass, with distance from the downwelling (Vervier and Naiman, 1992; Findlay *et al.*, 1993; Vervier *et al.*, 1993).

In hyporheic zones under the influence of upwelling ground water that is low in DOC, the chemical and biological gradients should be very similar to those for downwelling DOC-rich surface waters (Fig. 3B). However when particulate organic carbon (POC) is carried into the surface layers of the hyporheic zone by downwelling baseflow water or is entrained during scour and deposition associated with storms, the POC can generate DOC through leaching and enzymatic hydrolysis. In fact, the production of DOC from buried POC is believed to be a major factor in supporting hyporheic zone metabolism in some streams (Pusch and Schwoerbel, 1994; Pusch, 1996). Under those circumstances, oxygen concentration will show a minimum where POC is buried, and DOC plus microbial biomass will exhibit a slight increase that alters an otherwise declining trajectory with increasing depth of the hyporheic zone. Data that suggest this pattern have been collected for oxygen, DOC, and bacterial biomass in a third-order piedmont stream with

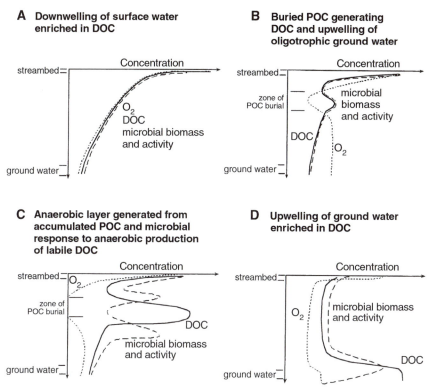

FIGURE 3 Profiles of DOC, oxygen, microbial biomass, and microbial activity generated within the hyporheic zones from (A) downwelling of surface waters enriched with DOC, (B) upwelling of low-DOC ground water and hydrolysis of buried POC to DOC, (C) upwelling of low-DOC ground water and isolated zones of anaerobiosis associated with accumulated POC, and (D) upwelling of ground water enriched with DOC.

a shallow hyporheic zone (T. J. Battin, L. A. Kaplan, and J. D. Newbold, unpublished data).

In circumstances where deposits of organic matter are particularly dense, such as burial of leaf packs or debris dams, organic matter could reduce local permeability and isolate the material from upwelling or downwelling waters. The associated microbial activity may be sufficient to create localized anoxic zones (Fig. 3C). Anoxic microsites, protected from upwelling or downwelling oxygenated water, could generate organic acids associated with fermentation. These organic acids would accumulate because of the lack of suitable electron acceptors but would diffuse into regions where aerobic metabolism would readily convert these molecules to carbon dioxide. Profiles of DOC, molecular weight, and methane reported for the hyporheic zone of Allequash Creek in northern Wisconsin are consistent with the development of an anaerobic zone associated with buried woody debris (Schindler and Krabbenhoft, 1998). The DOC concentration in the hyporheic zone of Allequash Creek was 7- to 50-fold higher than in the upwelling ground water, peaked in the upper sediments (10 to 15 cm below the sediment–water interface), and exhibited a minimum in molecular weight that coincided with the maximum in concentration.

Several studies have suggested that ground water enriched with DOC upwelling through the hyporheic zone provides an important source of carbon and energy for stream ecosystems (Hynes, 1983; Fiebig and Lock, 1991; Fiebig, 1995). Because this idea, commonly referred to as the "stripping hypothesis," seems to be gaining a foothold, it is worthwhile to examine it critically. The basic concept is that ground water rich in DOC is stripped of its organic load through adsorption during passage through the hyporheic zone, and then microorganisms metabolize the adsorbed molecules. As such, this is somewhat akin to the podzolization process responsible for the decline in DOC concentration with depth in some terrestrial systems (McDowell and Wood, 1984). Furthermore, the stripping idea has been extended to suggest that the accumulated organic matter in a biofilm polysaccharide matrix functions as a storehouse and maintains the activity of microorganisms when exogenous sources of DOC are low (Freeman and Lock, 1995).

To properly evaluate the stripping hypothesis, we must review what is known about DOC adsorption to inorganic and organic surfaces, precipitation of DOC, and the magnitude of the flux of water and carbon entering streams as ground water. Adsorption of DOC to inorganic surfaces has been recognized as an important phenomenon in streams (Madsen, 1972; Rounick and Winterborn, 1983), and the mechanisms of adsorption have been extensively studied, especially with humic substances, the dominant form of DOC in surface waters, and model metal oxides such as the iron oxyhydroxide, goethite. Adsorption of DOC to goethite is thought to involve ligand-exchange mechanisms wherein phenolic and carboxylic functional groups in humic substances interact with oxides on mineral surfaces (Tipping, 1990). The phenomenon shows characteristics of anionic sorbates with sorption enhanced at pH values below 4.0 and a dependence on the position of reactive functional groups (Evanko and Dzombak, 1998). Furthermore, DOC adsorption has been shown to be subject to saturation, behave in a manner consistent with the Langmuir adsorption model, and be consistent with a monolayer of DOC on mineral surfaces (Day *et al.*, 1994). Adsorption to organic surfaces is not as well understood (Tipping, 1990), but studies with humic substances and living cells also demonstrate a pH and concentration dependence seen with inorganic surfaces (Campbell *et al.*, 1997), and similar interactions with detrital surfaces are not only plausible but likely. Thus, through purely abiotic processes, the surfaces within the hyporheic zone of streams should equilibrate with the DOC concentration of perfusing waters, transiently altering the total amount of adsorbed DOC through adsorption or desorption as the bulk water DOC concentration changes.

Some of the molecules in solution can potentially exchange with the adsorbed molecules, but that would influence the quality of the adsorbed carbon, not the quantity.

Extensive precipitation or self-association of groundwater DOC passing through a hyporheic zone is unlikely. Even though ground water is often more acidic than surface waters from supersaturation with carbon dioxide, pH values generally do not fall below 5.5 (Harvey and Fuller, 1998), so most of the DOC will be ionized and soluble (Tipping, 1990). Complexation of DOC with metal cations can lead to precipitation at neutral pH values (Tipping, 1990), but data reported for the organic content of sediments from the continental shelf (Keil *et al.*, 1994), including sediments associated with river deltas (Mayer, 1994), and ranging from <1% to approximately 10% organic carbon content on a mass basis, also are consistent with an organic monolayer coating all mineral surfaces (Keil *et al.*, 1994; Mayer, 1994). No data suggest that DOC adsorbed to surfaces generates sites for additional DOC sorption or self-association resulting in a thick layer of deposited organic matter around sediments or detritus. Therefore, the continual immobilization of DOC from groundwater upwelling within the hyporheic zone of streams requires microbial metabolism of the adsorbed materials to regenerate adsorption sites on inorganic or organic surfaces. Alternatively, the biological uptake of DOC directly from the upwelling ground water and incorporation into biomass or a mucopolysaccharide matrix can "strip" DOC from ground water. Either way, ultimately the ability to "strip" DOC from ground water rests with biological activity, catabolism of DOC at rates equivalent to the rates of supply.

The plausibility of the stripping hypothesis has been argued by analogy to DOC removal from soil columns. However, the volumes of water perfusing through the hyporheic zone are so large that comparisons to soil column processes are inappropriate. Flow rates of upwelling ground water through hyporheic zones range widely, both within and between streams. For example, in the Breitenbach, a first-order stream in the uplands of central Germany, groundwater discharge rates range from 0.4 to 31.4 ml m^{-2} s^{-1} (Fiebig, 1995). Expressed as upwelling Darcian velocities, these discharge rates range from 13 to 990 m y^{-1}, which is vastly greater than percolation rates of precipitation through terrestrial soils. Indeed, because streambed areas typically constitute <1% of watershed area, one would expect average upwelling velocities to exceed downward soil percolation by a factor of 100 or more. Under these circumstances, especially with high DOC concentration in the upwelling ground water, we suggest that the capacity for immobilization and metabolism within the hyporheic zone is so overwhelmed by the flux of ground water and much of the DOC in upwelling ground waters passes through to the water column. In fact, empirical measurements from perfusion experiments with hyporheic zone sediments have demonstrated that only 10–25% of the DOC in ground waters enriched with DOC is removed by hyporheic zone sediment cores (Fiebig *et al.*, 1990; Fiebig, 1995), and these values overlap the estimates of biodegradable DOC in stream water (Volk *et al.*, 1997). Lastly, under circumstances where groundwater DOC is purportedly stripped from solution (immobilized) and then nourishes metabolic activity within the hyporheic zone, there should be a pattern of high DOC concentration, metabolic activity, and biomass, and an oxygen minimum associated with the bottom of the hyporheic zone (Fig. 3D). Such a pattern has not been documented for streams.

E. Impact on Energy Flow and Carbon Cycling

From the perspective of the whole stream, the hyporheic zone may be viewed as entrapping, transforming, and mineralizing organic carbon that would otherwise be transported farther downstream. The effect would be to shift the locus of utilization of organic carbon upstream and closer to the point of origin (i.e., to shorten the spiraling length of organic carbon), thereby increasing the processing efficiency of the ecosystem (sensu Fisher and Likens, 1973) and decreasing the downstream transport of organic carbon. The decreased transport of organic matter could have major impacts on downstream ecosystems such as lakes, reservoirs, and estuaries and must also be felt within the stream ecosystem itself. Such reductions in transport also may diminish the metabolism in surficial sediments. In this section, we consider the magnitude and prevalence of these effects with a focus on DOC dynamics.

We can gain some perspective on the potential influence of hyporheic zone processes by examining organic carbon budgets of stream ecosystems. Webster and Meyer (1997) calculated processing efficiencies from 17 streams for which whole-stream ecosystem budgets have been reported. Efficiency was calculated as the ratio of ecosystem respiration to total organic matter inputs calculated from the point of export upstream to include all headwaters. The efficiencies of all but three streams fell in the range of 10–80%, with a median efficiency of 31%. If we assume that 80% represents a general upper limit for stream processing efficiencies and that streams with 10% processing efficiency lack significant hyporheic metabolism, then the maximal impact that hyporheic-zone metabolism might have on the ecosystem is to respire an additional 70% of the organic inputs. At this theoretical limit, the hyporheic zone would account for 88% of the total ecosystem respiration, a proportion consistent with reported values (Jones *et al.*, 1995; Fuss and Smock, 1996; Naegeli and Uehlinger, 1997). However, there are two significant implications associated with this theoretical limit that make achieving an 80% ecosystem efficiency unlikely: an increase in ecosystem efficiency from 10 to 80% translates to a 78% decline in downstream transport and a 97% reduction in spiraling length. Thus, the question remains: can we observe cases in which hyporheic zone respiration results in a significant ecosystem effect, one that would be evidenced by a large reduction in organic carbon transport? Answering this question would require comparative ecosystem budgets that partition respiration between surficial and hyporheic zones. Few budgets report such partitioning, and many budgets do not include all components of inputs and outputs (Webster and Meyer, 1997) making firm conclusions difficult.

In streams where a high proportion of ecosystem respiration has been ascribed to the hyporheic zone, it appears that the primary carbon sources are buried POC or downwelling algal exudates. Stream sediments are effective in retaining POC, especially in the aftermath of storms. Long retention times make possible the degradation at very slow rates of even highly recalcitrant POC (Hedin, 1990; Fuss and Smock, 1996). In contrast to its ability to store POC, the hyporheic zone has a limited capacity to store DOC. And because stream-water DOC is dominated by refractory forms, hyporheic zones are unlikely to be as effective in metabolizing the transient DOC as compared to the stored POC. As POC degrades, it produces DOC through enzymatic processes and dissolution. To the extent that labile DOC is released from buried POC or enters the hyporheic zone from either downwelling surface water or upwelling ground water, it is likely to be selectively metabolized, contributing to ecosystem efficiency. But refractory DOC that accumulates in the hyporheic zone when bacterial enzymes cleave labile moieties from POC or DOC can upwell from the hyporheic zone and be transported downstream with a corresponding loss of ecosystem efficiency. Indeed, a further test of whether hyporheic zone metabolism results in a significant ecosystem-level impact would involve cross-system comparisons to determine whether the ratio of DOC:POC in transport correlates with rates of hyporheic zone metabolism.

A hyporheic zone of sufficient volume and exchange would likely have a significant effect on the utilization and concentration of bioavailable DOC, which in turn would have substantial implications to downstream ecosystems. In the White Clay Creek, bioavailable forms constitute typically 30% of the stream-water DOC (Volk *et al.*, 1997), and approximately one-third of that pool is highly labile molecules from various terrestrial sources plus algal exudates (Gremm and Kaplan, 1998). From chamber studies involving ^{14}C-labeled algal exudates, we have estimated that the highly labile DOC has a turnover length of about 2 km, with about 60% being reutilized within the third-order reach. The remaining two-thirds of the biodegradable DOC (or about 20% of the total DOC, and averaging about 0.4 mgC L^{-1}) are of intermediate lability. This fraction can be degraded in laboratory perfusion columns within about 2 hours of residence time, but this rate is only a few percent of the degradation rate for the most labile forms in the same columns. Thus, in the White Clay Creek, which has a very limited hyporheic volume (<0.1 of channel cross section), much of this DOC may be exported downstream without being used. Where hyporheic volume and exchange in White Clay Creek are large enough, however, to function in a manner similar to the perfusion columns and to metabolize all the biodegradable DOC that is not exported, areal organic matter respiration would increase by about 60 gC m^{-2} y^{-1}. This represents a 24% increase over reported ecosystem respiration (Newbold *et al.*, 1997) with a corresponding increase in efficiency. This increase in respiration of 24%, however, represents 100% of the bioavailable DOC that would otherwise be available to communities in downstream ecosystems. Last, labile DOC that enters the hyporheic zone may additionally augment ecosystem respiration by exerting a priming effect on the metabolism of refractory carbon that might not otherwise be oxidized, but little is known about the magnitude of this effect.

We conclude by addressing the question of the physical conditions of a stream under which a significant hyporheic influence on ecosystem efficiency might be expected. Findlay (1995) offered a conceptual model in which the influence of the hyporheic zone is an increasing function of two variables—the rate of hydrologic exchange between sediments and water column and the rate of biogeochemical processes within the hyporheic zone. The largest influences occur when both exchange and activity are high. This is an appropriate framework for understanding the contribution of DOC processes within the hyporheic zone. It accommodates the kinetics of adsorption and the role of DOC quality as it pertains to biodegradability (on the biogeochemical process axis) and the quantitative aspect of hydrologic fluxes that influence the supply of DOC (on the hydrologic exchange axis).

The model might usefully be carried a step farther by asking: what physical factors influence the rate of biogeochemical processing? With respect to physical factors, we suggest that DOC metabolism will depend on temperature, the supply rate of nutrients and oxygen, and the relative size of the hyporheic zone. The supply rate should be strongly influenced by hydrologic exchange rate acting in combination with biogeochemical inputs to the stream ecosystem, whereas the relative size of the hyporheic zone, in relation to the hydrologic exchange rate, will establish the hyporheic residence time of water (Morrice *et al.*, 1997) within the hyporheic zone. This leaves four relatively independent factors: temperature, inputs, exchange, and size, with the latter two being related to physical structure of the stream channel.

Figure 4 adapts Findlay's model to show the effect of size and exchange rate on stream ecosystem efficiency. We express exchange as the hydrologic exchange velocity, or the effective vertical velocity at which water mass is transferred from the water column into the sediments. We would expect the total utilization of DOC and therefore the ecosystem efficiency to increase as the water exchange rate increases up to some threshold at which either all of the bioavailable DOC is consumed within the hyporheic zone, or the capacity of the hyporheic zone to degrade DOC is reached. The capacity to degrade DOC, in turn, should increase with the size of the hyporheic zone,

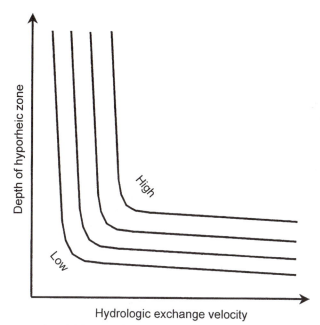

FIGURE 4   Proposed model for the influence of hyporheic zone DOC processing showing iso-clines of stream ecosystem metabolism (or stream ecosystem efficiency) as affected by depth and exchange velocity.

whereas for a given hydrologic exchange, the total DOC utilization will increase with the size of the hyporheic zone until all available DOC is used. Thus, either size or hydrologic exchange could act as the primary limiting factor, but we speculate that limitation by hydrologic exchange is the more common. Perhaps the major implication of the model as represented in Fig. 4, and a significant way in which it differs from Findlay's conceptualization, is that the role of both size and hydrologic exchange is limited. Beyond certain levels of either variable, DOC processing reaches a plateau at which virtually all of the bioavailable DOC is utilized. Whether such extensive utilization of the DOC is actually achieved in nature, however, remains a question to be investigated.

REFERENCES

Aiken, G., and E. Cotsaris. 1995. Soil and hydrology: Their effect on NOM. *Journal of the American Water Works Association* **87**:36–45.

Baker, C. D., P. D. Bartlett, I. S. Farr, and G. I. Williams. 1974. Improved methods for the measurement of dissolved and particulate organic carbon in fresh water and their application to chalk streams. *Freshwater Biology* **4**:467–481.

Boyer, E. W., G. M. Hornberger, K. E. Bencala, and D. M. McKnight. 1995. Variation of dissolved organic carbon during snowmelt in soil and stream waters of two headwater catchments, Summit County, Colorado. *IAHS Publication* **228**, 303–312.

Campbell, P. G. C., M. R. Twiss, and K. J. Wilkinson. 1997. Accumulation of natural organic matter on the surfaces of living cells: Implications for the interaction of toxic solutes with aquatic biota. *Canadian Journal of Fisheries and Aquatic Sciences* **54**:2543–2554.

Casey, H., and I. S. Farr. 1982. The influence of within-stream disturbance on dissolved nutrient levels during spates. *Hydrobiologia* **92**:44–462.

Castro, M. N., and G. M. Hornberger. 1991. Surface-subsurface water interactions in an alluviated mountain stream channel. *Water Resources Research* **27**:1613–1621.

Crocker, M. T., and J. L. Meyer. 1987. Interstitial dissolved organic carbon in sediments of a southern Appalachian headwater stream. *Journal of the North American Benthological Society* **63**:159–167.

Cronan, C. S. 1990. Patterns of organic acid transport from forested watersheds to aquatic ecosystems. *In* "Organic Acids in Aquatic Ecosystems" (E. M. Perdue and E. T. Gjessing, eds.), Dahlem Konferenzen, pp. 245–260. Wiley, New York.

Cronan, C. S., and G. R. Aiken. 1985. Chemistry and transport of soluble humic substances in forested watersheds of the Adirondack Park, New York. *Geochimica et Cosmochimica Acta* **49**:1697–1705.

Dahm, C. N. 1981. Pathways and mechanisms for removal of dissolved organic carbon from leaf leachate in streams. *Canadian Journal of Fisheries and Aquatic Sciences* **38**:68–76.

Dahm, C. N., E. H. Trotter, and J. R. Sedell. 1987. Role of anaerobic zones and processes in stream ecosystem productivity. *In* "Chemical Quality of Water and the Hydrologic Cycle" (R. C. Averett and D. M. McKnight, eds.), pp. 157–178. Lewis Publishers, Chelsea, MI.

Dahm, C. N., D. I. Carr, and R. L. Coleman. 1991. Anaerobic carbon cycling in stream ecosystems. *Verhandlungen Internationale Vereinigung für Theoretische und Angewandte Limnologie* **24**:1600–1604.

David, M. B., G. F. Vance, and J. S. Kahl. 1992. Chemistry of dissolved organic carbon and organic acids in two streams draining forested watersheds. *Water Resources Research* **28**:389–396.

Day, G. McD., B. T. Hart, I. D. McKelvie, and R. Beckett. 1994. Adsorption of natural organic matter onto goethite. *Colloids and Surfaces A: Physicochemical and Engineering Aspects* **89**:1–13.

Dillon, P. J., and L. A. Molot. 1997. Effect of landscape form on export of dissolved organic carbon, iron, and phosphorus from forested stream catchments. *Water Resources Research* **33**:2591–2600.

Dosskey, M. G., and P. M. Bertsch. 1994. Forest sources and pathways of organic matter transport to a blackwater stream: A hydrologic approach. *Biogeochemistry* **24**:1–19.

Duff, J. H., and F. J. Triska. 1990. Denitrification in sediments from the hyporheic zone adjacent to a small forested stream. *Canadian Journal of Fisheries and Aquatic Sciences* **47**:1140–1147.

Easthouse, K. B., J. Mulder, N. Christophersen, and H. M. Seip. 1992. Dissolved organic carbon fractions in soil and stream water during variable hydrological conditions at Birkenes, Southern Norway. *Water Resources Research* **28**:1585–1596.

Evanko, C. R., and D. A. Dzombak. 1998. Influence of structural features on sorption of NOM-analogue organic acids to goethite. *Environmental Science and Technology* **32**:2846–2855.

Fiebig, D. M. 1995. Groundwater discharge and its contribution of dissolved organic carbon to an upland stream. *Archiv für Hydrobiologie* **134**:129–155.

Fiebig, D. M. 1997. Microbiological turnover of amino acids immobilized from groundwater discharged through hyporheic sediments. *Limnology and Oceanography* **42**:763–768.

Fiebig, D. M., and M. A. Lock. 1991. Immobilization of dissolved organic matter from groundwater discharging through the stream bed. *Freshwater Biology* **26**:45–55.

Fiebig, D. M., M. A. Lock, and C. Neal. 1990. Soil water in the riparian zone as a source of carbon for a headwater stream. *Journal of Hydrology* **116**:217–237.

Findlay, S. 1995. Importance of surface-subsurface exchange in stream ecosystems: The hyporheic zone. *Limnology and Oceanography* **40**:159–164.

Findlay, S., D. Strayer, C. Goumbala, and K. Gould. 1993. Metabolism of streamwater dissolved organic carbon in the shallow hyporheic zone. *Limnology and Oceanography* **38**:1493–1499.

Fisher, S. G., and G. E. Likens. 1973. Energy flow in Bear Brook, New Hampshire: An integrative approach to stream ecosystem metabolism. *Ecological Monographs* **43**:421–439.

Foster, I. D., and I. C. Grieve. 1982. Short term fluctuations in dissolved organic matter concentrations in streamflow draining a forested watershed and their relation to the catchment budget. *Earth Surface Processes and Landforms* **7**:417–425.

Freeman, C., and M. A. Lock. 1995. The biofilm polysaccharide matrix: A buffer against changing organic substrate supply? *Limnology and Oceanography* **40**:273–278.

Fuss, C. L., and L. A. Smock. 1996. Spatial and temporal variation of microbial respiration rates in a blackwater stream. *Freshwater Biology* **36**:339–349.

Gremm, Th. J., and L. A. Kaplan. 1998. Dissolved carbohydrate concentration, composition, and bioavailability to microbial heterotrophs in stream water. *Acta Hydrochimica Hydrobiologica* **26**:167–171.

Grieve, I. C. 1984. Concentrations and annual loading of dissolved organic matter in a small moorland stream. *Freshwater Biology* **14**:533–537.

Grimm, N. B., and S. G. Fisher. 1984. Exchange between interstitial and surface water: Implications for stream metabolism and nutrient cycling. *Hydrobiologia* **111**:219–228.

Harvey, J. W., and C. C. Fuller. 1998. Effect of enhanced manganese oxidation in the hyporheic zone on basin-scale geochemical mass balance. *Water Resources Research* **34**:623–636.

Hedin, L. O. 1990. Factors controlling sediment community respiration in woodland stream ecosystems. *Oikos* **57**:94–105.

Hemond, H. F. 1990. Wetlands as the source of dissolved organic carbon to surface waters. *In* "Organic Acids in Aquatic Ecosystems" (E. M. Perdue and E. T. Gjessing, eds.), Dahlem Konferenzen pp. 301–313. Wiley, New York.

Hendricks, S. P. 1996. Bacterial biomass, activity, and production within the hyporheic zone of a north-temperate stream. *Archiv für Hydrobiologie* **136**:467–487.

Hendricks, S. P., and D. S. White. 1991. Physicochemical patterns within a hyporheic zone of a northern Michigan River, with comments on surface water patterns. *Canadian Journal of Fisheries and Aquatic Sciences* **48**:1645–1654.

Hinton, M. J., S. L. Schiff, and M. C. English. 1997. The significance of storms for the concentration and export of dissolved organic carbon from two Precambrian Shield catchments. *Biogeochemistry* **36**:67–88.

Hinton, M. J., S. L. Schiff, and M. C. English. 1998. Sources and flowpaths of dissolved organic carbon during storms in two forested watersheds of the Precambrian Shield. *Biogeochemistry* **41**:175–197.

Holmes, R. M., S. G. Fisher, N. B. Grimm, and B. J. Harper. 1998. The impact of flash floods on microbial distribution and biogeochemistry in the parafluvial zone of a desert stream. *Freshwater Biology* **40**:641–654.

Hornberger, G. M., K. E. Bencala, and D. M. McKnight. 1994. Hydrological controls on dissolved carbon during snowmelt in the Snake River near Montezuma, Colorado. *Biogeochemistry* **15**:147–165.

Hynes, H. B. N. 1983. Groundwater and stream ecology. *Hydrobiologia* **100**:93–99.

Ivarsson, H., and M. Jansson. 1994. Regional variation of dissolved organic matter in running waters in central northern Sweden. *Hydrobiologia* **286**:37–51.

Jones, J. B. 1995. Factors controlling hyporheic respiration in a desert stream. *Freshwater Biology* **34**:91–99.

Jones, J. B., S. G. Fisher, and N. B. Grimm. 1995. Vertical hydrologic exchange and ecosystem metabolism in a Sonoran Desert stream. *Ecology* **76**:942–952.

Jones, J. B., S. G. Fisher, and N. B. Grimm. 1996. A long-term perspective of dissolved organic carbon transport in Sycamore Creek, Arizona, USA. *Hydrobiologia* **317**:183–188.

Kaplan, L. A., and J. D. Newbold. 1993. Biogeochemistry of dissolved organic carbon entering streams. *In* "Aquatic Microbiology: An Ecological Approach" (T. E. Ford, ed.), pp. 139–165. Blackwell, Boston.

Kaplan, L. A., and J. D. Newbold. 1995. Measurement of streamwater biodegradable dissolved organic carbon with a plug-flow bioreactor. *Water Research* **29**:2696–2706.

Kaplan, L. A., R. A. Larson, and T. L. Bott. 1980. Patterns of dissolved organic carbon in transport. *Limnology and Oceanography* **15**:1034–1043.

Keil, R. G., D. B. Montlucon, F. G. Prahl, and J. I. Hedges. 1994. Sorptive preservation of labile organic matter in marine sediments. *Nature (London)* **370**:549–552.

Kirk, T. K., H. H. Yang, and P. Keyser. 1978. The chemistry and physiology of the fungal degradation of lignin. *Developments in Industrial Microbiology* **19**:51–61.

Leenheer, J. A., R. L. Malcolm, P. W. McKinley, and L. A. Eccles. 1974. Occurrence of dissolved organic carbon in selected ground-water samples in the United States. *Journal of Research of the U.S. Geological Survey* **2**:361–369.

Lewis, W. M., and J. F. Saunders. 1989. Concentration and transport of dissolved and suspended substances in the Orinoco River. *Biogeochemistry* **7**:203–240.

Lord, D. G., J. L. Barringer, P. A. Johnson, P. F. Schuster, R. L. Walker, J. E. Fairchild, B. N. Sroka, and E. Jacobsen. 1990. Hydrogeochemical data from an acidic deposition study at McDonalds Branch basin in the New Jersey Pinelands. *Geological Survey Open-File Report (U.S.) 88–500.*

Lush, D. L., and H. B. N. Hynes. 1978. Particulate and dissolved organic matter in a small partly forested Ontario stream. *Hydrobiologia* **60**:177–185.

Madsen, B. L. 1972. Detritus on stones in small streams. *Memorie Istituto Italiano di Idrobiologia Supplement* **29**:385–403.

Mayer, L. M. 1994. Surface area control of organic carbon accumulation in continental shelf sediments. *Geochimica et Cosmochimica Acta* **58**:1271–1284.

McDowell, W. H., and S. G. Fisher. 1976. Autumnal processing of dissolved organic matter in a small woodland stream ecosystem. *Ecology* **57**:561–569.

McDowell, W. H., and G. E. Likens. 1988. Origin, composition, and flux of dissolved organic carbon in the Hubbard Brook Valley. *Ecological Monographs* **58**:177–195.

McDowell, W. H., and T. Wood. 1984. Podzolization: Soil processes control dissolved organic carbon concentrations in stream water. *Soil Science* **137**:23–32.

Metzler, G. M., and L. A. Smock. 1990. Storage and dynamics of subsurface detritus in a sand-bottomed stream. *Canadian Journal of Fisheries and Aquatic Sciences* **47**:588–594.

Meybeck, M. 1982. Carbon, nitrogen, and phosphorus transport by world rivers. *American Journal of Science* **282**:401–450.

Meyer, J. L. 1986. Dissolved organic carbon dynamics in two subtropical blackwater rivers. *Archiv für Hydrobiologie* **108**:119–134.

Meyer, J. L., and C. M. Tate. 1983. The effects of watershed disturbance on dissolved organic carbon dynamics of a stream. *Ecology* **64**:33–44.

Michaelson, G. J., C. L. Ping, G. W. Kling, and J. E. Hobbie. 1998. The character and bioactivity of dissolved organic matter at thaw and in the spring runoff waters of the arctic tundra north slope, Alaska. *Journal of Geophysical Research* **103**:28,939–28,946.

Moeller, J. R., G. W. Minshall, K. W. Cummins, R. C. Petersen, C. E. Cushing, J. R. Sedell, R. A. Larson, and R. L. Vannote. 1979. Transport of dissolved organic carbon in streams of differing physiographic characteristics. *Organic Geochemistry* **1**:139–150.

Moore, T. R., and R. J. Jackson. 1989. Dynamics of dissolved organic carbon in forested and disturbed catchments, Westland, New Zealand 2. Larry River. *Water Resources Research* **25**:1331–1339.

Morrice, J. A., H. M. Valett, C. N. Dahm, and M. E. Campana. 1997. Alluvial characteristics, groundwater-surface water exchange and hydrologic retention in headwater streams. *Hydrologic Processes* **11**:253–267.

Mulholland, P. J. 1981. Organic carbon flow in a swamp-stream ecosystem. *Ecological Monographs* **51**:307–322.

Mulholland, P. J. 1993. Hydrometric and stream chemistry evidence of three storm flowpaths in Walker Branch Watershed. *Journal of Hydrology* **151**:291–316.

Mulholland, P. J., and W. R. Hill. 1997. Seasonal patterns in streamwater nutrient and dissolved organic carbon concentrations: Separating catchment flow path and in-stream effects. *Water Resources Research* **33**:1297–1306.

Mulholland, P. J., E. R. Marzolf, J. R. Webster, D. R. Hart, and S. P. Hendricks. 1997. Evidence that hyporheic zones increase heterotrophic metabolism and phosphorus uptake in forest streams. *Limnology and Oceanography* **42**:443–451.

Naegeli, M. W., and U. Uehlinger. 1997. Contribution of the hyporheic zone to ecosystem metabolism in a prealpine gravel-bed river. *Journal of the North American Benthological Society* **16**:794–804.

Nelson, P. N., E. Cotsaris, J. M. Oades, and D. B. Bursill. 1990. Influence of soil clay content on dissolved organic matter in stream waters. *Australian Journal of Marine and Freshwater Research* **41**:761–774.

Nelson, P. N., J. A. Baldock, and J. M. Oades. 1993. Concentration and composition of dissolved organic carbon in streams in relation to catchment soil properties. *Biogeochemistry* **19**:27–50.

Newbold, J. D., B. W. Sweeney, J. K. Jackson, and L. A. Kaplan. 1995. Concentrations and export of solutes from six mountain streams in northwestern Costa Rica. *Journal of the North American Benthological Society* **14**:21–37.

Newbold, J. D., T. L. Bott, L. A. Kaplan, B. W. Sweeney, and R. L. Vannote. 1997. Organic matter dynamics in White Clay Creek, Pennsylvania, USA. *Journal of the North American Benthological Society* **16**:46–54.

Pusch, M. 1996. The metabolism of organic matter in the hyporheic zone of a mountain stream, and its spatial distribution. *Hydrobiologia* **323**:107–118.

Pusch, M., and J. Schwoerbel. 1994. Community respiration in hyporheic sediments of a mountain stream (Steina, Black Forest). *Archiv für Hydrobiologie* **130**:35–52.

Reve, R., and I. F. Fergus. 1982. Black and white waters and their possible relationship to the podzolization process. *Journal of Soil Research* **21**:59–66.

Rounick, J. S., and M. J. Winterbourn. 1983. The formation, structure and utilization of stone surface organic layers in two New Zealand streams. *Freshwater Biology* **13**:57–72.

Rutherford, J. E., and H. B. N. Hynes. 1987. Dissolved organic carbon in streams and groundwater. *Hydrobiologia* **154**:33–48.

Schiff, S. L., R. Aravena, S. E. Trumbore, M. J. Hinton, R. Elgood, and P. J. Dillon. 1997. Export of DOC from forested catchments on the Precambrian Shield of Central Ontario: Clues from ^{13}C and ^{14}C. *Biogeochemistry* **36**:43–65.

Schindler, J. E., and D. P. Krabbenhoft. 1998. The hyporheic zone as a source of dissolved organic carbon and carbon gases to a temperate forested stream. *Biogeochemistry* **43**:157–174.

Tipping, E. 1990. Interactions of organic acids with inorganic and organic surfaces. *In* "Organic Acids in Aquatic Ecosystems" (E. M. Perdue and E. T. Gjessing, eds.), Dahlem Konferenzen, pp. 209–221. Wiley, New York.

Trumbore, S. E., S. L. Schiff, R. Aravena, and R. Elgood. 1992. Sources and transformation of dissolved organic carbon in the Harp Lake forested catchment: The role of soils. *Radiocarbon* **34**:626–635.

Vervier, P., and R. J. Naiman. 1992. Spatial and temporal fluctuations of dissolved organic carbon in subsurface flow of the Stillaguamish River (Washington, USA). *Archiv für Hydrobiologie* **123**:401–412.

Vervier, P., M. Dobson, and G. Pinay. 1993. Role of interaction zones between surface and ground waters in DOC transport and processing: considerations for river restoration. *Freshwater Biology* **29**:275–284.

Volk, C. J., C. B. Volk, and L. A. Kaplan. 1997. Chemical composition of biodegradable dissolved organic matter in streamwater. *Limnology and Oceanography* **42**:39–44.

Wallace, J. B., D. H. Ross, and J. L. Meyer. 1982. Seston and dissolved organic carbon dynamics in a southern Appalachian stream. *Ecology* **63**:824–838.

Wallis, P. M., H. B. N. Hynes, and S. A. Telang. 1981. The importance of groundwater in the transportation of allochthonous dissolved organic matter to the streams draining a small mountain basin. *Hydrobiologia* **79**:77–90.

Wassenaar, L. I., R. Aravena, P. Fritz, and J. F. Barker. 1991. Controls on the transport and carbon isotopic composition of dissolved organic carbon in a shallow groundwater system, Central Ontario, Canada. *Chemical Geology* **87**:39–57.

Webster, J. R., and J. L. Meyer. 1997. Organic matter budgets for streams: A synthesis. *Journal of the North American Benthological Society* **16**:141–161.

Wehr, J. D., S. P. Lonergan, and J. H. Thorp. 1997. Concentrations and controls of dissolved organic matter in a constricted-channel region of the Ohio River. *Biogeochemistry* **38**:41–65.

White, D. S. 1983. Perspectives on defining and delineating hyporheic zones. *Journal of the North American Benthological Society* **12**:61–69.

Williams, M. R., T. R. Fisher, and J. M. Melack. 1997. Solute dynamics in soil water and groundwater in a central Amazon catchment undergoing deforestation. *Biogeochemistry* **38**:303–335.

Wroblicky, G. J., M. E. Campana, H. M. Valett, and C. N. Dahm. 1998. Seasonal variation in surface-subsurface water exchange and lateral hyporheic area of two stream-aquifer systems. *Water Resources Research* **34**:317–328.

11

Anoxia, Anaerobic Metabolism, and Biogeochemistry of the Stream-water–Ground-water Interface

Michelle A. Baker[1], Clifford N. Dahm*,
and H. Maurice Valett[†]

*Department of Biology
University of New Mexico
Albuquerque, New Mexico

[†]Virginia Polytechnic Institute and State University
Department of Biology
Blacksburg, Virginia

I. Introduction   260
II. Methods Used in Studies of Anaerobic Metabolism   264
    A. Dissolved Gases   264
    B. Solutes   265
    C. Metabolism   266
III. Controls on the Establishment of Anoxia   266
IV. Biogeochemistry and Evidence for Anaerobic Metabolism   269
    A. Spatial Distributions   269
    B. Temporal Variation   272
    C. Rates of Anaerobic Metabolism   273
V. Influence of Subsurface Anaerobic Metabolism on Stream
    Ecosystem Processes   276
VI. Conclusions and Future Research Directions   278
    References   280

[1]Present address: Department of Biology, Utah State University, Logan, Utah.

I. INTRODUCTION

Oxygen constitutes 21% of the Earth's atmosphere and dissolves into surface waters via diffusion and turbulent mixing. Unless contaminated by organic pollution, stream surface waters are usually well oxygenated owing to equilibration with the atmosphere (Allan, 1995). The major biological processes influencing dissolved oxygen concentration in streams are photosynthesis, where oxygen is produced as a by-product, and aerobic respiration, where oxygen is consumed during organic matter oxidation.

Anoxia, operationally defined here as dissolved oxygen levels below 0.5 $mgO_2 \ L^{-1}$, exists in regions of streams where advective velocity of water is slowed, allowing for accumulation of sediments and organic matter. In streams, these environments include pools behind beaver and debris dams, backwaters, side channels, and algal mat interiors at night (Dahm *et al.*, 1987; Yavitt *et al.*, 1992; Devito and Dillon, 1993; Naiman *et al.*, 1994). Anoxia also occurs in alluvial sediments within the stream-water–groundwater interface, when contact time with metabolically active alluvial sediments is long (Valett *et al.*, 1990, 1996; Dahm *et al.*, 1991).

Anoxia in the stream-water–groundwater interface is regulated by hydrologic exchange with the stream surface. Surface stream water enters the subsurface at locations of groundwater recharge (downwelling zones), supplying oxygen to sediments, which is consumed during aerobic respiration (Pusch and Schwoerbel, 1994; Jones *et al.*, 1995a). Depending on metabolic rates and contact time between water and sediments, oxygen can be removed from solution and organic matter mineralization then proceeds via a suite of microbially mediated alternative terminal electron-accepting processes (Vroblesky and Chapelle, 1994), which strongly influence biogeochemical constituents and metabolic rates in the subsurface (Table I, Fig. 1).

The types of terminal electron-accepting processes occurring in the stream-water–groundwater interface can be predicted based on thermodynamic (Champ *et al.*, 1979; Stumm and Morgan, 1996; Hedin *et al.*, 1998), geochemical (Potsma and Jakobsen, 1996), and ecological (Lovley and Klug, 1983), constraints (Table I, Fig. 1). Once oxygen is consumed, respiration is predicted to occur first via denitrification, with nitrate (NO_3^-) used as the terminal electron acceptor (Fig 1). This occurs both because this process has the highest free energy change (Table I) and because most denitrifying bacteria are facultative anaerobes able to use a broad range of organic substrates (Korom, 1992). The occurrence of the remaining terminal electron-accepting processes is regulated in part by fermentation. Fermentation is an anaerobic metabolic process where organic molecules are used both as electron donors and electron acceptors (Chapelle, 1993). Although fermentation is the least energetically favorable terminal electron-accepting processes (Table I), fermentation must co-occur with metal and sulfur reduction and methanogenesis because fermentation products, such as hydrogen gas (H_2) and acetate,

TABLE I   Stoichiometry of Possible Redox Reactions in the Stream-water–Groundwater Interface

Reaction	Reductive processes equation	Free energy $\Delta G^\circ(w)$ (kcal)
Aerobic respiration	$CH_2O + O_2 = CO_2 + H_2O$	−120.0
Denitrification	$CH_2O + (4/5)NO_3^- + (4/5)H^+ = (7/5)H_2O + (2/5) N_2 + CO_2$	−113.9
Mn(IV) reduction	$CH_2O + 2MnO_2 + 4H^+ = 2Mn^{2+} + 3H_2O + CO_2$	−81.3
Fe(III) reduction	$CH_2O + 8H^+ + 4Fe(OH)_3 = 4Fe^{2+} + 11H_2O + CO_2$	−27.7
Sulfate reduction	$CH_2O + (½)SO_4^{2-} + (½)H^+ = (½)HS^- + H_2O + CO_2$	−25.0
Methanogenesis	$CH_2O + (½)CO_2 = (½)CH_4 + CO_2$	−19.2
Fermentation	$CH_2O + (1/3)H_2O = (2/3)CH_3OH + (1/3)CO_2$	−8.6

Reaction	Oxidative processes equation	Free energy $\Delta G^\circ(w)$ (kcal)
Methane oxidation	$O_2 + (½)CH_4 = (½)CO_2 + H_2O$	−97.7
Sulfur oxidation	$O_2 + (½)HS^- = (½)SO_4^{2-} + (½)H^+$	−95.0
Fe(II) oxidation	$O_2 + 4Fe^{2+} + 10H_2O = 4Fe(OH)_3 + 8H^+$	−92.3
Nitrification	$O_2 + (½)NH_4^+ = (½)NO_3^- + H^+ + (½)H_2O$	−41.7
Mn(II) oxidation	$O_2 + 2Mn^{2+} + 2H_2O = 2MnO_2 + 4H^+$	−38.6

Note. $\Delta G^\circ(w)$ is the change in Gibbs free energy calculated as $\Delta G^\circ(w) = \Delta G^\circ - RT \ln[H^+]^p$, where $[H^+] = 1 \times 10^{-7}$ M and p is the stoichiometric coefficient for H^+ in a given reaction. For reductive reactions it is important to note that the actual $\Delta G^\circ(w)$ for a given system will depend on the type of organic matter being oxidized, and that alternative terminal electron accepting processes often involve oxidation of low molecular weight fermentation products. Adapted from Champ *et al.* (1979), Stumm and Morgan (1996), and Hedin *et al.* (1998).

serve as electron donors. Metals that exist in more than one redox state (e.g., Fe, Mn, Se, U, Mo, and others) are used as electron acceptors by anaerobic bacteria (Oremland *et al.,* 1989; Lovley, 1991; Lovley *et al.,* 1991a,b; Chapelle, 1993; Ghani *et al.,* 1993). These organisms can use a variety of organic molecules as substrates, including aromatic compounds as well as fermentation products (Chapelle, 1993). After nitrate is depleted and fermentation products accumulate, anaerobic respiration in the stream-water–groundwater interface is likely to occur via metal reduction (Table I, Fig. 1). Nonetheless, when fermentation rates are low, the metabolic rate of metal reduction is limited by the availability of electron donors (Lovley and Klug, 1983). Where suitable electron donor concentrations exist, metal-reducing bacteria can frequently out-compete sulfate-reducing and methanogenic bacteria for fermentation products (Chapelle, 1993). Once available oxidized

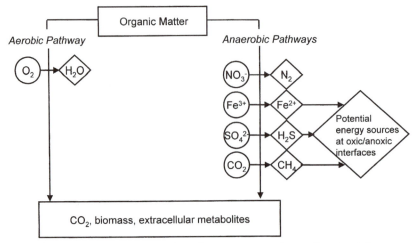

FIGURE 1 Model of how anoxia in the surface water–ground water interface influences metabolism and biogeochemistry. As stream water enters subsurface zones, organic matter oxidation may strip available dissolved oxygen (O_2). When dissolved oxygen is absent, a sequence of alternative terminal electron-acceptors (represented in circles) can be used by bacteria in anaerobic metabolism of organic matter. Because of geochemical, thermodynamic, and ecological constraints, these anaerobic processes may be spatially and/or temporally segregated along flowpaths in the surface water–ground water interface. When anoxic subsurface water returns to the stream, reduced products of anaerobic metabolism (represented in diamonds) may be re-oxidized, influencing the biology and ecology at the surface water-ground water and benthic interfaces.

metals have been reduced and fermentation products have accumulated, sulfate reduction is predicted to be the dominant catabolic process (Table I, Fig. 1). Like metal reducers, sulfate reducers use fermentation products as electron donors; however, sulfate reducers are more restricted in the types of organic substrates that can be used (Postgate, 1984). In addition, they are known to out-compete methanogens for fermentation products (Lovley and Klug, 1983). The last terminal electron-accepting processes predicted to occur along flowpaths in the stream-water–groundwater interface is methanogenesis, where methane (CH_4) is produced by acetate fermentation or by the reduction of carbon dioxide (CO_2; Table I, Fig. 1). Due to the combined effects of local geochemistry, thermodynamics, and competition for electron donors, individual terminal electron-accepting processes likely dominate discrete zones along flowpaths (Champ *et al.,* 1979; Vroblesky and Chapelle, 1994) in the stream-water–groundwater interface.

Terminal electron-accepting processes occurring in the stream-water–groundwater interface may be clearly segregated in space (Fig. 1); however, because of the spatial heterogeneity associated with deposition of alluvial sediments, segregation of dominant terminal electron-accepting processes may occur at much smaller scales than is commonly sampled with wells (see Fig. 2). Wells screened over vertical distances from a few centimeters to many tens of centimeters are likely to intercept several anoxic zones or multiple flowpaths with different terminal electron-accepting processes (Fig. 2). This may explain the observance of multiple anaerobic metabolic end products in well-oxygenated subsurface water (Holmes *et al.*, 1994).

The sequence of terminal electron-accepting processes described earlier can result in accumulation of metabolic products that may influence surface-water biogeochemistry and ecosystem functioning when interstitial water enters the stream at upwelling zones, or regions of groundwater discharge (Fig. 1). Most of these metabolic end products are chemically reduced and can be reoxidized at oxic–anoxic interfaces. Oxidative processes likely to occur at

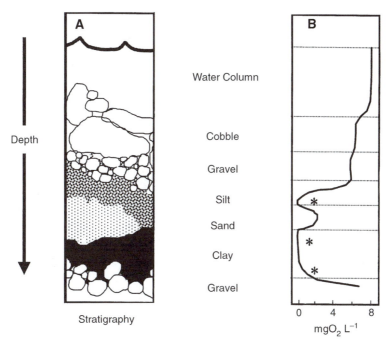

FIGURE 2 Diagram of how alluvial sediment stratigraphy at the surface water–ground water interface may generate multiple oxic–anoxic interfaces. (A) Stratigraphic layers through the surface water–ground water interface. (B) Hypothetical relationship between oxygen concentration (mg O_2 L^{-1}) and depth below the streambed. Multiple oxic–anoxic interfaces (indicated by *) may result from water moving through different stratigraphic layers with differing residence times and rates of respiration.

these interfaces include nitrification, metal oxidation, sulfur oxidation, and methane oxidation (Table I, Fig. 1).

As stream ecologists have looked in greater detail at the subsurface in studies of carbon and nutrient cycling, they have gained new insights on the importance of anaerobic metabolism in this environment. Considerable research has been directed at the process of denitrification at the stream-water–groundwater interface and is addressed elsewhere in this volume (see also Duff and Triska, Chapter 8). This chapter focuses on metal reduction, sulfate reduction, and methanogenesis at the stream-water–groundwater interface. We summarize current information on these anaerobic metabolic processes at the stream-water–groundwater interface, examine distributions of the end products of these processes, and discuss implications of these end products on surface-water biogeochemistry and ecosystem functioning.

II. METHODS USED IN STUDIES OF ANAEROBIC METABOLISM

Before discussing anaerobic metabolism and its influences on biogeochemistry at the stream-water–groundwater interface, a description of how redox-sensitive constituents and anaerobic processes are measured is warranted. Many of these measurements are not part of routine analyses by lotic ecologists but are commonly used in lake and marine studies. Brief descriptions of commonly used techniques to determine redox-sensitive gases solutes and metabolic rates are presented.

A. Dissolved Gases

Sampling must be conducted in a manner that prevents bubble formation and exposure of samples to the atmosphere. Sampling hyporheic and groundwater wells for analysis of dissolved gases should be conducted with a peristaltic pump. This minimizes bubble formation and degassing of supersaturated samples which can occur if a manual bailer, vacuum pump, or syringe is used to obtain water (M. A. Baker, unpublished data).

Dissolved oxygen is frequently measured in the field using Clark-type polarographic electrodes. Although these types of electrodes are common, relatively inexpensive, and provide rapid measurements, accuracy declines at low dissolved oxygen levels (ca. <1 mgO_2 L^{-1}) because products of anaerobic sulfur metabolism compete with oxygen for reaction at the electrode surface (Hale, 1983). This effect can be minimized with the use of high-sensitivity, 12.7-μm (0.5-mil) Teflon membranes in place of standard 25.4-μm (1-mil) membranes. In all cases, electrode readings should be compared to occasional Winkler titrations (Wetzel and Likens, 1991) in waters of low dissolved oxygen concentration.

The carbon dioxide in water can be calculated from pH, alkalinity, or total dissolved inorganic carbon (Wetzel and Likens, 1991). The value obtained when calculating P_{CO_2} in this way is strongly dependent on the quality of the pH measurement, so accurate pH measurement is critical. In addition, this method assumes that all of the alkalinity is derived from inorganic carbon (bicarbonate and carbonate), which may be an inappropriate assumption in some cases, especially in waters with low alkalinity, high dissolved organic carbon, or high ionic strength (Wetzel and Likens, 1991). Alternatively, carbon dioxide can be measured directly by allowing the water sample to equilibrate with a headspace of known volume and gas composition. The resultant carbon dioxide concentration in the headspace can then be measured using a gas chromatograph with a thermal conductivity detector (Kling *et al.*, 1992) or with an infrared gas analyzer (Hope *et al.*, 1995).

Gaseous methane is typically analyzed using a gas chromatograph with a flame ionization detector. Samples are transferred to gas-tight syringes or bottles in the field. The dissolved methane in the known volume of sample is partitioned into headspace of known volume and gas composition by shaking. A subsample of the headspace is then analyzed on the gas chromatograph, and the dissolved methane concentration is calculated (Dahm *et al.*, 1991; Baker *et al.*, 1994). Hydrogen gas is a product of fermentation and is obtained by passing a stream of water through a gas-sampling bulb and stripping H_2 from the sample using nitrogen (N_2) gas (Vroblesky and Chapelle, 1994; Chapelle *et al.*, 1995). Hydrogen concentration is then quantified using a gas chromatograph with a reduction gas detector. Because hydrogen gas is extremely volatile and can diffuse through glass, it is best measured in the field (Chapelle *et al.*, 1995).

B. Solutes

Dissolved reduced iron (Fe^{2+}) can be quantified colorimetrically by reacting filtered water samples with ferrozine (Stookey, 1970). Aqueous iron and manganese are measured using atomic absorption spectroscopy [American Public Health Association (APHA), 1992]. Samples are acidified to prevent precipitation of metal hydroxides (APHA, 1992) however, atomic absorption spectroscopy measures total concentration. Determining metal speciation and redox state requires further analysis using geochemical models such as PHREEQE or MINTEQ (Drever, 1988).

Numerous sulfur species can occur in solution; the most commonly measured forms are sulfate (SO_4^{2-}) and sulfide (S^{2-}). Sulfate is usually measured using ion chromatography (APHA, 1992), whereas reduced sulfur species (H_2S, S^{2-} and HS^-) are more difficult to quantify because they are extremely reactive chemically. Samples for sulfide analyses need to be isolated from the atmosphere to prevent chemical oxidation and out of contact with metallic surfaces to minimize precipitation. A standard technique for the measurement of dissolved H_2S, S^{2-}, and HS^- uses colorimetric analysis (Cline, 1969).

C. Metabolism

Anaerobic respiration can be quantified from the rate of change in dissolved carbon dioxide and methane in the same way that change in dissolved oxygen concentration is used to measure aerobic respiration (Chapelle *et al.*, 1996). Measurements of rates of specific terminal electron-accepting processes can be performed by quantifying loss of oxidized species or increase of reduced species with time in a microcosm or along a known flowpath (Jones, 1995; Chapelle *et al.*, 1996; Baker, 1998). Killed controls or assays to determine chemical reactivity of the solutes with the sediment matrix are needed to separate abiotic from biotic processes. Rates of terminal electron-accepting processes also can be measured by coupling results from subsurface tracer injections to groundwater solute transport models (Smith *et al.*, 1991, 1996). In addition, *in situ* quantification of hydrogen gas concentration has been used in groundwater ecosystems as a predictor of the dominant terminal electron-accepting processes because microbially produced hydrogen can be an electron donor for which microorganisms compete in anoxic environments (Lovley and Goodwin, 1988; Vroblesky and Chapelle, 1994; Chapelle *et al.*, 1995, 1996). The efficiency of dissolved hydrogen uptake varies among microbes that use different terminal electron-accepting processes. Hydrogen concentration therefore can be used as an indicator of the dominant terminal electron-acceptor process.

Measurement of metabolic rates also can be determined using electron acceptors labeled with radioactive isotopes in microcosms or sediment slurries. These techniques are appropriate for some microbial processes but not others. Roden and Lovley (1993) reported that ^{55}Fe-labeled iron oxides were not an effective tracer of iron reduction since the labeled ferric iron (Fe^{3+}) exchanged with unlabeled ferrous iron (Fe^{2+}) in the sediments, reducing recovery of the tracer. In ocean and lake sediments, $^{35}SO_4^{2-}$ is commonly used to trace sulfate reduction (Urban *et al.*, 1994), and this method could be applied to alluvial sediments. Finally, radiolabeled carbon dioxide or acetate can be used to measure methanogenic rates in sediments (Schulz *et al.*, 1997). To date, these techniques have not been applied to sediments from the stream-water–groundwater interface.

III. CONTROLS ON THE ESTABLISHMENT OF ANOXIA

Consumption of dissolved oxygen at the stream-water–groundwater interface is controlled by water residence time and respiration rate. An important physical variable that controls the development and persistence of anoxic conditions in the subsurface is hydrologic residence time. For the stream-water–groundwater interface, this can be expressed as the time it takes for a parcel of stream surface water to reach a sampling well (i.e., nominal travel time; sensu Triska *et al.,* 1989), or contact time with sediment (Findlay, 1995; Valett *et al.,* 1996). The longer a parcel of water remains in contact with biologically or chemically reactive sediment, the more oxygen will be consumed. Using solute injection techniques, interstitial dissolved oxygen concentration was inversely related to nominal travel time in the stream-water–groundwater interface of three headwater streams (Valett *et al.,* 1996; Fig. 3). Similarly, dissolved oxygen concentration in the stream-water–groundwater interface (expressed as a percent of surface water concentration) is inversely related with the amount of time water was in contact ($\log_{10}$ contact time in hours) with sediments for eight stream ecosystems throughout the world (Findlay, 1995).

Physical properties of the alluvium also influence dissolved oxygen in the stream-water–groundwater interface by affecting hydrologic exchange. Hydraulic conductivity is a measure of the propensity for aquifer material to allow water flow (Fetter, 1988). Fine-grained sediments, such as clay and silt, have lower hydraulic conductivity than coarse sediments like sand and gravel (Kelson and Wells, 1989). In a comparison among streams with differing

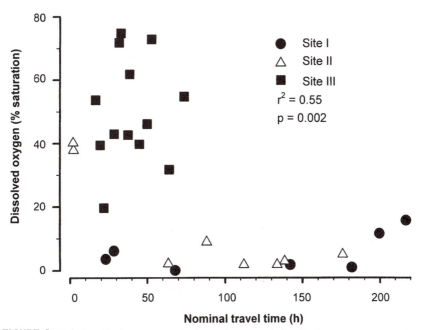

FIGURE 3 Relationship between nominal travel time and dissolved oxygen concentration from three hyporheic zones in geologically distinct [sandstone/siltstone (Site I), volcanic tuff (Site II), granite/gneiss (Site III)] catchments in New Mexico (Valett *et al.,*1996). Linear regression was performed on arcsin $(x)^{1/2}$ to transformed dissolved oxygen data to produce statistics given in figure.

alluvial hydraulic conductivity, dissolved oxygen in the stream-water–groundwater interface with low hydraulic conductivity (sandstone/siltstone) was significantly lower (mean = 5.7% saturation) than in a granitic (high hydraulic conductivity) stream (50.2% saturation; Valett *et al.*, 1996). A site with intermediate hydraulic conductivity (volcanic tuff) had a mean oxygen saturation of 12.6% (Valett *et al.*, 1996). This pattern was attributed to greater subsurface water residence time due to slower flow rate and greater oxygen consumption in the lower hydraulic conductivity systems. Similarly, in a study using groundwater microcosms filled with sediments of differing hydraulic conductivity, lower dissolved oxygen concentration was recorded in microcosms filled with silt compared to those filled with sand or gravel (Dodds *et al.*, 1996). Because microbial activity (measured as ^{3}H-incorporation) was highest in the gravelly sediment, lower oxygen concentration associated with low hydraulic conductivity sediments was attributed to a lower rate of dissolved oxygen transport (i.e., water transport; Dodds *et al.*, 1996).

The rate of respiration by organisms in the stream-water–groundwater interface also can influence dissolved oxygen availability and the distribution of terminal electron-accepting processes. Respiration rate, as measured in sediment slurries, in microcosms, or along flowpaths is influenced by the abundance of respiring bacteria, which in turn can be influenced by sediment grain size. For example, microbial colonization is associated with sediment grain size, with smaller grain sizes having greater surface area and greater microbial density (Lock *et al.*, 1984). Further, respiration rates in small-grained hyporheic sediment were higher than those observed for larger particles (Jones, 1995). The difference in respiration rate was attributed in part to epilithic community size.

Transport of metabolic substrates, including electron donors (e.g., organic matter) and electron acceptors (e.g., dissolved oxygen) also influence respiration rate. Smaller sediment grain sizes may restrict transport of metabolic substrates as discussed earlier (Dodds *et al.*, 1996). In addition, sediments may adsorb dissolved organic matter, thereby locally increasing its availability to microbes. For example, organic matter derived from leaf leachates can sorb to stream benthic sediments (Dahm, 1981), dissolved organic matter can sorb to iron and aluminum hydroxides (McDowell and Wood, 1984), and the polysaccharide matrix of biofilms on sediment can retain dissolved organic matter (Freeman and Lock, 1995). Respiration in subsurface sediments is often limited by the availability of organic matter (Jones, 1995; Pusch, 1996) and higher metabolic rates in smaller grained sediments may be partly attributed to a higher availability of nutrients and organic matter as a result of sorption.

It is important to note that surface-water dissolved oxygen concentration also can vary on diel and seasonal timescales as a result of differing photosynthesis and respiration rates. These variations can directly influence the establishment of anoxia in the stream-water–groundwater interface by changing initial dissolved oxygen availability (Brick and Moore, 1996). Anoxic conditions in the stream-water–groundwater interface may be more extensive at night when dissolved oxygen levels are lowest in the surface stream.

In general, the combination of hydrologic residence time and sediment respiration rate determines the extent, location, and rate of onset of anoxia at the stream-water–groundwater interface. Sediment grain size is perhaps the key parameter, influencing oxygen consumption and establishment of anoxia in several ways, including (1) surface area available for microbial colonization, with smaller particles having more surface area; (2) residence time in the subsurface, with water passing through smaller particles having longer residence time; (3) rate of transport of electron donors and acceptors with larger sediment particles resulting in faster transport rates; and (4) abiotic retention of dissolved organic carbon. Therefore, sediment grain size and the stratigraphy of alluvial sediments are important features that affect dissolved oxygen availability, sediment redox potential, and distribution of terminal electron-accepting processes in the subsurface.

IV. BIOGEOCHEMISTRY AND EVIDENCE FOR ANAEROBIC METABOLISM

Under anoxic conditions, metabolic products of anaerobic metabolism accumulate. The relative concentrations and spatial distribution of redox-sensitive solutes and dissolved gases such as Fe^{3+}/Fe^{2+}, SO_4^{2-}/H_2S, and CO_2/CH_4 are indicative of the relative importance of different anaerobic microbial processes.

A. Spatial Distributions

We measured dissolved ferrous iron concentration concurrently with conservative tracer additions in three geologically distinct sites in New Mexico (Valett *et al.*, 1996; Morrice *et al.*, 1997). Plots of ferrous iron concentration in the stream-water–groundwater interface versus nominal subsurface travel time show a positive linear relationship between concentration and time for sites with intermediate and high hydraulic conductivity (Fig. 4; Baker, 1998). This relationship also is seen at the site with intermediate hydraulic conductivity along a 2-m flowpath (Fig. 5; Baker, 1998) as identified from conservative tracer injection experiments (Morrice *et al.*, 1997) and hydrologic modeling (Wroblicky *et al.*, 1998). The site with the lowest average hydraulic conductivity showed no relationship between ferrous iron concentration and nominal subsurface travel time (Fig. 4; Baker, 1998), and peak ferrous iron concentration was an order of magnitude lower than at the

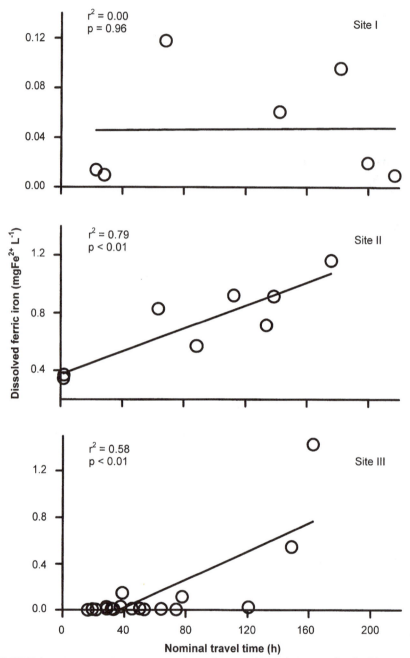

FIGURE 4 Relationship between nominal travel time and hyporheic zone dissolved iron concentration for three sites in New Mexico. Average hydraulic conductivity was lowest for Site I, intermediate for Site II, and highest for Site III (Morrice *et al.*, 1997; data are from Baker, 1998).

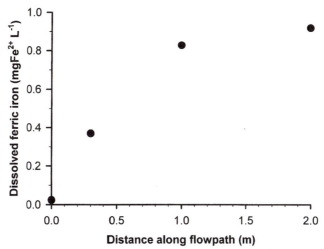

FIGURE 5   Relationship between dissolved iron concentration (mg L^{-1}) and distance (meters) along a hyporheic flowpath at a headwater stream (Site II) in New Mexico (data are from Baker, 1998).

higher conductivity sites. Anaerobic metabolism at this site likely included a high rate of sulfate reduction, which may have decreased soluble ferrous iron concentration by precipitation of iron sulfides (e.g., pyrite, Potsma and Jakobsen, 1996; Stumm and Morgan, 1996).

Sulfate exhibits a general trend of higher concentration in surface waters relative to subsurface waters at sites in New Mexico (Valett *et al.*, 1996). Lower sulfate concentration in anoxic subsurface water is likely the result of microbial sulfate reduction, but direct measurements of sulfate reduction rates at the stream-water–groundwater interface have not been made.

Methane supersaturation relative to the atmosphere routinely occurs within stream surface water and waters of the stream-water–groundwater interface (deAngelis and Lilley, 1987; Carr, 1989; Dahm *et al.*, 1991; Baker *et al.*, 1994; Jones *et al.*, 1995b). Subsurface methane concentrations at the three sites in New Mexico showed no statistically significant relationship with nominal travel time (Valett *et al.*, 1996; Baker, 1998). However, subsurface methane concentrations were greater than surface-water concentrations at all sites (Table II; Baker *et al.*, 1994; Baker, 1998). Methane concentration in the stream-water–groundwater interface at these sites is highly variable (coefficient of variation ranges from 69 to 158%; Table II; Baker, 1998). Similar variability in methane concentration was reported for several desert streams in Arizona during the winter (Jones *et al.*, 1995b). It is unknown whether methane observed in the stream-water–groundwater interface is produced *in situ* or transported from other subsurface environments.

TABLE II   Methane Concentrations [mean + SE (CV)] in Surface, Hyporheic and Ground Water in New Mexico Streams with Differing Geology

| | Methane concentration [$\mu gC\ L^{-1}$, mean $\pm$ SE (CV)] | | |
Site	Surface water	Hyporheic zone[a]	Ground water
Aspen Creek	11.1 $\pm$ 1.4 (26.7)	33.7 $\pm$ 6.6 (69.1)	76.1 $\pm$ 17.3 (93.3)
Rio Calaveras	2.6 $\pm$ 0.3 (27.9)	69.5 $\pm$ 98.3 (189.0)	45.0 $\pm$ 12.6 (119.1)
Gallina Creek	5.5 $\pm$ 2.9 (80.1)	7.3 $\pm$ 4.0 (157.9)	—[b]

[a]Hyporheic zone water was defined as having >10% surface water (Triska *et al.* 1989) as determined using a conservative tracer (Bromide; Valett *et al.* 1996; Morrice *et al.* 1997).
[b]All wells were connected to the surface stream at Gallina Creek.
From Baker (1998).

B. Temporal Variation

Changes in dissolved oxygen availability on diel and seasonal timescales also may influence anaerobic metabolism and the accumulation of chemically reduced solutes in the stream-water–groundwater interface. For example, in the Upper Clark Fork River, Montana, dissolved and acid-soluble particulate metals exhibit a diel cycle (Brick and Moore, 1996). Dissolved manganese concentration in the surface water of the river was higher during the night. The increased manganese was attributed to diel variation in dissolved oxygen availability and consequently anaerobic metabolism (metal reduction) in interstitial sediments, as well as to a change in flux of these constituents from the stream-water–groundwater interface as a result of decreased evapotranspiration at night. Higher levels of acid-soluble particulate iron were also observed in the stream at night, and these were associated with increases in total suspended solids that may have been related to increased invertebrate activity in the dark (Brick and Moore, 1996).

In the River Glatt, Switzerland, where river water penetrates into the surrounding aquifer, subsurface dissolved iron and manganese concentrations vary seasonally (von Gunten *et al.*, 1991, 1994; von Gunten and Lienert, 1993). Dissolved oxygen concentration was lowest during the summer when heterotrophic metabolism was greatest due to higher temperatures. During this time, manganese and iron oxides dissolved due to metal reduction by anaerobic bacteria. Dissolved oxygen concentration was highest and dissolved metal concentrations were lowest during winter, presumably because of decreased biological activity due to cold temperatures. Dissolved manganese concentration in the River Glatt also decreased over a period of 12 years (von Gunten and Lienert, 1993). This decline was attributed to decreased surface-water phosphate concentrations which, in turn, lowered the supply of algal-derived labile organic matter to the stream-water–ground-water interface, and, consequently, decreased the rate of respiration and the relative importance of anaerobic processes in the subsurface.

Seasonal variation in stream hydrographs can be an important factor in determining the extent and location of anaerobic metabolism in alluvial ground water. Hydrographs of temperate mountain streams are often dominated by spring snowmelt. In these systems, snowmelt supplies well-oxygenated water to the subsurface (Fig. 6; Baker, 1998), which, in turn, can influence anaerobic metabolism and biogeochemically important end products. For example, the proportion of anoxic wells at 3 montane sites in New Mexico decreased during snowmelt (Fig. 6; Baker, 1998). The prevailing oxic conditions during snowmelt were associated with significant decreases in interstitial reduced iron and methane concentrations (Fig. 6; Baker, 1998).

C. Rates of Anaerobic Metabolism

Measurements of the rates of specific anaerobic processes that cycle metals, sulfur, and methane in streams are uncommon. However, from biogeochemical evidence presented earlier, it is clear that these processes occur in the stream-water–groundwater interface of many catchments. In this section, we summarize the few studies that have investigated metal reduction, sulfate reduction, and methanogenesis. Because data on these processes in the stream-water–groundwater interface are rare, some of the data we present are from other aquatic ecosystems.

Microbial processes that involve metal reduction in sediments have moved to the forefront of research in microbial ecology in the past decade. Organisms capable of reducing iron were first cultured in 1987, and those capable of reducing other metals like manganese have only recently been isolated (Lovley, 1991). In culture, the type of carbon source influences the rate of iron reduction, with acetate oxidation yielding the highest rate, followed by glucose and benzoate oxidation (Lovley *et al.*, 1991b). Iron-reducing bacteria have recently been isolated from sediments of the stream-water–groundwater interface at Rio Calaveras, New Mexico (Markwiese *et al.*, 1998). This ongoing research establishes the presence of iron-reducing bacteria in near-stream alluvial sediments, but their role in organic matter mineralization at this site remains to be assessed (Baker, 1998).

The importance of sulfate reduction has been demonstrated in anoxic sediments from a variety of aquatic systems, including wetlands, rivers, lakes, and estuaries. An extremely high rate of sulfate reduction (up to 33.9 mgS kg sediment^{-1} d^{-1}) was reported for sediments from Lake Kinneret, Israel (Table III; Haadas and Pinkas, 1995). Rate of sulfate reduction in sediment from the Pregolya River ranged from 0.02 to 8.26 mgS kg sediment^{-1}

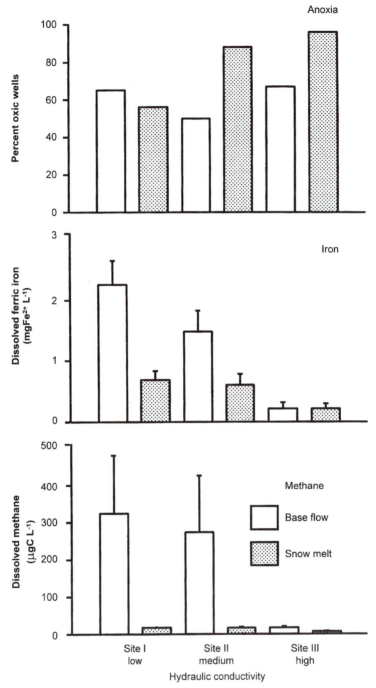

FIGURE 6 Changes in extent of anoxia, dissolved iron, and methane concentration in the sub-surface waters of three geologically distinct catchments in New Mexico during baseflow and snowmelt conditions. Statistical analysis was performed using repeated measures ANOVA (Baker, 1998).

TABLE III   Rates of Sulfate Reduction in Sediments from Several
Aquatic Ecosystems

Site	Sulfate reduction rate (mgS kg sediment^{-1} d^{-1})	Reference
Little Rock Lake, Wisconsin	0–13.6	Urban *et al.* (1994)[a]
Pregolya River, Kalingrad	0.02–8.26	Ivanov *et al.* (1995)
Lake Kinneret, Israel	0.11–33.9	Haadas and Pinkas (1995)[a]
Congo River Estuary, West Africa	0.39–3.34	Pimenov *et al.* (1994)

[a]Assumes 1 ml sediment = 2 g.

d^{-1} (Table III; Ivanov *et al.*,1995) and is within the same range as those reported for estuarine and lake sediments. Rates of sulfate reduction in sediments from the Congo River Estuary range from 0.39 to 3.34 mgS kg sediment^{-1} d^{-1}, and sediments from Little Rock Lake in Wisconsin range from 0 to 13.56 mgS kg sediment^{-1} d^{-1} (Table III; Pimenov *et al.*, 1994; Urban *et al.*, 1994). Comparison of these rates with those reported for subsurface ecosystems is difficult because different units are used, and sulfate reduction in the stream-water–groundwater interface of lotic ecosystems has not been extensively studied. Based on an injection of sulfate into a known hyporheic flowpath at Rio Calaveras, New Mexico, the estimated rate of sulfate retention was on the order of 0.05–0.10 mgS L^{-1} h^{-1} (Morrice, 1997). Using geochemical models, the estimated sulfate reduction rate ranged from 6.3 × 10^{-9} to 2.5 × 10^{-8} mgS L^{-1} h^{-1} for the Black Creek coastal plain aquifer in South Carolina (Chapelle and McMahon, 1991). This groundwater ecosystem differs from the stream-water–groundwater interface in that it is thousands of meters deep and extremely oligotrophic, thus the low rate of sulfate reduction.

Methanogenesis has become of interest in recent years because methane is an important greenhouse gas. In Sycamore Creek, Arizona, methanogenesis rates in sediments ranged from 10 to 604 mgC kg sediment^{-1} d^{-1} (Jones *et al.*, 1995b). Additionally, potential rates of methanogenesis appear to vary widely in the stream-water–groundwater interface of headwater streams across the United States and among habitats within the same stream (Table IV; Carr, 1989). The highest potential rate of methanogenesis (24.5 mgC kg sediment^{-1} d^{-1}) was recorded for pool sediments from a Zuni Mountain stream in New Mexico. Low rates were measured for streams at Coweeta in North Carolina and H.J. Andrews in Oregon (Table IV; Carr, 1989). Differences in hydrologic residence time among the different geologic settings and habitat likely contribute to the variation in methanogenesis rate. Although surface waters of streams are almost always supersaturated with methane

TABLE IV Rates of Potential Methanogenesis and Methane Oxidation
in the Sediments of Pools and Riffles in Small Streams (mean ± SE)

Site	Methanogenesis (μgC g sediment^{-1} h^{-1})	Methane oxidation (μgC g sediment^{-1} h^{-1})
Zuni Mountains, New Mexico		
Pools	25 ± 23	34 ± 16
Riffles	0.48 ± 0.24	27 ± 8.4
Konza Prairie, Kansas		
Pools	1.9 ± 1.2	8.9 ± 1.7
Riffles	0.48 ± 0.24	11 ± 3.8
Coweeta, North Carolina		
Pools	0.24 ± 0.00	8.4 ± 1.9
Riffles	<0.24 ± 0.00	4.1 ± 0.48
H.J. Andrews, Oregon		
Pools	0.24 ± 0.00	5.0 ± 0.48
Riffles	0.24 ± 0.00	5.0 ± 0.96

From Carr (1989).

relative to the atmosphere (DeAngelis and Lilley, 1987; Dahm *et al.*, 1991; Baker *et al.*, 1994; Jones *et al.*, 1995b), the methane sources are generally not known. Methanogenesis in alluvial sediments is a likely source.

V. INFLUENCE OF SUBSURFACE ANAEROBIC METABOLISM ON STREAM ECOSYSTEM PROCESSES

Water and solutes from alluvial ground water and the stream-water–groundwater interface can enter surface water in upwelling or groundwater discharge zones. If metabolic products of anaerobic processes have accumulated in the subsurface, these solutes can affect in-stream processes at the benthic interface and perhaps for some distance downstream. At anoxic–oxic interfaces, chemically reduced solutes resulting from anaerobic metabolism may be re-oxidized via chemical or biological pathways (Table I, Fig. 1). Chemoautotrophic organisms capable of catalyzing the oxidation of reduced solutes can use the chemical energy of the oxidation reaction to fix CO_2 and thus contribute to stream primary production (Table I).

The overall contribution of chemoautotrophy to stream energetics has not been assessed for many lotic ecosystems. In Sycamore Creek, Arizona sediments, chemoautotrophy (measured as $^{14}CO_2$ uptake) was only 1–1.3% of aerobic respiration and was orders of magnitude lower than carbon dioxide fixation via photosynthesis (Jones *et al.*, 1994). However, in heavily shaded streams with lower rates of algal production, chemoautotrophy at oxic–anoxic interfaces may be more important to the overall energy budget.

Nitrification is an important chemoautotrophic process in freshwater because it oxidizes ammonium (NH_4^+) to nitrate, a more mobile form of nitrogen. In many catchments, subsurface waters are enriched in ammonium compared to surface waters, suggesting that ammonium is oxidized to nitrate at oxic–anoxic interfaces (Triska *et al.*, 1989, 1993; Valett *et al.*, 1994, 1996). Nitrification enhances nitrogen availability to surface algal communities, locally increasing primary production in nitrogen-limited streams (Valett *et al.*, 1994). This chemoautotrophic process is discussed in greater detail in Chapter 8 (Duff and Triska).

Because it represents a sink for an important greenhouse gas, methane oxidation is another chemoautotrophic process of great interest (Fig. 1, Table I). Methane can account for a sizable fraction (4–18%) of the total dissolved organic carbon pool at the stream-water–groundwater interface (Dahm *et al.*, 1991; Baker *et al.*, 1994; Jones *et al.*, 1995b). This energy source can be consumed by methanotrophic bacteria that utilize methane for energy as well as biomass accrual (Table I). By comparing the rate of methane evasion to methanogenesis, methane oxidation was estimated to consume more than 95% of the methane produced (Jones *et al.*, 1995b). Similarly, sediments from the hyporheic zone of small streams at Konza Prairie (Kansas), H.J. Andrews (Oregon), and Coweeta (North Carolina) Long-Term Ecological Research sites, as well as headwater streams in the Zuni Mountains of New Mexico, had higher methane oxidation potentials than potential rates of methanogenesis (Table IV; Carr, 1989). Methane oxidation at the stream-water–groundwater interface may be a major control on rates of methane efflux to the atmosphere from stream corridors.

The contribution of methane cycling to the overall energy budget of stream ecosystems is poorly understood. However, in Sycamore Creek, Arizona, methanogenesis rates (measured as methane evasion) in anoxic sediments (3500 mgC m^{-2} d^{-1}) were nearly equal to rates of aerobic respiration in oxic sediments (Jones *et al.*, 1995b). Most of the methane was consumed (3330 mgC m^{-2} d^{-1}) as it was transported across the oxic–anoxic interface (Jones *et al.*, 1995b). Methane represents a labile energy source, which may be an important substrate for aerobic respiration in the stream-water–groundwater interface where methane and dissolved oxygen co-occur.

The role of metal and sulfur oxidation has received limited consideration by stream ecologists. Visual examples of the occurrence of these biogeochemical processes abound. Iron-rich flocculent oxyhydroxide precipitates with a yellow to orange coloration commonly can be seen at seepage areas in streams during baseflow conditions. Black manganese oxide deposits are routinely encountered coating the undersides of stream gravel and rocks. Mats of filamentous sulfur-oxidizing bacteria may be present where sulfide-rich ground waters discharge into surface stream water.

A few studies have addressed the impact of oxidative processes on stream biota where products of anaerobic metal reduction are oxidized at an aerobic interface. For example, deposition of metal oxides in streambeds alters the substrate available for epilithic algal growth either by preventing algal attachment due to the flocculent texture of the oxyhydroxide deposits (Sode, 1983), by competing for space and nutrients by metal oxidizing bacteria (Sheldon and Skelly, 1990), or by increasing metal concentrations to toxic levels. These hypotheses were tested by measuring diatom colonization using *in situ* experimental chambers in which iron and manganese concentrations were increased in the absence of ferromanganese-depositing bacteria (Wellnitz and Sheldon, 1995). Results from chambers showed that the presence of iron oxides alone decreased diatom abundance by at least an order of magnitude, whereas the presence of manganese oxides did not influence algal growth. In addition, the presence of metal-oxidizing bacterial blooms has been shown to decrease stream macroinvertebrate diversity and abundance, perhaps via the loss of food resources for algal grazers (Wellnitz *et al.*, 1994).

The presence of iron oxides in streambeds resulting from upwelling of iron-rich anoxic water also can influence algal productivity by sorbing macro- and micronutrients, including PO_4^-, Cu, Mo, Co, and Zn (see also Hendricks and White, Chapter 9 in this volume). Sorption of phosphate (PO_4^-) by iron oxides is of particular interest since algal production in many streams is limited by phosphorus availability. In some streams, iron oxides may serve as nutrient reservoirs whereby the limiting nutrient is released upon reduction of the iron oxide by bacteria or light (McKnight *et al.*, 1988; Tate *et al.*, 1995). Similarly, anaerobic processes in the subsurface may liberate sorbed phosphate, increasing phosphorus availability for algal production in upwelling zones (Hendricks and White, Chapter 9). The role of metal oxidation and reduction (particularly of iron) in biogeochemical phosphorus cycling within the subsurface and the subsequent influence on benthic primary production in streams has been explored very little compared to sediment phosphorus cycling in lentic ecosystems. The significance of these oxidative processes as energy sources and as sites for adsorption of nutrients (e.g., PO_4-P, NH_4-N) in lotic ecosystems is worth considering beyond the extensive work examining these processes in acid mine drainage catchments (e.g., Lind and Hem, 1993; McKnight *et al.*, 1992; Tate *et al.*, 1995).

VI. CONCLUSIONS AND FUTURE RESEARCH DIRECTIONS

The biogeochemical and ecological effects of anoxia in the stream-water–groundwater interface have rarely been addressed. The limited data showing measurable changes in concentrations of redox-sensitive chemical species such as ferrous iron, manganous manganese, sulfate, and methane indicate that anaerobic metabolism is not uncommon in the stream-water–groundwater interface of many lotic ecosystems. Rates of metal reduction, sulfate reduction, and methanogenesis are likely to be high in finer-grained alluvial sediments, where water residence time is increased, organic carbon availability is often higher, and microbial densities and activities are high.

Terminal electron-accepting processes in bottom sediments of lentic ecosystems are distributed over relatively short distances (e.g., millimeter to centimeter scales). Because transport in these systems primarily is via diffusion, the spatial segregation of terminal electron-accepting processes is relatively stable over time (Carlton *et al.*, 1989; Sweerts *et al.*, 1991). It may be the case that redox reactions in the stream-water–groundwater interface also will be spatially segregated along flowpaths (e.g., Fig. 1; Champ *et al.*, 1979). This zonation is likely to occur on a larger spatial scale relative to lentic ecosystems and to be quite responsive to discharge variability owing to advection of water and solutes through alluvial sediments. Furthermore, multiple oxic–anoxic interfaces may occur within the same flowpath because alluvial sediments are deposited in layers with different characteristics (Fig. 2). Determinations of the extent, scale, and temporal dynamics of terminal electron-accepting process zonation and oxic–anoxic interfaces constitute a major challenge in defining the three-dimensional structure of stream ecosystems.

Given that anaerobic metabolic processes occur the stream-water–groundwater interface of many streams, and that much of the respiration at reach and catchment scales may be attributed to subsurface organisms (Grimm and Fisher, 1984; Mulholland *et al.*, 1997), an important question to be addressed is: what is the contribution of these anaerobic processes to total respiration and energy flow in streams? Anaerobic processes may be an important component of organic matter oxidation in stream ecosystems when the underlying alluvium is considered. Measurements of ecosystem metabolism based on changes in dissolved oxygen concentration may underestimate actual respiration rates when anaerobic processes are important. To date, total anaerobic metabolism and the importance of individual anaerobic terminal electron-accepting processes have not been compared to total stream ecosystem metabolism. Anaerobic metabolism of organic carbon in alluvial sediments may well be an important component of total ecosystem metabolism for some streams.

Finally, the contribution of chemoautotrophic processes involving the oxidation of reduced products of anaerobic metabolism (e.g., nitrification, metal and sulfur oxidation, and methanotrophy) at oxic–anoxic interfaces in streams should be quantified in stream energy budgets. These processes function to recapture chemical energy that would otherwise be lost from the ecosystem and may be an important component of total production in some streams.

REFERENCES

Allan, J. D. 1995. "Stream Ecology." Chapman & Hall, London.

American Public Health Association (APHA). 1992. "Standard Methods for the Examination of Water and Wastewater," 18th ed. American Water Works Association, Washington, DC.

Baker, M. A. 1998. Organic carbon retention and metabolism in near-stream groundwater. Ph. D. Dissertation, University of New Mexico, Albuquerque.

Baker, M. A., C. N. Dahm, H. M. Valett, J. A. Morrice, K. S. Henry, M. E. Campana, and G. J. Wroblicky. 1994. Spatial and temporal variation in methane distribution at the ground water–surface water interface in headwater catchments. *In* "Proceedings of the Second International Conference on Ground Water Ecology." (J. A. Stanford and H. M. Valett, eds.), pp. 29–37. American Water Resources Association, Herndon, MD.

Brick, C. M., and J. N. Moore. 1996. Diel variation of trace metals in the Upper Clark Fork River, Montana. *Environmental Science and Technology* 30:1953–1960.

Carlton, R. G., G. S. Walker, M. J. Klug, and R. G. Wetzel. 1989. Relative values of oxygen, nitrate, and sulfate to terminal microbial processes in the sediments of Lake Superior. *Journal of Great Lakes Research* 15:133–140.

Carr, D. L. 1989. Nutrient dynamics of stream and interstitial waters of three first-order streams in New Mexico. M.S. Thesis, University of New Mexico, Albuquerque.

Champ, D. R., J. Gulens, and R. E. Jackson. 1979. Oxidation-reduction sequences in ground water flow systems. *Canadian Journal of Earth Science* 16:12–23.

Chapelle, F. H. 1993. "Ground-Water Microbiology and Geochemistry." Wiley, New York.

Chapelle, F. H., and P. B. McMahon. 1991. Geochemistry of dissolved inorganic carbon in a coastal plain aquifer. 1. Sulfate from confining beds as an oxidant in microbial CO_2 production. *Journal of Hydrology* 127:85–108.

Chapelle, F. H., P. B. McMahon, and N. M. Dubrovsky. 1995. Deducing the distribution of terminal electron-accepting processes in hydrologically diverse groundwater systems. *Water Resources Research* 31:359–371.

Chapelle, F. H., P. M. Bradley, D. R. Lovley, and D. A. Vroblesky. 1996. Measuring rates of biodegradation in a contaminated aquifer using field and laboratory methods. *Ground Water* 34:691–698.

Cline, J. D. 1969. Spectrophotometric determination of hydrogen sulfide in natural waters. *Limnology and Oceanography* 14:454–458.

Dahm, C. N. 1981. Pathways and mechanisms for removal of dissolved organic carbon from leaf leachate in streams. *Canadian Journal of Fisheries and Aquatic Sciences* 38:68–76.

Dahm, C. N., E. H. Trotter, and J. R. Sedell. 1987. Role of anaerobic zones in stream ecosystem productivity. *In* "Chemical Quality of Water and the Hydrologic Cycle" (R. C. Averett and D. M. McKnight, eds.), pp. 157–178. Lewis Publishers, Chelsea, MI.

Dahm, C. N., D. L. Carr, and R. L. Coleman. 1991. Anaerobic carbon cycling in stream ecosystems. *Vereinigung für theoretische und angewandte Limnologie* 24:1600–1604.

deAngelis, M. A., and M. D. Lilley. 1987. Methane in surface waters of Oregon estuaries and rivers. *Limnology and Oceanography* 32:716–722.

Devito, K. J., and P. J. Dillon. 1993. Importance of runoff and winter anoxia to the P and N dynamics of a beaver pond. *Canadian Journal of Fisheries and Aquatic Sciences* 50:2222–2234.

Dodds, W. K., C. A. Randel, and C. C. Edler. 1996. Microcosms for aquifer research: Application to colonization of various sized particles by ground-water microorganisms. *Ground Water* 34(4):756–759.

Drever, J. I. 1988. "The Geochemistry of Natural Waters," 2nd ed. Prentice Hall, Englewood Cliffs, NJ.

Fetter, C. W. 1988. "Applied Hydrogeology," 2nd ed. Macmillan, New York.

Findlay, S. 1995. Importance of surface-subsurface exchange in stream ecosystems: The hyporheic zone. *Limnology and Oceanography* **40**:159–164.

Freeman, C., and M. A. Lock. 1995. The biofilm polysaccharide matrix: A buffer against changing organic substrate supply? *Limnology and Oceanography* **40**:273–278.

Ghani, B., M. Takai, N. Z. Hisham, N. Kishimoto, A. K. Ismail, T. Tano, and T. Sugio. 1993. Isolation and characterization of a Mo^{6+}-reducing bacterium. *Applied and Environmental Microbiology* **59**:1176–1180.

Grimm, N. B., and S. G. Fisher. 1984. Exchange between interstitial and surface water: Implications for stream metabolism and nutrient cycling. *Hydrobiologia* **111**:219–228.

Haadas, O., and R. Pinkas. 1995. Sulfate reduction processes in sediments at different sites in Lake Kinneret, Israel. *Microbial Ecology* **30**:55–66.

Hale, J. M. 1983. The action of hydrogen sulfide on polarographic oxygen sensors. *In* "Polarographic Oxygen Sensors, Aquatic and Physiological Applications" (E. Gnaiger and H. Forstner, eds.), 1st ed., pp. 73–75. Springer-Verlag, New York.

Hedin, L. O., J. C. von Fischer, N. E. Ostrom, B. P. Kennedy, M. G. Brown, and G. P. Robertson. 1998. Thermodynamic constraints on nitrogen transformations and other biogeochemical processes at soil-stream interfaces. *Ecology* **79**:684–703.

Holmes, R. M., S. G. Fisher, and N. B. Grimm. 1994. Parafluvial nitrogen dynamics in a desert stream ecosystem. *Journal of the North American Benthological Society* **13**:469–478.

Hope, D., J. J. C. Dawson, M. S. Cresser, and M. F. Billet. 1995. A method for measuring free CO_2 in upland streamwater using headspace analysis. *Journal of Hydrology* **166**:1–14.

Ivanov, M. V., N. V. Pimenov, A. S. Savvichev, A. Y. Opekunov, and M. E. Bart. 1995. Microbial processes of sulfide production in the Pregolya River (Kaliningrad). *Microbiology* **64**:112–118.

Jones, J. B. 1995. Factors controlling hyporheic respiration in a desert stream. *Freshwater Biology* **34**:91–99.

Jones, J. B., R. M. Holmes, S. G. Fisher, and N. B. Grimm. 1994. Chemoautotrophic production and respiration in the hyporheic zone of a Sonoran Desert stream. *In* "Proceedings of the Second International Conference on Ground Water Ecology" (J. A. Stanford and H. M. Valett, eds.), pp. 329–338. American Water Resources Association, Herndon, MD.

Jones, J. B., S. G. Fisher, and N. B. Grimm. 1995a. Vertical hydrologic exchange and ecosystem metabolism in a Sonoran Desert stream. *Ecology* **76**:942–952.

Jones, J. B., R. M. Holmes, S. G. Fisher, N. B. Grimm, and D. M. Greene. 1995b. Methanogenesis in Arizona, USA dryland streams. *Biogeochemistry* **31**:155–173.

Kelson, K. I., and S. G. Wells. 1989. Geologic influences on fluvial hydrology and bedload transport in small mountainous watersheds, New Mexico, USA. *Earth Surface Processes and Landforms* **14**:671–690.

Kling, G. W., G. W. Kipphut, and M. C. Miller. 1992. The flux of CO_2 and CH_4 from lakes and rivers in arctic Alaska. *Hydrobiologia* **240**:23–36.

Korom, S. F. 1992. Natural denitrification in the saturated zone: A review. *Water Resources Research* **28**:1657–1668.

Lind, C. J., and J. D. Hem. 1993. Manganese minerals and associated fine particles in the streambed of Pinal Creek, Arizona, USA: a mining-related acid drainage problem. *Applied Geochemistry* **8**:67–80.

Lock M. A., R. R. Wallace, J. W. Costerson, R. M. Ventullo, and S. E. Charlton. 1984. River epilithon: Toward a structural-functional model. *Oikos* **42**:10–22.

Lovley, D. R. 1991. Dissimilatory Fe(III) and Mn(IV) reduction. *Microbiological Reviews* **55**:259–287.

Lovley, D. R., and S. Goodwin. 1988. Hydrogen concentrations as an indicator of the predominant terminal electron-accepting reactions in aquatic sediments. *Geochimica et Cosmochimica Acta* **52**:2993–3003.

Lovley, D. R., and M. J. Klug. 1983. Sulfate reducers can outcompete methanogens at freshwater sulfate concentrations. *Applied and Environmental Microbiology* **45**:187–192.

Lovley, D. R., E. J. P. Phillips, Y. A. Gorby, and E. R. Landa. 1991a. Microbial reduction of uranium. *Nature (London)* **350**:413–416.

Lovley, D. R., E. J. P. Phillips, and D. J. Lonergan. 1991b. Enzymatic versus nonenzymatic mechanisms for Fe(III) reduction in aquatic sediments. *Environmental Science and Technology* **25**:1062–1067.

Markwiese, J. T., J. S. Meyer, and P. J. S. Colberg. 1998. Copper tolerance in iron reducing bacteria: implications for copper mobilization in aquatic sediments. *Environmental Toxicology and Chemistry* **17**:675–678.

McDowell, W. H., and T. Wood. 1984. Podzolization: Soil properties control DOC concentrations in stream water. *Soil Science* **37**:23–32.

McKnight, D. M., B. A. Kimball, and K. E. Bencala. 1988. Iron photoreduction and oxidation in an acidic mountain stream. *Science* **240**:365–375.

McKnight, D. M., R. L. Wershaw, K. E. Bencala, G. W. Zellweger, and G. C. Feder. 1992. Humic substances and trace metals associated with Fe and Al oxides deposited in an acidic mountain stream. *Science of the Total Environment* **118**:485–498.

Morrice, J. A., H. M. Valett, C. N. Dahm, and M. E. Campana. 1997. Alluvial characteristics, groundwater-surface water exchange, and hydrologic retention in headwater streams. *Hydrological Processes* **11**:253–267.

Morrice, J. A. 1997. Influences of stream-aquifer interactions on nutrient cycling in streams. Ph.D. Dissertation. University of New Mexico, Albuquerque.

Mulholland, P. J., E. R. Marzolf, J. R. Webster, and D. R. Hart. 1997. Evidence that hyporheic zones increase heterotrophic metabolism and phosphorus uptake in forest streams. *Limnology and Oceanography* **42**:443–451.

Naiman, R. J., G. Pinay, C. A. Johnston, and J. Pastor. 1994. Beaver influences on the long-term biogeochemical characteristics of boreal forest drainage networks. *Ecology* **75**:905–921.

Oremland, R. S., J. T. Hollibaugh, A. S. Maest, T. S. Presser, L. G. Miller, and C. W. Culbertson. 1989. Selenate reduction to elemental selenium by anaerobic bacteria in sediments and culture: biogeochemical significance of a novel, sulfate independent respiration. *Applied and Environmental Microbiology* **55**:2333–2345.

Pimenov, N. V., I. A. Davydova, I. I. Rusanov, A. Yu. Lein, S. S. Belyaev, and M. V. Ivanov. 1994. Microbial processes of the carbon and sulfur cycles in bottom sediments of the Congo River discharge zone. *Microbiology* **63**:198–206.

Postgate, J. R. 1984. "The Sulphate-Reducing Bacteria," 2nd ed. Cambridge University Press, New York.

Potsma, D., and R. Jakobsen. 1996. Redox zonation: Constraints on the Fe(III)/SO_4 reduction interface. *Geochimica et Cosmochimica Acta* **60**:3169–3175.

Pusch, M. 1996. The metabolism of organic matter in the hyporheic zone of a mountain stream, and its spatial distribution. *Hydrobiologia* **323**:107–118.

Pusch, M., and J. Schwoerbel. 1994. Community respiration in hyporheic sediments of a mountain stream (Steina, Black Forest). *Archiv für Hydrobiologie* **130**:35–52.

Roden, E. E., and D. R. Lovley. 1993. Evaluation of ^{55}Fe as a tracer of Fe(III) reduction in aquatic sediments. *Geomicrobiology Journal* **11**:49–56.

Schulz, S., H. Matsuyama, and R. Conrad. 1997. Temperature dependence of methane production from different precursors in a profundal sediment (Lake Constance). *FEMS Microbiology Ecology* **22**:207–213.

Sheldon, S. P., and D. K. Skelly. 1990. Differential colonization and growth of algae and ferro-manganese-depositing bacteria in a mountain stream. *Journal of Freshwater Ecology* **5**:475–485.

Smith, R. L., B. L. Howes, and S. P. Garabedian. 1991. In situ measurement of methane oxidation in groundwater by using natural-gradient tracer tests. *Applied and Environmental Microbiology* **57**:1997–2004.

Smith, R. L., S. P. Garabedian, and M. H. Brooks. 1996. Comparison of denitrification activity measurements in groundwater using cores and natural-gradient tracer tests. *Environmental Science and Technology* **30**:3448–3456.

Sode, A. 1983. Effect of ferric hydroxide on algae and oxygen consumption by sediment in a Danish stream. *Archiv für Hydrobiologie, Supplementband* **65**:134–162.

Stookey, L. L. 1970. Ferrozine—A new spectrophotometric reagent for iron. *Analytical Chemistry* **42**:779–781.

Stumm, W., and J. J. Morgan. 1996. "Aquatic Chemistry: Chemical Equilibria and Rates in Natural Waters," 3rd ed. Wiley, New York.

Sweerts, J. -P. R. A., M. -J. Bar-Gilissen, A. A. Cornelese, and T. E. Cappenberg. 1991. Oxygen consuming processes at the profundal and littoral sediment-water interface of a small meso-eutrophic lake (Lake Vechten, The Netherlands). *Limnology and Oceanography* **36**:1124–1133.

Tate, C. M., R. E. Broshears, and D. M. McKnight. 1995. Phosphate dynamics in an acidic mountain stream: Interactions involving algal uptake, sorption by iron oxide, and photoreduction. *Limnology and Oceanography* **40**:938–946.

Triska, F. J., V. C. Kennedy, R. J. Avanzino, G. W. Zellweger, and K. E. Bencala. 1989. Retention and transport of nutrients in a third-order stream in northwestern California: Hyporheic processes. *Ecology* **70**:1893–1903.

Triska, F. J., J. H. Duff, and R. J. Avanzino. 1993. Patterns of hydrological exchange and nutrient transformation in the hyporheic zone of a gravel-bottom stream: Examining terrestrial-aquatic linkages. *Freshwater Biology* **29**:259–274.

Urban, N. R., P. L. Brezonik, L. A. Baker, and L. A. Sherman. 1994. Sulfate reduction and diffusion in sediments of Little Rock Lake, Wisconsin. *Limnology and Oceanography* **39**:797–815.

Valett, H. M., S. G. Fisher, and E. H. Stanley. 1990. Physical and chemical characteristics of the hyporheic zone of a Sonoran Desert stream. *Journal of the North American Benthological Society* **9**:201–215.

Valett, H. M., S. G. Fisher, N. B. Grimm, and P. Camill. 1994. Vertical hydrologic exchange and ecological stability of a desert stream ecosystem. *Ecology* **75**:548–560.

Valett, H. M., J. A. Morrice, C. N. Dahm, and M. E. Campana. 1996. Parent lithology, surface-groundwater exchange, and nitrate retention in headwater streams. *Limnology and Oceanography* **41**:333–345.

von Gunten, H. R., and C. Lienert. 1993. Decreased metal concentrations in ground water caused by controls of phosphate emissions. *Nature (London)* **364**:220–222.

von Gunten, H. R., G. Karametaxas, U. Krähenbül, M Kuslys, R. Giovanoli, E. Hoehn, and R. Keil. 1991. Seasonal biogeochemical cycles in riverborne groundwater. *Geochimica et Cosmochimica Acta* **55**:3597–3609.

von Gunten, H. R., G. Karametaxas, and R. Keil. 1994. Chemical processes in infiltrated riverbed sediments. *Environmental Science and Technology* **28**:2087–2093.

Vroblesky, D. A., and F. H. Chapelle. 1994. Temporal and spatial changes of terminal electron-accepting processes in a petroleum hydrocarbon contaminated aquifer and the significance for contaminant biodegradation. *Water Resources Research* **30**:1561–1570.

Wellnitz, T. A., and S. P. Sheldon. 1995. The effects of iron and manganese on diatom colonization in a Vermont stream. *Freshwater Biology* **34**:465–470.

Wellnitz, T. A., K. A. Grief, and S. P. Sheldon. 1994. Response of macroinvertebrates to blooms of iron-depositing bacteria. *Hydrobiologia* **281**:1–17.

Wetzel, R. G., and G. E. Likens. 1991. "Limnological Analyses." Springer-Verlag, New York.

Wroblicky, G. J., M. E. Campana, H. M. Valett, and C. N. Dahm. 1998. Seasonal variation in surface-subsurface water exchange and lateral hyporheic area of two stream-aquifer systems. *Water Resources Research* **34**:317–328.

Yavitt, J. B., L. L. Angell, T. J. Fahey, C. P. Cirmo, and C. T. Driscoll. 1992. Methane fluxes, concentrations and production in two Adirondack beaver impoundments. *Limnology and Oceanography* **37**:1057–1066.

SECTION *THREE*

*ORGANISMAL
ECOLOGY*

Microbial Communities in Hyporheic Sediments

Stuart Findlay and William V. Sobczak

Institute of Ecosystem Studies
Millbrook, New York

I. Introduction   287
II. Physical and Chemical Environment   288
III. The Organisms   290
    A. Fungi   290
    B. Protozoa   291
    C. Bacteria   292
IV. Respiration   293
V. Carbon Supply   295
VI. Alternate Controls on Hyporheic Bacteria   300
    A. Sediment Size   300
    B. Inorganic Chemistry   300
    C. Grazing   301
VII. Community Composition   301
VIII. Conclusions and Research Needs   302
    References   304

I. INTRODUCTION

Microbial communities catalyze an array of biogeochemical processes integral to carbon fluxes and nutrient cycling in sediments of aquatic ecosystems (Chapelle, 1993; Ford, 1993). The microbial loop (Pomeroy, 1974) suggests that bacteria in planktonic systems assimilate dissolved organic carbon (POC) making it available to particle feeders at higher trophic levels. In sediments, the direct contribution of microbial biomass to higher trophic levels depends on the rate of new biomass production, as well as the efficiency of the consumer (Findlay *et al.,* 1986; Hall, 1995). Heterotrophic microorganisms, such as bacteria and fungi, play a fundamental role in decomposition of particulate and dissolved organic matter in sediments because of their high metabolic rates and diverse array of carbon-degrading enzymes. Given that microbes regulate key processes in surface sediments of streams, it seems almost certain that they will also dominate many aspects of material and energy flow in hyporheic sediments. It is also likely that bacteria and fungi represent the vast majority of hyporheic biodiversity. Knowledge of hyporheic microbial communities and the processes they control remains limited, although there has been considerable progress over the past decade (Hendricks, 1993; Findlay, 1995; Jones and Holmes, 1996).

We define the hyporheic zone to be the saturated sediments found beneath and lateral to the stream channel in which there is active two-way exchange of surface and interstitial water. The relatively small number of studies which specifically examine the microbial ecology of hyporheic zones warrants a more inclusive scope of review, so many of our comparisons are based upon information from surface sediments of streams and other aquatic ecosystems.

Microbial ecology is the science that examines the relationships between microorganisms and their natural biotic and abiotic environments (Atlas and Bartha, 1998). Microbial ecology usually includes all organisms less than about 100 μm including bacteria, fungi, protozoans, viruses, and sometimes algae and micrometazoans, although criteria vary (see also Hakencamp and Palmer, Chapter 13 in this volume). Microbial ecology is considered a subdiscipline of microbiology, although most practitioners maintain strong links to both general ecology and biogeochemistry. Unfortunately, microbial ecology is commonly limited by suitable and comparable techniques, particularly in sediments. Recent methods-oriented textbooks (Kemp *et al.*, 1993; Hauer and Lamberti, 1996; Hurst *et al.*, 1997) have addressed this concern, yet methods for sampling and describing microbes vary among environmental disciplines for historical, conceptual, and pragmatic reasons. In general, less is known about the microbial ecology of stream ecosystems compared to the plankton of lentic aquatic ecosystems (Leff, 1994), and the microbial ecology of hyporheic zones is poorly understood relative to other stream habitats.

In this chapter, we (1) review the available information on microbial abundances and activities, (2) discuss environmental controls on hyporheic bacterial abundances and activities since there are more data for bacteria than other microbes, (3) compare the microbial activities of hyporheic sediments with general relationships derived from broad-scale studies of aquatic sediments, (4) examine carbon supply to hyporheic microbes, and (5) synthesize our knowledge and highlight areas that seem ripe for progress.

II. PHYSICAL AND CHEMICAL ENVIRONMENT

Most microbes in hyporheic sediments are associated with organic particles or biofilms coating inorganic sediment particles (as shown in Fig. 1).

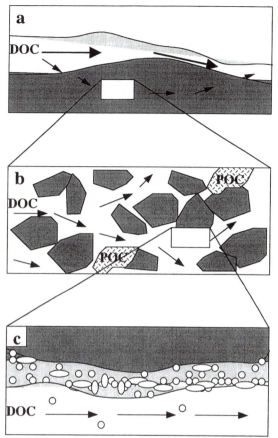

FIGURE 1   Conceptual diagram of hyporheic biofilms. (a) Cross section of stream riffle with hyporheic flowpath highlighted. Stream surface water is the source of dissolved organic carbon (DOC). (b) Close-up of hyporheic flowpath at the scale of interstitial flow. Buried particulate organic carbon (POC) and DOC are highlighted. (c) Close-up of hyporheic flowpath at the scale of an interstitial biofilm. Unshaded shapes represent bacteria imbedded in a biofilm matrix.

By associating with surfaces rather than moving freely with interstitial water, these microbes can elaborate exoenzymes to release assimilable compounds from complex organic matter dissolved in interstitial water or associated with particles. Biofilms can be highly organized with distinct vertical layering of diverse microbial communities (Stevenson, 1997) and may temporarily buffer embedded bacteria from fluctuating resource supply (Freeman and Lock, 1995). Relatively large pores may penetrate all sections of biofilms and influence the movement of dissolved constituents in and out of biofilms (Lock, 1993). Bacteria and other microorganisms in solution may not be representative of the organisms embedded in interstitial biofilms (Madsen and Ghiorse, 1993).

Although life in a biofilm confers many advantages to its microbial constituents (Meyer-Reil, 1994), it also means that the external milieu is controlled by the rate of water flow (or diffusion) resupplying or removing dissolved gases and solutes. Factors affecting hydraulic conductivity and interstitial velocities in hyporheic sediments will consequently influence the renewal of oxygen and alternate terminal electron acceptors (Baker *et al.*, Chapter 11 in this volume). In addition, heterotrophic biofilms rely on external energy sources, such as reduced carbon substrates (Kaplan and Newbold, Chapter 10 in this volume), and essential inorganic nutrients (Duff and Triska, Chapter 8 in this volume; Hendricks and White, Chapter 9 in this volume). Besides affecting resupply of energy and materials, the grain size of particles and porosity at a site will also influence accessibility to meiofauna and macrofauna, which may feed upon accumulated biofilms. Hence, physical characteristics govern surface area for biofilm formation and greatly influence both resource availability and predation pressure to biofilms.

III. THE ORGANISMS

In many ecological studies, microorganisms are separated into morphologically distinct groups which represent different steps in the trophic transfer of energy. These groups frequently include heterotrophic bacteria, fungi, protozoa (e.g., ciliates and flagellates), and autotrophs (e.g., cyanobacteria, diatoms, and other algae). It is important to recognize that such "ecological" groupings have considerable phylogenetic overlap, but these classifications have proven useful in studies of planktonic marine and freshwater environments.

A. Fungi

Fungi are known to be a significant and critical component of the microbial community in leaf packs, wood surfaces, surface sediments, and epilithic biofilms (Lock, 1993; Sinsabaugh and Findlay, 1995; Suberkropp, 1997), but little is known about their role in hyporheic zones. To date, microbial studies of hyporheic zones have largely excluded fungi even though fungi were included in one of the earliest investigations of hyporheic microbes (Bärlocher and Murdoch, 1989). Bärlocher and Murdoch (1989) found fungal mycelium to be prominent on glass beads incubated in hyporheic sediments. Length of fungal mycelium was greatest in the winter, possibly due to low water temperatures. Improved methods for determining fungal biomass and turnover should help to further research in this area (Suberkropp and Weyers, 1996). Fungal biomass may be significant in hyporheic zones that have been recently disturbed and contain freshly buried leaf or wood material. It seems likely that the aquatic hyphomycetes so prevalent on surface leaf litter may not tolerate stagnant conditions and low dissolved oxygen concentrations often found in hyporheic sediments; however, the paucity of hyporheic studies that examine fungal biomass make this claim speculative.

B. Protozoa

Protozoa, unicellular eukaryotic microorganisms that lack cell walls, are generally thought to be an important component of microbial food webs and a potential intermediate between bacteria and larger invertebrates in the trophic transfer of energy. However, very little is known about the composition, abundance, and activity of protozoa in hyporheic zones. Protozoa are routinely partitioned into ciliates and flagellates (Bott, 1996), although more refined functional fractionation is possible (Pratt and Cairns, 1985). Unfortunately, few studies have attempted to quantify the abundance and production of protozoans in stream sediments. In one of the few studies, flagellate, ciliate and amoebae abundance, biomass, and production were measured in chalk stream sediments in the United Kingdom (Sleigh *et al.*, 1992). Ciliates contributed 75% of the production (flagellates 15%, amoebae 10%). Flagellate abundance was 8×10^3 cm^{-3}, ciliate abundance was 3.5×10^3 cm^{-3}, and amoebae abundance was 2.5×10^3 cm^{-3}. In sediments of White Clay Creek, PA, ciliate abundance was 9.1×10^3 cm^{-3}, and microflagellate abundance was 1.3×10^6 cm^{-3} (at 20°C; Bott and Kaplan, 1989). In a recent review of microbes in food webs, Bott (1996) summarized protozoan abundance estimates in White Clay Creek. Microflagellate densities are routinely two orders of magnitude greater than ciliate densities (10^6 versus 10^4 organisms cm^{-3}; Bott, 1996). More broadly, meiofauna in stream sediments are poorly characterized, even though significant ecological importance is speculated (Hakencamp and Palmer, Chapter 13 in this volume; Bott, 1995).

Few studies have examined bacterivory by protozoa (i.e., grazing) in hyporheic sediments. In a chalk stream in the United Kingdom, protozoan grazing removed 20% of bacterial production (Sleigh *et al.*, 1992). In White Clay Creek, ciliates and flagellates in surface sediments were estimated to remove 21–48 g bacterial C m^{-2} y^{-1} (Bott and Kaplan, 1990); ciliates and flagellates consumed an estimated 213 and 12 bacteria h^{-1}, respectively, in grazing studies conducted at 20°C, resulting in consumption of 80–183% of estimated annual bacterial productivity. Hence, under certain conditions, protozoa can be an important consumer of bacteria and may be a trophic intermediate between bacteria and macrofauna. Research by Bott and Kaplan (1989) on protozoan abundance in the field coupled with laboratory grazing rates (Bott and Kaplan 1990) is the model for arriving at accurate estimates of bacterivory by protozoa.

C. Bacteria

Bacterial abundance has become a routine measure in investigations of aquatic ecosystems (Sander and Kalff, 1993; Schallenberg and Kalff, 1993), but estimates of bacterial abundance in hyporheic zones are rare. In Ontario, Canada, bacterial densities were between 2.7×10^5 and 2.4×10^7 cm^{-2} on glass beads incubated in hyporheic sediments (Bärlocker and Murdoch, 1989). Decreases in bacterial abundance have been observed in porewater along hyporheic flowpaths, presumably in response to decreases in dissolved organic matter (Findlay *et al.*, 1993). Findlay *et al.* (1993) reported bacterial abundance in interstitial water declines from 1.2×10^5 to 0.5×10^5 cells cm^{-3}, although these abundances do not account for the attached bacteria. More recent research has examined bacterial abundance at multiple depths and locations within streams (Fischer et al., 1996; Hendricks, 1996). In the Steina, a third-order mountain stream in southern Germany, bacterial abundance was heterogeneous but averaged 7.2×10^8 cm^{-3} (Fischer *et al.*, 1996). Abundance data were converted to biomass and averaged 7.2 μgC cm^{-3} at 40 cm in depth. Overall, bacterial abundance and biomass declined with sediment depth (from 10 to 40 cm) in the Steina. Similarly, bacterial abundance and biomass declined with depth (from 10 to 50 cm) in a riffle in the Maple River, Michigan (Hendricks, 1996). If values from all cores are pooled; bacterial biomass averaged 21 μgC cm^{-3} in the shallow sediments of the downwelling zone, 0.3 μgC cm^{-3} in the deep sediments of the downwelling zone, 1.3 μgC cm^{-3} in the shallow sediments of the upwelling zone, and 0.2 μgC cm^{-3} in the deep sediments of the upwelling zone (Hendricks, 1996). In an interstitial seep of a first-order stream (Dryman Fork) at the Coweeta Hydrologic Laboratory, North Carolina, benthic bacterial biomass ranged from 4 to 24 μgC cm^{-3} (Crocker and Meyer, 1987). Bacterial biomass in unmanipulated sediments averaged 24 μgC cm^{-3} at the upstream site and 18 μgC cm^{-3} at the downstream site. Bacterial abundance in the hyporheic zone of a mountain stream in Austria was 7.2×10^8 cm^{-3} (Bretschko, 1991). Lastly, bacteria have been examined at the interface between the Rhône River and its alluvial aquifer (Lyon, France; Marmonier *et al.*, 1995). This is not a natural hyporheic zone because the downwelling gradient results from mechanical pumping of interstitial water 80 m from the river. However, bacterial abundance decreased from approximately 40×10^7 cells cm^{-3} in shallow sediments (20 cm in depth) to approximately 10×10^7 cells cm^{-3} in deeper sediments (100 cm in depth) (assuming 1 cm^3 sediment weighs 2 g dry mass; Marmonier *et al.*, 1995). Overall, bacterial abundance and biomass decline with sediment depth and distance from downwelling zone.

Relationships between bacterial biomass and environmental parameters have been examined in marine and freshwater ecosystems (Sander and Kalff, 1993; Schallenberg and Kalff, 1993). Sander and Kalff (1993) review bacterial biomass in sediments from numerous aquatic ecosystems and report areal means for different surface sediments (marine: 380 μgC cm^{-2}, lake: 210 μgC cm^{-2}, river: 40 μgC cm^{-2}). In an examination of 22 lakes in Ontario, Canada, mean bacterial biomass was 99 μgC cm^{-3} (Schallenberg and Kalff, 1993). Hyporheic bacterial biomass ranges from approximately 1 to 50 μgC cm^{-3} and appears lower than sediments from other aquatic ecosystems, possibly due to lower sediment organic content. However, additional factors may contribute to this observation: (1) inadequate number of estimates, (2) larger average grain size, thus less surface area, (3) greater grazing pressure, and (4) lower production.

Bacterial production in hyporheic sediments has been measured in only a handful of studies. Rates vary broadly, from undetectable (Hendricks, 1996) to 800 ngC cm^{-3} h^{-1} (Marxsen, 1996). In the Maple River, Michigan, bacterial production was greater in a downwelling area relative to an upwelling area (Hendricks, 1996), although "within zone" variability was large (range: 0–0.33 ngC cm^{-3} h^{-1}). In addition, bacterial production was generally greater in the shallow sediments relative to the deeper sediments. In the Breitenbach, a first-order stream with sandy sediments in central Germany, bacterial production was estimated at approximately 230 ngC cm^{-3} h^{-1} (Marxsen, 1996). Although Marxsen (1996) does not specify these sediments as "hyporheic," he does acknowledge the influence of perfusion velocity and the vertical distribution of bacteria. Production was greater in recirculating cores with slower perfusion velocities, although production estimates are based on cores that had natural velocity. Production decreased markedly with both depth and organic content (Marxsen, 1996); surface sediments had production estimates that averaged approximately 800 ngC cm^{-3} h^{-1}, whereas deeper sediments averaged approximately 200 ngC cm^{-3} h^{-1}. Bacterial production correlated with sediment organic matter. In surface sediments in White Clay Creek, Pennsylvania, bacterial production has been estimated in a variety of microbial food web studies and averages approximately 430 ngC cm^{-3} h^{-1} (Bott, 1995). In general, hyporheic productivity appears lower than productivity in other aquatic sediments (Fig. 2), probably because these sediments have lower organic content than other environments in the data set. Interestingly, production per unit organic matter is not lower than predicted by the general regression suggesting that subsurface organic matter is not necessarily of poor quality. These findings suggest hyporheic bacterial productivity is at least a partial function of the amount of sediment organic matter; however, other factors may contribute to a relationship between bacterial productivity and organic matter.

IV. RESPIRATION

The potential importance of hyporheic respiration to whole-stream respiration was first recognized in Sycamore Creek, Arizona (Grimm and Fish-

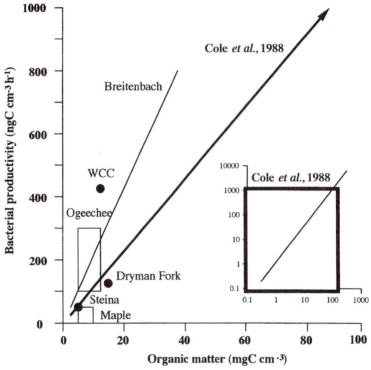

FIGURE 2   Hyporheic bacterial productivity from different streams related to sediment organic matter. The regression from Cole *et al.* (1988) covering a wide range of aquatic sediments is shown for comparison (see insert). We focus on the lower end of the Cole et al. (1988) regression line in order to compare hyporheic sediments. Hyporheic data are from the following sources: Breitenbach (Marxsen, 1996), Maple River (Hendricks, 1996), Steina (Fischer *et al.*, 1996), Dryman Fork (Crocker and Meyer, 1987), and the Ogeechee River (Marxsen, 1988a). In cases where an average value could not be derived, a range is indicated with a box.

er, 1984). Respiration rates in Sycamore Creek sediments have been reported in several studies (Grimm and Fisher, 1984; Valett *et al.*, 1990; Jones *et al.*, 1995). Hyporheic respiration in Sycamore Creek, measured *in situ* at three independent downwelling and upwelling sites, ranged between 50 and 4400 μgO_2 L of sediment^{-1} h^{-1} (Jones *et al.*, 1995). Respiration at downwelling sites was much higher than at upwelling sites (1112 versus 466 μgO_2 L^{-1} h^{-1}), presumably due to greater dissolved organic matter availability from algal exudates. Buried particulate organic matter accounted for only a small portion of total respiration (15%) even though Sycamore Creek experiences frequent flash floods which are capable of depositing freshly buried particulate organic matter several times a year. Dissolved organic carbon (DOC) in hyporheic water was higher than that found in surface waters, although upwelling-zone DOC was lower than downwelling-zone DOC. In the hyporheic zone of the Steina, a mountain stream in southern Germany, mean respiration was estimated at $71 \mu gO_2 \ L^{-1} \ h^{-1}$ along a riffle-pool reach, and spatial variation in respiration correlated with sediment organic matter (Pusch, 1996). Respiration in the riffle exceeded that in the pool by approximately fivefold and the upper 10 cm in the riffle had the highest respiration rate. Earlier work in the Steina demonstrated that respiration was strongly temperature-dependent, related to protein content of organic matter, and dominated by organisms <100 μm (Pusch and Schwoerbel, 1994). This study reports a higher respiration rate of 330 $\mu gO_2 \ L^{-1} \ h^{-1}$ (Pusch and Schwoerbel, 1994; Pusch, 1996). In low-gradient blackwater streams in the southeastern United States, variation in benthic respiration correlates with sediment organic matter as well (Fuss and Smock, 1996). In the East Branch Wappinger Creek, a gravel bottom stream in the northeastern United States, respiration correlates with DOC loss along subsurface flowpaths (Findlay *et al.,* 1993; Findlay and Sobczak, 1996). In surface sediments in first- and second-order streams in the Hubbard Brook Experimental Forest, New Hampshire, respiration ranged between 29 and 379 $\mu gO_2 \ L^{-1} \ h^{-1}$ and was related to sediment organic matter (Hedin, 1990).

Data on hyporheic respiration are sparse, and the lack of standardized methodology and temporal coverage weaken, but do not preclude, general comparisons among hyporheic zones and between hyporheic sediments and surface sediments. For comparisons, we use the general relationship between respiration and sediment organic matter content in surface sediments presented by Hedin (1990) covering a range of stream types. Based on six separate studies, the respiration rates in hyporheic sediments are generally at least as high as those for surface sediments within the same range of sediment organic carbon (Fig. 3). Tentatively, we conclude that respiration in hyporheic sediments is not markedly different from respiration in surface sediments despite the presumed lower quality of older buried particulate organic carbon (POC) and lack of subsurface algal production. Similar to bacterial production in hyporheic sediments, respiration appears to be influenced by the supply of organic carbon.

V. CARBON SUPPLY

In surface sediments from a wide range of environments, significant positive relationships have been described between sediment organic matter and bacterial abundance (Schallenberg and Kalff, 1993), production (Cole *et al.,* 1988), and respiration (Hedin, 1990). Because organic matter supply appears to be a common control on bacteria in surface sediments, we first examine factors influencing carbon supply to hyporheic microbes. Alternative controlling factors are discussed later.

Supply of organic carbon to hyporheic sediments can occur through two

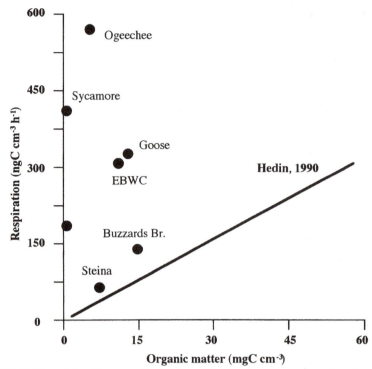

FIGURE 3 Hyporheic sediment respiration from different streams plotted as a function of sediment organic matter. The general regression from Hedin (1990) is shown for comparison. Data sources are: Ogeechee River (Marxsen, 1988a), Goose Creek (Hakencamp unpublished), East Branch Wappinger Creek (EBWC; Findlay and Sobczak, 1996), Sycamore Creek (downwelling and upwelling zone means; Jones *et al.*, 1995), Buzzards Branch (Fuss and Smock, 1996), and Steina (Pusch and Schwoerbel, 1994).

distinct (although by no means independent) mechanisms. (1) The DOC enters subsurface sediments at downwelling points (Findlay, 1995) or from groundwater inputs (Fiebig, 1995), and several investigators have documented its importance in affecting abundance (Hendricks, 1996), respiration (Jones *et al.*, 1995), and production (Findlay *et al.*, 1993). (2) The POC can be carried into sediments by water currents, traveling many centimeters into coarse sediments when transported at relatively rapid velocities (Huettel *et al.*, 1996). Resupply of freshly buried POC during floods (Metzler and Smock, 1990) is a more sporadic but probably quantitatively larger input of organic carbon.

Although it is tempting to view dissolved and particulate organic carbon as dichotomous (i.e., DOC versus POC) when considering the carbon supply to hyporheic microbes, in reality such a dichotomy is unlikely for several reasons: (1) the demarcation between DOC and POC is arbitrary (e.g.,

0.45 μm), (2) transformations between DOC and POC may occur (e.g., adsorbed or assimilated DOC may be considered POC), and (3) both DOC and POC are used by bacteria. Hence, the relevant question is not which is more important? but what controls the relative contribution of both?

We highlight factors affecting the contribution of DOC and POC to hyporheic microbes with an illustrative exercise. To draw quantitative comparisons between DOC and POC requires simplifying assumptions. For purposes of this exercise we assume that (1) DOC and POC are independent, (2) DOC is supplied as a unidirectional continuous input carried by downwelling stream water, and (3) POC is buried, measured as a standing stock, and not found in downwelling stream water. The supply of DOC to a given volume of subsurface sediment can be estimated as the flux of downwelling water multiplied by its DOC concentration. The flux of downwelling water through a given volume of sediment is a function of interstitial water velocity and porosity (i.e., volume water:volume sediment). Thus, interstitial water velocity multiplied by porosity and the amount of bioavailable DOC represents the maximal contribution or input of DOC to hyporheic microbes for a given volume of sediment per time (Fig. 4). For example, downwelling water carrying 10 mgC L^{-1} (50% of which is bioavailable) at a velocity of 1 m d^{-1} through a sediment cross section with 20% porosity provides 1 g C m^{-3} d^{-1} (Fig. 5a).

To estimate the potential contribution of POC, we can use turnover rates (i.e., percent loss per time) as a surrogate for rate of supply of POC. Thus percent loss per time multiplied by standing stock POC represents the "in-

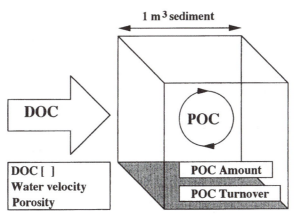

FIGURE 4 Conceptual diagram illustrating factors affecting carbon "supply" to a designated volume of hyporheic sediments. These factors are used in an exercise that examines the relative contributions of DOC and POC to hyporheic microbes. Simplifying assumptions are discussed in the text (see Section V).

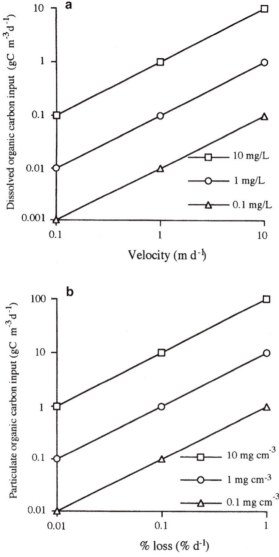

FIGURE 5 Numerical exercise illustrating relative contribution of DOC and POC supply to hyporheic microbes. (a) Input of DOC to a cubic meter of sediment estimated as a function of velocity and concentration assuming a fixed interstitial volume of 200 L (i.e., 20% porosity) and 50% bioavailability. (b) Input of POC within a cubic meter of sediment estimated as a function of standing stock in the sediment and turnover times (% loss d^{-1}).

put" of POC to hyporheic microbes for a given volume of sediment per time (Fig. 4). For example, a POC standing stock of 10^4 gC m^{-3} (= 10 mgC cm^{-3}, ~4% OM) decomposing at 0.1% d^{-1} provides 10 gC m^{-3} d^{-1} (Fig. 5). More complex models could fractionate DOC and POC pools (e.g., labile versus refractory), include transformations between DOC and POC (e.g., adsorbed DOC becoming POC), and include additional DOC vectors (e.g., ground water).

Using the simplifying assumptions discussed earlier, we compare the DOC carried into a cubic meter of sediment to the POC being degraded in that same volume using a realistic range of controlling factors such as water velocity and POC standing stock. These comparisons reveal a disparity between DOC and POC inputs (Fig. 5) suggesting that, based strictly on relative mass, POC should be the predominant source of carbon for hyporheic sediment microbial communities. In reality, qualitative factors seem to be important because correlations between respiration and bulk sediment organic matter are not always strong (Fischer *et al.,* 1996; Pusch, 1996), and correlations between respiration and DOC supply have been described (Jones *et al.,* 1995; Findlay and Sobczak, 1996). In addition, DOC adsorption and assimilation may account for a large fraction of POC in hyporheic sediments that rarely experience episodic burial of POC. Moreover, it is possible that POC and DOC supply will be negatively correlated because high-organic sediments will tend to have low hydraulic conductivity thereby minimizing inward fluxes of DOC.

To provide a context for these estimated carbon inputs, we can compare the supply terms to one of the loss terms, hyporheic respiration. The reported range in hyporheic respiration rates (expressed as C) is approximately 1–15 $gC\,m^{-3}\,d^{-1}$ (Fig. 3). These rates of carbon respiration could be supported either by relatively high DOC supply or almost any of the estimated rates of POC supply (Fig. 5). We have estimated previously that approximately half the respiration in subsurface sediments of the East Branch of the Wappinger Creek, New York, is supported by DOC, leaving half to be supported by POC (Findlay and Sobczak, 1996). One possible scenario is that a large fraction of baseline metabolism is supported by buried POC with variations in DOC triggering short-term fluctuations in excess of baseline (Vervier *et al.,* 1993). It should be kept in mind that DOC supply can be "accumulated" over some period of time by adsorption to sediment surfaces and biofilms. Hence, accumulated DOC may be metabolized sometime after the original input (Fiebig, 1995; Freeman and Lock, 1995).

Although we have focused on carbon supply, additional factors influence carbon availability to hyporheic microbes (Kaplan and Newbold, Chapter 10 in this volume). It is becoming increasingly clear that rates of respiration and production are influenced by carbon quality, as well as quantity. Bulk chemical characteristics of organic matter such as the degree of reduction (Vallino *et al.,* 1996) and elemental composition (Sun *et al.,* 1997) are useful predictors of absolute rates of growth as well as growth efficiency. In addition, compositional differences in stream DOC (Volk *et al.,* 1997) and groundwater DOC (Fiebig, 1997) pools can greatly influence bioavailability to sediment bacteria.

VI. ALTERNATE CONTROLS ON HYPORHEIC BACTERIA

Evidence to date suggests carbon supply frequently correlates with hyporheic bacterial biomass, production, and respiration; hence it appears likely that carbon supply is a major control on hyporheic bacteria. Although we have focused on carbon supply, numerous physical (e.g., sediment-size distribution), chemical (e.g., inorganic nutrients), and biological factors (e.g., grazing) may have varying direct influence on hyporheic bacteria in specific habitats. At a broader scale, these same physical, chemical, and biological factors may influence carbon supply and availability, suggesting that attempts to designate a single controlling factor on hyporheic bacteria will fail. Even though carbon supply appears to be an important control on hyporheic bacteria, it is useful to highlight several alternate controlling factors recognizing that in most systems hyporheic microbes will be affected by several factors acting concurrently.

A. Sediment Size

At larger spatial scales, hyporheic bacteria are likely to vary in response to differences in sediment composition among stream habitats (e.g., depositional versus erosional regions). Stream hydraulics and geomorphology will affect the stability of downwelling locations thereby influencing the predictability of resource supply from overlying waters. If downwelling is controlled by relatively permanent features such as bedrock or debris dams, then hyporheic microbial communities will exist in a fairly predictable chemical environment. If downwelling varies with streamflow, water table elevation, or ephemeral features such a sandwaves, then the supply of materials from surface water will be highly variable. Paradoxically, sediment attributes (e.g., large particle size, high hydraulic conductivity) that favor the resupply of resources by surface water infiltration may constrain organic matter accumulation and/or surface area for attached bacteria (Leichtfried, 1996).

B. Inorganic Chemistry

There is accumulating evidence from planktonic systems that the supply of inorganic nutrients can directly and indirectly influence bacterial growth (Pace and Cole, 1996; Rivkin and Anderson, 1997). Phosphorus limitation has been ascribed to stream bacteria on occasion (Miettinen *et al.*, 1997). Also, fungal production has been related to nutrient availability in streams differing in nutrient concentrations (Suberkropp and Chauvet, 1995). Hence, the rate of nutrient supply or stoichiometry of organic material (e.g., C:N:P) may influence hyporheic bacteria (see Duff and Triska, Chapter 8 in this volume; Hendricks and White, Chapter 9 in this volume). In addition, the supply of dissolved oxygen may constrain rates of carbon use, and thus limit hyporheic microbes. For example, in anoxic sediments, alternate terminal electron acceptors (e.g., nitrate) are required for anaerobic respiration, yet theoretical rates cannot exceed aerobic respiration rates and are usually several fold to several orders of magnitude less (see Baker *et al.*, Chapter 11 in this volume).

C. Grazing

A major tenet of the microbial loop concept so widely applied in planktonic systems is that micrograzers efficiently harvest bacterial biomass produced from otherwise unavailable forms of organic carbon. Grazing loss has long been considered an important mechanism for the control of bacterial abundance in numerous aquatic ecosystems (Gasol, 1994; Bott, 1996). Evidence that grazing by protozoa, meiofauna, or macrofauna could control bacterial abundance is limited to a few detailed studies, and there is limited information on this subject for almost all benthic environments (Kemp, 1990). Further, grazing pressure has not been assessed quantitatively along a hyporheic flowpath. There are occasional reports that micro-metazoans or protozoans can remove significant quantities of bacterial biomass from leaf surfaces or sediments (Bott and Kaplan, 1990; Perlmutter and Meyer, 1991). However, only under maximal densities of protozoans is removal via grazing a significant quantity. Other studies show grazing pressure on sediment microbes to be low with removal of only a small to modest fraction of standing stocks or daily production (Epstein and Shiaris, 1992; Borchardt and Bott, 1995). In addition, many bacteria in hyporheic sediments appear inactive (Fischer *et al.*, 1996; Marxsen, 1988a), suggesting that abundances would not be maintained under intense grazing pressure. Per capita bacterial consumption may be lower for protozoans in sediments compared to the water column because of alternate food resources in sediments or difficulty in harvesting particle-associated microbes. Alternatively, bacteria:protozoa ratios may simply be much higher in sediments compared to the water column. The tentative conclusion at this point is that grazing may occasionally be heavy but does not generally appear significant in controlling hyporheic bacteria.

VII. COMMUNITY COMPOSITION

Our current understanding of the functional diversity of hyporheic microbes is based largely on inferential, biogeochemical evidence. Efforts to examine specific bacterial functional guilds (e.g., denitrifiers or methanogens) or to characterize or compare bacterial community composition (e.g., phylogenetic diversity) are rare. Microbial ecologists have had increased success at quantifying microbial community processes, yet they have been limited in their ability to thoroughly characterize the composition of natural microbial communities. Hence, microbial communities have been relegated to a nebulous, generic grouping (or black box) of species in most conceptual models of specific aquatic ecosystems. This is generally true for stream ecosystems (Cummins, 1974; Leff, 1994; Meyer, 1994), and certainly true for hyporheic zones. The degree to which a bulk classification of microbes limits our understanding of the trophic structure of ecological communities or specific ecosystem-level processes is uncertain.

Hyporheic sediments appear to be ideal systems for examining the relationship between microbial community structure and function. Gradients in microbial function frequently occur along discreet flowpaths (Findlay *et al.*, 1993; Jones *et al.*, 1995) in response to sharp chemical gradients (Baker *et al.*, Chapter 11 in this volume; Duff and Triska, Chapter 8 in this volume). At present, we do not know whether microbial community composition varies along chemical gradients. Likewise, we do not know whether gradients in microbial function (e.g., aerobic respiration) are accompanied by alterations in community composition. Functional differences may be the result of shifts in the relative abundances of various taxa within a microbial community, or they may be the result of functional plasticity in which the same taxa express alternate phenotypes. In reality, both processes are likely to occur; hence, the challenge is to assess their relative strength. The issue is relevant to general ecology where there is increasing interest in whether species diversity is connected to variance in nutrient supply and other system-level processes (Wedin and Tilman, 1996; McGrady-Steed *et al.*, 1997).

There is an applied side to this issue as well because if hyporheic sediments are important sites of material retention and transformation, then pollutants will also be accumulated and degraded in subsurface sediments (Madsen and Ghiorse, 1993). If microbial diversity is high, then it becomes more likely that taxa capable of degrading novel compounds will be present in these sediments. In addition, the practice of pumping drinking water from alluvial sediments connected to surface waters can infuse pollutants and possibly pathogens into the hyporheic zone. The likelihood that these compounds and organisms will survive passage through hyporheic sediments may be a function of endemic community diversity or the presence of particular taxa.

VIII. CONCLUSIONS AND RESEARCH NEEDS

Hyporheic zones have received increased attention from a variety of researchers in the last decade; however, our understanding of the microbial ecology of this habitat remains rudimentary. Hyporheic zones are known to harbor a diverse assemblage of macroinvertebrates and meiofauna and likely contain an even greater diversity of microorganisms. Microbes are thought to account for the bulk of hyporheic biomass. In most hyporheic zones, microbial biomass and abundance are likely dominated by heterotrophic bacteria. The wide array of physical and chemical conditions suggests that hyporheic bacterial communities are functionally diverse, yet empirical evidence is limited. Overall, bacterial abundance and activity among samples within a typical stream study site (e.g., riffle or pool reach) are heterogeneous, yet some patterns emerge. Bacterial biomass, production, and respiration generally decrease with depth and distance from downwelling zone. In addition, hyporheic bacterial abundance, production, and respiration appear to be closely related to carbon supply, yet it is difficult to resolve the relative contribution of DOC and POC. In anoxic zones, the supply of alternate terminal electron acceptors may be an important control. Sediment-size distribution invariably influences resource supply and dictates surface area for biofilm formation. Empirical evidence for inorganic nutrient limitation and biotic control (i.e., intense grazing pressure) in hyporheic zones is limited.

Synthesis of the microbial ecology of hyporheic zones is inherently limited by the small number of studies. Bacteria probably dominate microbial biomass and are sensitive to organic carbon supply, yet studies that explicitly examine factors that control hyporheic bacteria are rare (Jones, 1995). In addition, studies that explicitly examine bacterial community composition are absent. An array of nucleic acid-based techniques for revealing the structure of microbial communities could be applied in hyporheic research. Non-bacterial components of the hyporheic microbial community (e.g., fungi and protozoa) are poorly understood as well. Although important trophic linkages (i.e., carbon to bacteria/fungi to consumer) have been hypothesized (Marxsen, 1988b; Bärlocher and Murdoch, 1989; Bott and Kaplan, 1990), empirical evidence is limited. Hyporheic bacteria are known to perform numerous important biogeochemical transformations, yet factors that limit or regulate these transformations have not been identified in a general way. We as yet have no conceptual or quantitative framework for predicting where and when hyporheic metabolism will play a major role in overall stream material fluxes and transformations. The difficulty in predicting the importance of the hyporheic zone is partly due to our inability to scale-up from "plot-scale" or individual flowpaths to the reach or catchment scale (Meyer, 1988). Comparative and experimental studies are certainly required in order to construct a general and predictive mechanistic model of the microbial ecology of hyporheic zones. In addition, microbial ecology needs to be better integrated into biogeochemical and hydrological models of hyporheic zones. For instance, water quality models can now routinely describe patterns of planktonic chlorophyll, nutrients, and dissolved oxygen in surface waters, but the same level of integration has not been attained in studies of the hyporheic zone.

REFERENCES

Atlas, R. M., and R. Bartha. 1998. "Microbial Ecology: Fundamentals and Applications, 4th ed. Benjamin/Cummins, Menlo Park, CA.

Bärlocher, F., and J. H. Murdoch. 1989. Hyporheic biofilms—a potential food source for interstitial animals. *Hydrobiologia* 184:61–67.

Borchardt, M. A., and T. L. Bott. 1995. Meiofaunal grazing of bacteria and algae in a Piedmont stream. *Journal of the North American Benthological Society* 14:278–298.

Bott, T. L. 1995. Microbes in food webs. *American Society of Microbiology News* 61:580–585.

Bott, T. L. 1996. Algae in microscopic foodwebs. *In* "Algal Ecology: Freshwater Benthic Ecosystems" (R. J. Stevenson, M. L. Bothwell, and R. L. Lowe, eds.), pp. 573–608. Academic Press, San Diego, CA.

Bott, T. L., and L. A. Kaplan. 1989. Densities of benthic protozoa and nematodes in a Piedmont stream. *Journal of the North American Benthological Society* 8:187–196.

Bott, T. L., and L. A. Kaplan. 1990. Potential for protozoan grazing of bacteria in streambed sediments. *Journal of the North American Benthological Society* 9:336–345.

Bretschko, G. 1991. The limnology of a low order alpine gravel stream (Ritrodat-Lunz study area, Austria). *Verhandlungen Internationale Vereinigung für Theoretische ünd Angewandte Limnologie* 24:1908–1912.

Chapelle, F. H. 1993. "Ground-water Microbiology and Geochemistry." Wiley, New York.

Cole, J. J., S. Findlay, and M. L. Pace. 1988. Bacterial production in fresh and saltwater ecosystems: A cross-system overview. *Marine Ecology: Progress Series* 43:1–10.

Crocker, M. T., and J. L. Meyer. 1987. Interstitial dissolved organic carbon in sediments of a southern Appalachian headwater stream. *Journal of the North American Benthological Society* 6:159–167.

Cummins, K. W. 1974. Structure and function of stream ecosystems. *BioScience* 24:631–641.

Epstein, S. S., and M. P. Shiaris. 1992. Rates of microbenthic and meiobenthic bacterivory in a temperate muddy tidal flat community. *Applied and Environmental Microbiology* 58:2426–2431.

Fiebig, D. M. 1995. Groundwater discharge and its contribution of dissolved organic carbon. *Archiv für Hydrobiologie* 134:129–155.

Fiebig, D. M. 1997. Microbiological turnover of amino acids immobilized from groundwater discharged through hyporheic sediments. *Limnology and Oceanography* 42:763–768.

Findlay, S. 1995. Importance of surface-subsurface exchange in stream ecosystems: The hyporheic zone. *Limnology and Oceanography* 40:159–164.

Findlay, S., and W. V. Sobczak. 1996. Variability in removal of dissolved organic carbon in hyporheic sediments. *Journal of the North American Benthological Society* 15:35–41.

Findlay, S., J. L. Meyer, and P. J. Smith. 1986. Incorporation of microbial biomass by *Peltoperla* sp. (Plecoptera) and *Tipula* sp. (Diptera). *Journal of the North American Benthological Society* 5:306–310.

Findlay, S., D. Strayer, C. Goumbala, and K. Gould. 1993. Metabolism of streamwater dissolved organic carbon in the shallow hyporheic zone. *Limnology and Oceanography* 38:1493–1499.

Fischer, H., M. Pusch, and J. Schwoerbel. 1996. Spatial distribution and respiration of bacteria in stream-bed sediments. *Archiv für Hydrobiologie* 137:281–300.

Ford, T. E., ed. 1993. "Aquatic Microbiology: An Ecological Approach." Blackwell, Boston.

Freeman, C., and M. Lock. 1995. The biofilm polysaccharide matrix: A buffer against changing organic substrate supply? *Limnology and Oceanography* 40:273–278.

Fuss, C. L., and L. A. Smock. 1996. Spatial and temporal variation of microbial respiration rates in a blackwater stream. *Freshwater Biology* 36:339–349.

Gasol, J. M. 1994. A framework for the assessment of top-down vs. bottom-up control of heterotrophic nanoflagellate abundance. *Marine Ecology: Progress Series* 113:291–300.

Grimm, N. B., and S. G. Fisher. 1984. Exchange between interstitial and surface water: Implications for stream metabolism and nutrient cycling. *Hydrobiologia* **111**:219–228.

Hall, R. O. 1995. Use of a stable carbon isotope addition to trace bacterial carbon through a stream food web. *Journal of the North American Benthological Society* **14**:269–277.

Hauer, F. R., and G. A. Lamberti, eds. 1996. "Methods in Stream Ecology." Academic Press, San Diego, CA.

Hedin, L. O. 1990. Factors controlling sediment community respiration in woodland stream ecosystems. *Oikos* **57**:94–105.

Hendricks, S. P. 1993. Microbial ecology of the hyporheic zone: A perspective integrating hydrology and biology. *Journal of the North American Benthological Society* **12**:70–78.

Hendricks, S. P. 1996. Bacterial biomass, activity, and production within the hyporheic zone of a north-temperate stream. *Archiv für Hydrobiologie* **136**:467–487.

Huettel, M., W. Ziebis, and S. Forster. 1996. Flow-induced uptake of particulate matter in permeable sediments. *Limnology and Oceanography* **41**:309–322.

Hurst, C. J., G. R. Knudsen, M. J. McInerney, L. D. Stetzenbach, and M. V. Walter, eds. 1997. "Manual of Environmental Microbiology." American Society of Microbiology Press, Washington, DC.

Jones, J. B. 1995. Factors controlling hyporheic respiration in a desert stream. *Freshwater Biology* **34**:91–99.

Jones, J. B., and R. M. Holmes. 1996. Surface-subsurface interactions in stream ecology. *Trends in Ecology and Evolution* **11**:239–242.

Jones, J. B., S. G. Fisher, and N. B. Grimm. 1995. Vertical hydrologic exchange and ecosystem metabolism in a Sonoran Desert stream. *Ecology* **76**:942–952.

Kemp, P. F. 1990. The fate of benthic bacterial biomass. *Reviews in Aquatic Sciences* **2**:109–124.

Kemp, P. F., B. F. Sherr, E. B. Sherr, and J. J. Cole, eds. 1993. "Handbook of Methods in Aquatic Microbial Ecology." CRC Press, Boca Raton, FL.

Leff , L. G. 1994. Stream bacterial ecology: A neglected field? *American Society of Microbiology News* **60**:135–138.

Leichtfried, M. 1996. Organic matter in bed-sediments of the River Danube and a small unpolluted stream, the Ober Seebach. *Archiv für Hydrobiologie, Supplementband* **113**:87–98.

Lock, M. A. 1993. Attached microbial communities in rivers. *In* "Aquatic Microbiology: An Ecological Approach" (T. E. Ford, ed.), pp. 113–138. Blackwell Scientific Publications, Boston.

Madsen, E. L., and W. C. Ghiorse. 1993. Groundwater microbiology: Subsurface ecosystem processes. *In* "Aquatic Microbiology: An Ecological Approach" (T. E. Ford, ed.), pp. 167–213. Blackwell, Boston.

Marmonier, P., D. Fontvieille, J. Gibert, and V. Vanek. 1995. Distribution of dissolved organic carbon and bacteria at the interface between the Rhone River and its alluvial aquifer. *Journal of the North American Benthological Society* **14**:382–392.

Marxsen, J. 1988a. Evaluation of the importance of bacteria in the carbon flow of a small open grassland stream the Breitenbach Europe. *Archiv für Hydrobiologie* **111**:339–350.

Marxsen, J. 1988b. Investigations into the number of respiring bacteria in groundwater from sandy and gravelly deposits. *Microbial Ecology* **16**:65–72.

Marxsen, J. 1996. Measurement of bacterial production in stream-bed sediments via leucine incorporation. *FEMS Microbial Ecology* **21**:313–325.

McGrady-Steed, J., P. M. Harris, and P. J. Morin. 1997. Biodiversity regulates ecosystem predictability. *Nature (London)* **390**:162–165.

Metzler, G. M., and L. A. Smock. 1990. Storage and dynamics of subsurface detritus in a sand-bottomed stream. *Canadian Journal of Fisheries and Aquatic Sciences* **47**:588–594.

Meyer, J. L. 1988. Benthic bacterial biomass and production in a blackwater river. *Verhandlungen Internationale Vereinigung für Theoretische ünd Angewandte Limnologie* **23**:1832–1838.

Meyer, J. L. 1994. The microbial loop in flowing waters. *Microbial Ecology* **28**:195–200.

Meyer-Reil, L. A. 1994. Microbial life in sedimentary biofilms—the challenge to microbial ecologists. *Marine Ecology: Progress Series* **112**:303–311.

Miettinen, I. T., T. Vartiainen, and P. J. Martikainen. 1997. Phosphorus and bacterial growth in drinking water. *Applied and Environmental Microbiology* **63**:3242–3245.

Pace, M. L., and J. J. Cole. 1996. Regulation of bacteria by resources and predation tested in whole-lake experiments. *Limnology and Oceanograpy* **41**:1448–1460.

Perlmutter, D. G., and J. L. Meyer. 1991. The impact of a stream-dwelling harpacticoid copepod upon detritally associated bacteria. *Ecology* **72**:2170–2180.

Pomeroy, L. R. 1974. The ocean's food web, a changing paradigm. *BioScience* **24**:499–504.

Pratt, J. R., and J. Cairns. 1985. Functional groups in the protozoa: Roles in the differing ecosystems. *Journal of Protozoology* **32**:415–423.

Pusch, M. 1996. The metabolism of organic matter in the hyporheic zone of a mountain stream, and its spatial distribution. *Hydrobiologia* **323**:107–118.

Pusch, M., and J. Schwoerbel. 1994. Community respiration in hyporheic sediments of a mountain stream (Steina, Black Forest). *Archiv für Hydrobiologie* **130**:35–52.

Rivkin, R. B., and M. R. Anderson. 1997. Inorganic nutrient limitation of oceanic bacterioplankton. *Limnology and Oceanography* **42**:730–740.

Sander, B. C., and J. Kalff. 1993. Factors controlling bacterial production in marine and freshwater sediments. *Microbial Ecology* **26**:79–99.

Schallenberg, M., and J. Kalff. 1993. The ecology of sediment bacteria in lakes and comparisons with other aquatic ecosystems. *Ecology* **74**:919–934.

Sinsabaugh, R. L., and S. Findlay. 1995. Microbial production, enzyme activity, and carbon turnover in surface sediments of the Hudson River Estuary. *Microbial Ecology* **30**:127–141.

Sleigh, M. A., B. M. Baldock, and J. H. Baker. 1992. Protozoan communities in chalk streams. *Hydrobiologia* **248**:53–64.

Stevenson, R. J. 1997. Scale-dependent causal frameworks and the consequences of benthic algal heterogeneity. *Journal of the North American Benthological Society* **16**:248–262.

Suberkropp, K. 1997. Annual production of leaf-decaying fungi in a woodland stream. *Freshwater Biology* **38**:169–178.

Suberkropp, K., and E. Chauvet. 1995. Regulation of leaf breakdown in streams: Influences of water chemistry. *Ecology* **76**:1433–1445.

Suberkropp, K., and H. Weyers. 1996. Application of fungal and bacterial production methodologies to decomposing leaves in streams. *Applied and Environmental Microbiology* **62**:1610–1615.

Sun, L. E., M. Perdue, J. L. Meyer, and J. Weis. 1997. Using elemental composition to predict bioavailability of dissolved organic matter in a Georgia River. *Limnology and Oceanography* **42**:714–721.

Valett, H. M., S. G. Fisher, and E. H. Stanley. 1990. Physical and chemical characteristics of the hyporheic zone of a Sonoran desert stream. *Journal of the North American Benthological Society* **9**:201–205.

Vallino, J. J., C. S. Hopkinson, and J. E. Hobbie. 1996. Modeling bacterial utilization of dissolved organic matter: Optimization replaces Monod growth kinetics. *Limnology and Oceanography* **41**:1591–1609.

Vervier, P., M. Dobson, and G. Pinay. 1993. Role of interaction zones between surface and ground water in DOC transport and processing: Considerations for river restoration. *Freshwater Biology* **29**:275–284.

Volk, C. J., C. B. Volk, and L. A. Kaplan. 1997. Chemical composition of biodegradable dissolved organic matter in streamwater. *Limnology and Oceanography* **42**:39–44.

Wedin, D., and D. Tilman. 1996. Influence of nitrogen loading and species composition on the carbon balance of grasslands. *Science* **274**:1720–1723.

13

The Ecology
of Hyporheic Meiofauna

Christine C. Hakenkamp* and Margaret A. Palmer†

*Department of Biology
James Madison University
Harrisonburg, Virginia

†Department of Zoology
University of Maryland
College Park, Maryland

I. Introduction   307
II. Meiofaunal Taxa and Their Relative Abundances in the Hyporheic
     Zone   309
III. Spatial Distribution and Abundance   316
IV. Meiofauna and the Physical Environment   319
V. Tolerance to Anoxia, Body Size, and Biomass   321
VI. Trophic Roles, Dispersal Dynamics, and Response to Spatial–
     Temporal Heterogeneity   322
VII. Meiofaunal–Microbial Interactions   324
VIII. Meiofauna and Ecosystem Ecology   325
     A. Scenario A: High Interstitial Flow Rates   326
     B. Scenario B: Intermediate Interstitial Flow Rates   327
     C. Scenario C: Low Interstitial Flow Rates   327
IX. Conclusions and Future Research Needs   328
     References   329

I. INTRODUCTION

Meiofauna are sediment-associated organisms intermediate in size between the microbes and macrofauna (Palmer and Strayer, 1996). The term
meio is a Greek term for smaller (Mare, 1942) and meiofauna are generally
classified as protists and invertebrates between 50 and 1000 µm (Mare,
1942; although some researchers use a 500-µm upper size limit; Meyer,

1990). Invertebrates > 1000 μm would not be included as meiofauna, unless they spend part of their life as smaller interstitial organisms (temporary meiofauna). Meiofauna occur in all aquatic ecosystems (marine to freshwater) and include representatives from two-thirds of the known animal phyla (Coull, 1988). Thus, it should not be surprising that hyporheic meiofauna include a very diverse group of organisms from a wide range of taxa (Table I).

Some species of marine and hyporheic meiofauna share similar characteristics for adapting to interstitial habitats, including small size, wormlike shape, adhesive organs (to hold onto sediment grains), armor (protection from sediment movement), direct fertilization, and lack of free-swimming larvae (Giere, 1993; Palmer and Strayer, 1996). Because meiofauna are defined as interstitial organisms and spend most of their lives in the sediments, for the sake of simplicity we will consider all streambed meiofauna as members of the hyporheic community.

There are many advantages to studying meiofauna in both marine and freshwater systems. These include their short generation times, high abundances, intimate association with the sediments (they generally lack a planktonic or aerial stage), quick response to any change in the system, and ease of collection even with small-sized samples (Hummon *et al.*, 1978; Warwick, 1993). They can be ideal experimental organisms (Coull and Palmer, 1984).

Compared to marine meiofauna, much less is generally known about hyporheic meiofauna even though they have been studied in both systems for over 100 years (Coull and Giere, 1988). During the early 1900s, when marine ecologists were first documenting large abundances of these small fauna in marine systems, European stream ecologists also found numerous meiofauna in streams (for a review, see Pennak, 1988). Aquatic ecologists discovered that they were missing many taxa, as well as much of the total biomass, by sampling with sieves of large mesh sizes (i.e., 250 μm instead of 50 μm; Hummon, 1981). The study of marine meiofauna quickly entered a period of intense effort and rich discovery, especially when compared to equivalent research in hyporheic systems. There are at least two factors that probably influenced these different rates of research progress. First, many marine meiofaunal taxa also have macrofaunal constituents so that taxonomic expertise was in place quite early for many groups. This was in contrast to a lack of taxonomic experts on many freshwater meiofauna. Second, there were simply more researchers actively studying marine systems and their fauna. Recently (late 1970s to present), study of stream meiofauna has advanced more rapidly, aided by our vast knowledge of factors influencing marine meiofauna and the technology that marine researchers have developed to study the meiobenthos.

Here we summarize meiofaunal research in stream hyporheic zones. We also include what are generally considered to be the major results from marine studies because they are much more extensive (Table II, developed from Coull and Giere, 1988; Giere, 1993) and because this research is quite relevant to streams (Palmer, 1990b, 1992). Throughout this chapter, we ask whether questions addressed in marine systems have also been addressed in hyporheic systems. Further, we point out several studies in which streams have been shown to be ideal locations for testing new ideas in meiofaunal research. This chapter is not intended as a complete overview on meiofaunal ecology, and the reader is directed to other literature for more complete reviews (Table III).

II. MEIOFAUNAL TAXA AND THEIR RELATIVE ABUNDANCES IN THE HYPORHEIC ZONE

Hyporheic systems contain extremely large densities of meiofauna which may equal or even exceed those in marine systems (Palmer, 1990a; Borchardt and Bott, 1995). Hyporheic meiofauna are also very taxonomically diverse (see hyporheic references in Table I). For example, more than 145 species of meiofauna occur in Goose Creek, a sandy stream in Northern Virginia (Turner and Palmer, 1996; P. Silver, M. Shofner, and J. Reid, personal communication), and over 300 species occur at the Oberer Seebach, a gravel stream in Austria (Schmid-Araya and Schmid, 1995; Schmid and Schmid-Araya, 1997). Even though all hyporheic systems may not be this diverse, small invertebrates obviously contribute greatly to the biodiversity of streams (Palmer *et al.*, 1997).

The number of hyporheic studies using mesh sizes for organism sampling small enough to collect meiofauna has increased considerably. However, comparisons between taxa identified in these studies can be problematic given the large variability in sampling approaches and the range in sampling depths. Organisms such as crustaceans, insects, rotifers, oligochaetes, and nematodes appear to be important components of most hyporheic communities, but considerable differences exist in the percent composition of these taxa in different streams (Table IV). For example, a comparison of meiofauna in sandy versus coarse substrates (e.g., cobble, gravel) indicates that although most taxa occur in both sediment types, coarse sediments often contain more crustaceans and insects, and sandy sediments contain more rotifers, nematodes, and tardigrades (Fig. 1). This difference in taxa is perhaps not surprising given that crustaceans and insects tend to occur within the upper size range of meiofauna and might have more difficulty moving through the smaller interstitial spaces of sandy sediments. Another possibility, however, is that the use of different sampling methodologies greatly influenced the taxa collected (e.g., pumping versus hand coring; see also following discussion). Another important point is that even though an impressive amount

TABLE 1 Taxonomic Composition of Meiofauna[a]

Group	Freshwater	Marine	Hyporheic references
Meiofaunal-sized and smaller			
Protista (Ciliophora, Sarcomastigophora)	X	X	Schmid-Araya (1994a,b)
All meiofaunal-sized			
Gastrotricha	X	X	Schmid-Araya and Schmid (1995)
Tardigrada	X	X	Strayer *et al.* (1994); Schmid-Araya and Schmid (1995)
Kinorhyncha		X	
Loricifera		X	
Groups primarily freshwater			
Aelosomatidae and Potamodrilidae	XX	X	
Insecta (Collembola, Ephemeroptera, Odonata, Plecoptera, Trichoptera, Coleoptera, Diptera)	XX	X	Williams (1984); Pennak and Ward (1986); Strommer and Smock (1989); McElravy and Resh (1991); Boulton *et al.* (1992); Bretschko (1992); Williams (1993); Ward and Voelz (1994); Schmid-Araya and Schmid (1995); Wolz and Shiozawa (1995) Chironomidae: Schmid (1993); Ruse (1994) Collembola: Bretschko and Christian (1989)
Rotifera	XX	X	Braioni and Gottardi (1979); Schmid-Araya (1993, 1995); Turner and Palmer (1996)
Occur in both freshwater and salt water			
Acari	X	X	Schwoerbel (1961); Gledhill (1982); Strayer (1988a)
Crustacea (Copepoda, Ostracoda, Amphipoda, Isopoda, Cladocera, Syncarida, Thermobaenacea)	X	X	Cladocera: Vila (1989); Dumont and Negrea (1996) Copepods: Rouch (1968, 1991); Lescher-Moutoué (1973); Strayer (1988b); Kowarc (1990); Reid and Strayer (1994) Microcrustaceans: Rouch (1988, 1995); Rundle and Ormerod (1991); Shiozawa (1986, 1991) Ostracods: Danielopol (1980); Marmonier and Ward (1990); Marmonier and Creuzé des Châtelliers (1992) Isopods: Coireau (1971) Thermobaenaceans: Wagner (1994)

310

Taxon		Reference
Mollusca (Aplacophora, Bivalvia, Gastropoda)	×	
Oligochaeta	×	Schwank (1981–1982); Strayer and Bannon-O'Donnell (1988); Lafont et al. (1992)
Platyhelminthes (Turbellaria)	×	Schwank (1981–1982); Kolasa (1982, 1983); Kolasa et al. (1987)
Groups more diverse and abundant in marine systems		
Nematoda	××	Eder (1983); Strayer and O'Donnell (1992); Wolz and Shiozawa (1995)
Nemertina	××	
Polychaeta	××	
Marine groups		
Brachiopoda	×	
Echinodermata (Holothuroidea)	×	
Entoprocta	×	
Gnathostomulida	×	
Priapulida	×	
Pycnogonida	×	
Sipuncula	×	
Tunicata	×	

[a]Groups that occur preferentially in one habitat over another are indicated with a double ×. Terms in parentheses on left side of table are provided to indicate major types of meiobenthic organisms within that group. Annelids (Polychaeta, Oligochaeta, Aelosomatidae, and Potamodrilidae) have been broken into groups to represent different preferences for freshwater versus marine habitats. A nonexclusive list of hyporheic references is provided for relevant groups.

TABLE II Historical Overview[a]

Marine meiofauna research	Stream equivalent
1900–1950	
Surprisingly large numbers of small fauna found and studied systematically for first time. Research focused on:	Also considerable stream research, primarily in Russia, Poland, Switzerland, and France. See also Coull (1988) and Pennak (1988).
1. Taxonomy (who is there)	1. See Table I
2. Recognition of large abundances	2. Large abundances found in streams (e.g., Williams and Hynes, 1974; Hummon et al., 1978; Palmer, 1990a; Bretschko, 1992)
3. Recognition of distinct ecological assemblages (e.g., sandy vs. muddy beaches)	3. Distinct areas within hyporheic zone (Boulton et al., 1992; Williams, 1993)
1950s to mid 1960s	
Three main areas of research were:	These areas of research are still very active today. Examples of studies summarizing some of this work include:
1. Broadening taxonomic knowledge of various taxa including the description of many new taxa	1. Numerous taxanomic studies in streams (see Pennak, 1988)
2. Descriptive ecology and the importance of limiting factors such as granulometry and oxygen	2. Similar factors probably limiting in streams (Schwoerbel, 1967; Williams, 1993; Strayer, 1994b)
3. Documenting large biomass of meiofauna relative to other groups of organisms	3. Meiofaunal biomass in streams can exceed macrofaunal biomass (Poff et al., 1993)
Late 1960s and into 1970s	
Four general areas of interest:	Research along same lines as in marine studies, but fewer studies. Some interesting exceptions include:
1. Ecophysiology and behavior	1. Examining behavior in terms of active and passive drift components (Richardson, 1991; Palmer, 1992; Tokeshi, 1994)
2. Tolerance limits, meiofauna as bioassay	2. Using multivariate statistics to look at importance of such factors as pH (Rundle, 1993) and metals (Plenet and Gibert, 1994) with respect to tolerance limits and for using meiofauna for bioassay
3. Energetics	3. Comparison to size spectra different in freshwater systems compared to marine systems (Poff et al., 1993; Strayer, 1994a)
4. Species diversity and meiofauna as a bioassay	4. Importance of species diversity in the hyporheic zone as it relates to ecosystem functioning (Palmer et al., 1997)

1970s and 1980s

Experimental ecology and hypothesis testing (emphasis on using rigorous approaches and controls)—questions asked include:

1. Macro–meiofaunal interactions
2. Predator–prey interactions
3. Effect of structure
4. Recolonization after disturbance
5. Pollution effects

Late 1980s and into 1990s

Four general areas of interest:

1. Meiofauna feeding on microbes [results varied, some (not all) showed feeding by meiofauna on either diatoms or bacteria] using correlation studies, radioisotope studies, stable isotopes
2. Mineralization of organic matter (meiofauna appear to enhance rates of organic matter mineralization)
3. Bioturbation of sediment and effect on solute transport
4. Documenting and describing spatial variability in distribution

Much of stream work discussed previously already tied to hypothesis testing (marine work had shown rigorous approaches were critical to studying these factors). New topics explored in this section include:

1 & 2. Predator–prey interactions between macro- and meiofauna (Amores-Serrano, 1991; Shofner and Palmer, 1997)
3. Effect of structure in terms of influence on mitigating disturbance or providing habitat (Palmer et al., 1995, 1996a; Robertson et al., 1995, 1997)
4. Examining the hyporheic refuge hypothesis (Palmer et al., 1992; Dole-Olivier et al., 1997)

All currently "hot topics" in hyporheic ecology as stream ecologists try to tie fauna to function and explore spatially explicit ecology issues. In general, hyporheic studies lag behind marine studies. Exceptions include:

1. Meiofauna feeding on microbes (Meyer, 1990; Borchardt and Bott, 1995; Hall, 1995)
2. Documenting spatial variability (Schmid, 1992; Schmid-Araya, 1994a,b; Ward and Palmer, 1994; Palmer et al., 1996b)
3. Oligochaetes my make detritus available to microbes (Chauvet et al., 1993)

aMuch of meiofaunal research has historically been conducted in marine systems. Major research topics are presented (developed from Coull and Giere, 1988; Giere, 1993). For comparison, we ask: are there equivalent studies in streams systems?

TABLE III   Meiofaunal Reviews[a]

Meiofaunal texts
  Meiofaunal ecology (Giere, 1993)
  History of meiofauna, sampling methodology, taxonomy (Higgins and Thiel, 1988)
Various reviews
  Review of marine meiofauna research (Coull, 1988)
  Review of freshwater meiofauna research (Pennak, 1988)
  Comparison of fauna of freshwater and marine beaches (Pennak, 1951)
  Importance of marine meiofauna to benthic community (Gerlack, 1971)
  Ecology of microfauna and meiofauna in marine sediment (Fenchel, 1978)
  Hyporheic/phreatic invertebrate literature (Danielopol, 1982)
  Evolutionary modifications of freshwater fauna from marine ancestors (Pennak, 1985)
  Biomass and abundance estimates for an entire lake benthos (Strayer, 1985)
  Zoogeography of groundwater fauna (Botosaneanu, 1986)
  Evolutionary considerations of marine meiofauna (Warwick, 1989)
  Groundwater ecology text (Gibert *et al.*, 1994)
Distribution patterns
  Factors limiting distribution of fauna (Strayer, 1994b)
  Factors influencing spatial distribution (Fleeger and Decho, 1987; Decho and Fleeger, 1988)
Trophic studies with Meiofauna
  Blackwater river trophic interactions (Meyer, 1990)
  Role of bacteria and algae as food for meiofauna (Borchardt and Bott, 1995)
  Role of bacteria in aquatic systems, comparisons with protists and invertebrates (Kemp, 1990)
  Meiofaunal impacts on microbial communities (Tietjen, 1980)
  Meiofauna as food for fish (Gee, 1989; Coull, 1990)
Pollution research using meiofauna as bioindicators
  Pollution studies (primarily marine; Coull and Chandler, 1992)
  Toxicity studies (Traunspurger and Drews, 1996)
Streams
  Overview meiofaunal role in stream systems, laboratory and field exercises (Palmer and Strayer, 1996)
  Flow and physicochemical factors that may influence faunal distribution (Williams, 1989, 1993)
  Microhabitats in hyporheic zone, influence of abiotic factors (Boulton *et al.*, 1992)
Field experiments
  Marine field experiments (Coull and Palmer, 1984)
  Experimentation and research needs in hyporheic zones (Palmer, 1993)

[a]References are provided for those interested in more information regarding meiofaunal research, much of which has been completed in marine systems but has direct relevance to freshwater habitats.

of research on European hyporheic meiofauna exists (e.g., see Table I); much of that research has been conducted using mesh sizes >100 μm or has focused on specific taxonomic groups. We include a list of terms commonly encountered in the meiofaunal literature to encourage future cross-system comparisons (Table V).

TABLE IV Comparison of Hyporheic Taxa from Studies Using Small Sieve Mesh Sizes (≤ 100 μm)[a]

Stream/ reference group	Coarse sediments			Sandy sediments						
	Danube, Austria (Danielopol, 1984)	Multiple streams, NY (Strayer, 1988a)	Wappinger Ck, NY (Hakenkamp, 1997)	Multiple streams, OH (Hummon et al., 1978)	Headwater stream, MS (Hummon 1987)	Goose Creek, VA (Palmer, 1990a)	Praire stream, TX (Golladay and Hax, 1995)	Buzzards Branch, VA (Hakenkamp, 1997)	White Clay Ck, PA (Hakenkamp, 1997)	Goose Creek, VA (Hakenkamp, 1997)
Oligochaetes	5.9	19.2	*	7.8	4.8	19.1	*	*	*	*
Nematodes	8.3	10.4	2.2	38.5	11.2	3.1	32.2	12.1	17.0	2.8
Other worms	*	9.8	4.9	4.2	0.9	*	*	3.8	16.1	21.9
Crustaceans	81.8	37.8	1.7	3.9	2.9	12.0	18.3	2.4	8.8	10.2
Rotifers	*	*	26.0	31.3	58.6	47.7	30.9	47.9	13.3	23.4
Acari	*	6.3	*	*	*	*	*	*	*	*
Insects	4.0	14.8	62.2	9.0	4.4	12.9	*	16.4	37.9	18.2
Tardigrades	*	2.2	*	1.7	11.8	4.4	18.6	12.0	6.0	0
Others	*	0.3	3.0	3.1	5.3	0.8		5.4	0.9	23.5
Total	100.0	100.8	100.0	99.5	99.9	100.0	100.0	100.0	100.0	100.0
Sediment type	Coarse	Coarse	Coarse	Sand	Sand	Sand	Sand/ gravel	Sand	Sand	Sand
Depth	0.5–2.3 m	Variable	10 cm	7 cm	7 cm	50 cm	5–10 cm	10 cm	10 cm	10 cm
Sampler	BRP	KC	HC	HC	HC	SPC	HC	HC	HC	HC
Mesh size (μm)	100	100	63	38	37	48	45	63	63	63

[a]Numbers represent percent of total taxa, insufficient data indicated with a *. "Other worms" include taxa such as Aelosomatidae, Potamodrilidae, Gastrotricha, Platyhelminthes, and Nemertina. Samples include Bou-Rouch pumps (BRP), Karman-Chappuis method (KC), hand-held corer (HC), and a standpipe corer (SPC).

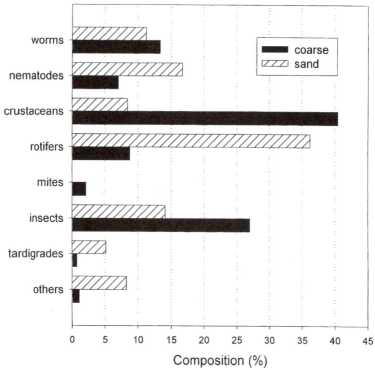

FIGURE 1 Comparison of hyporheic faunal composition in sandy versus coarse sediment streams. Values represent means for multiple streams and are taken from Table IV. Category "others" includes protists, gastrotrichs, and molluscs.

III. SPATIAL DISTRIBUTION AND ABUNDANCE

In many, if not most streams, meiofauna are abundant below the surface of the streambed (e.g., often 25 to >75% of total fauna occur below benthic sediments; Table VI). Most studies of stream meiofauna have been conducted at only a few locations within a streambed study site. Few studies have compared benthic (surface) versus hyporheic meiofaunal abundances between all sediment types within a stream (e.g., fine to coarse sediment zones, across more than one stream reach), longitudinally within a watershed (Ward and Voelz, 1990), or between watersheds (Strayer, 1988a; Boulton *et al.*, 1992; Ward and Palmer, 1994). Understandably, such studies would require extensive effort, but they would yield considerable information for generalizing across stream systems.

Our ability to make cross-system comparisons is also greatly hindered by an inability to quantitatively compare among different sampling methods (Palmer, 1993; Willliams, 1993). Hand-held corers cannot be used to sample very deeply into coarse sediments. Although pumping samples from coarse sediments is an alternative approach to quantifying abundances, determining from what depth samples are collected is problematic (Strayer and Bannon-O'Donnell, 1988; Creuzé des Châtelliers and Dole-Olivier, 1991; Bretschko,

---

TABLE V Terms Commonly Encountered in the Meiofaunal Literature[a]

Bedsediments: channel-forming sediments dominated by epigeic organisms, uppermost hyporheic zone

Benthos: organisms that live primarily on (not in) stream sediments

Epigeic: used to describe organisms that live in surface sediments or very shallow sediments (fauna are epigeam fauna)

Eupsammolittoral: habitat above the water line of sandy margins of streams

Ground water: term varies to describe fauna that occur below the hyporheic zone (when streambed is split into benthic, hyporheic, and ground water zones) to fauna that occur below depth of light penetration (when streambed is split into surface sediments and ground water sediments; thus term may actually include hyporheic fauna or may exclude hyporheic fauna), synonymous with phreatic

Hypogeic: used to describe organisms that live in the subsurface, *not* epibenthically or in very shallow sediment (fauna are hypogean faunam not synonymous with hyporheos)

Hyperbenthic: living at the surface–water interface or just above the sediments

Hyporheos: organisms that live in the hyporheic zone

Interstitial: organisms that live between sediment grains

Karstic: term used to describe organisms associated with limestone caves and streams

Meiobenthos: in marine systems, most meiofauna occur in upper few centimeters of sediments, so this term is often synonymous with meiofauna. In streams, it is probably more appropriate to use the term *meiofauna* because many species can occur in both shallow and deep sediments

Meiofauna: term generally used to describe organisms between 44 and 1000 μm in size; term has been used in both singular and plural sense

Occasional or temporary hyporheos: organisms that do not live entire life in hyporheic zone

Permanent hyporheos: organisms restricted to hyporheic zone through entire life cycle

Phreatic: organisms that live in the subsurface, below the level of light penetration, may or may not include hyporheos; synonymous with ground water

Psammobites: organisms restricted to interstitial habitats

Psammon: organisms associated with sand

Psammophiles: organisms that spend entire life interstitially

Psammoxenes: organisms whose presence in substrate is accidental

Rithrostygon: term applied equal to hyporheic

Stygobios: organisms that live in subsurface waters

Stygobites: organisms that only live interstitially and whose bodies are modified for subsurface life style

Stygophiles: organisms that can occur on or in sediments but whose bodies are not modified for interstitial way of life

Stygoxenes: organisms that generally live on surface of stream sediments but may be found interstitially

Troglobios: organisms that live in caves

---

[a]Terms have been gleaned from many sources including Williams and Hynes (1974), Pennak (1988), Giere (1993), and Marmonier *et al.* (1993).

TABLE VI Vertical Distribution of Fauna in Hyporheic Sediments[a]

Reference	Hyporheic system	Sediment type	Mesh size	Sampler	Total faunal found deep in hyporheic zone	
					Percentage	Indicated depth (cm)
Williams and Hynes (1974)	Speed-R, Ontario	Coarse	100	SPC&KS	97	5–75
Klemens (1991)	Oberer Seeback, Austria	Coarse	100	NFCE	75	10–70
Pehofer (in Klemens, 1991)	Alpbacher Ache, Austria	Coarse	100	NFCES	50	20–60
Bretschko (in Klemens, 1991)	Donau, Austria	Coarse	100	NFCES	25	20–60
Strommer and Smock (1989)	Buzzards Branch, VA	Sand	53	HSC	66	1–40
McElravy and Resh (1991)	Big Canyon Ck, CA	Coarse	63	PS	48	5–35
Pennak and Ward (1986)	Colorado River, CO	Coarse	48	BRP	95	30–50
Kowarc (1992)	Oberer Seebach	Coarse	50	FC	>90[b]	10–70
Bretschko (1992)	Hundsheimer Haufen	Coarse	63	FC	50	10–60

[a]Studies that show that a large percentage of total hyporheic fauna occur below surficial sediments are shown. Percentage of fauna found deep in hyporheic zone are shown; rest of fauna were found at shallower depths. Studies shown used a mesh with a size ≤ 100 μm and sampled at least two depths within the hyporheic zone. Selected references are reported in Klemens (1991). Samplers include standpipe corer (SPC), kick sampler (KS), freeze corer (FC), nitrogen freeze corer with electric field (NFCE), nitrogen freeze corer with electric field and box-type sampler (NFCES), hand-held sediment corer (HSC), Bou-Rouch pump (BRP), pot sampler (PS).

[b]Percentage for harpacticoid copepods only.

1992). Recent studies (Fraser and Williams, 1997; Mauclaire *et al.*, 1998) have compared different sampling methods (freeze-coring, Bretschko, 1990; Bou-Rouch pumping, Giere, 1993; Palmer and Strayer, 1996; Karman-Chappuis digging, Strayer and Bannon-O'Donnell, 1988; Giere, 1993; hyporheic pot sampling, McElravy and Resh, 1991; and hand-coring Palmer and Strayer, 1996); however, more information is needed, especially to compare between coarse and fine substrates. Additionally, a study is needed to compare subsurface sampling to surface sampling (e.g., Surber samples, McElravy and Resh, 1991). Such studies would reveal methodological biases and allow for generalizations between studies using different techniques.

Given the foregoing caveats, several generalizations can be made with respect to our knowledge of hyporheic taxa. First, many of the species found in the hyporheic zone are not endemic groundwater specialists but, in fact, are broad generalists that are surprisingly little modified for interstitial life beyond their general small size and often wormlike shape (Rouch and Danielopol, 1987; Strayer, 1988b; Reid and Strayer, 1994). Second, detailed taxonomic information is generally available for only two meiofaunal groups (crustaceans and insects). Third, widespread species-level data for most of the other meiofaunal groups (Table I) may never exist because of the difficulty in identifying these taxa. This is particularly likely for groups such as turbellarians and rotifers which are often identified live (Kolasa *et al.*, 1987; Schmid-Araya, 1993; Turner and Palmer, 1996), nematodes for which there are few trained freshwater taxonomists (especially in North America), and protists which are often too fragile to sample even using hand-held corers. Even though taxon-specific meiofaunal studies have been valuable in broadening our general understanding of hyporheic systems, as well as furthering the study of important basic ecological questions (Schmid, 1993; such as the impact of disturbance on the resilience and resistance in communities, Schmid, 1992; Schmid-Araya, 1994b, 1995; Palmer *et al.*, 1995, 1996a; Dole-Olivier *et al.*, 1997; and the growth of spatially explicit ecology, Robertson *et al.*, 1995), many basic ecology questions may be addressed using less precise taxonomic identification. For example, the impact of pollution, acidity, and nearness to open water channels on microcrustacean communities is obvious at relatively "simple" taxonomic levels for some streams (Ward and Palmer, 1994; Rundle and Attrill, 1995). However, clearly more taxonomic expertise and training are needed among hyporheic meiobenthologists (especially in North America), and the difficulty in identifying meiofauna to species is perhaps their greatest research liability.

IV. MEIOFAUNA AND THE PHYSICAL ENVIRONMENT

The distribution of meiofauna is greatly influenced by the direction and magnitude of water flow. Whether meiofauna are flushed from sediments or can settle from the water column is largely dependent on hydrologic characteristics of the stream (Richardson, 1991, 1992; Palmer *et al.*, 1992, 1995, 1996a). Water flow is very different in stream hyporheic zones compared to marine sediments that are exposed to fairly predictable and cyclic tidal fluctuations. In hyporheic zones the important flows influencing faunal distribution patterns are likely to be the magnitude of surface water flow (water velocity and sudden changes in discharge) and the amount and direction of vertical exchange (upwelling and downwelling; see Boulton, Chapter 14 in this volume). We will consider velocity and discharge later in this chapter; here we consider the importance of vertical exchange.

The vertical flux of surface and subsurface waters may influence the distribution and abundances of various meiofaunal taxa (e.g., distinct upwelling and downwelling assemblages seem to occur; see Boulton, Chapter 14 in this volume). Distinct species assemblages have been identified within the hyporheic zone, and more importantly, these assemblages have been correlated with physicochemical factors that should vary with the magnitude and direction of vertical water flux (Boulton *et al.*, 1992; Williams, 1993; Ruse, 1994). Thus, biota offer great potential for identifying the types and perhaps even the rates of biotic processes occurring within the hyporheic zone (see also Boulton, Chapter 14 in this volume). We must be cautious, though, because often whether meiofauna are responding to vertical hydrologic exchange or whether both are responding to other site-specific factors is not clear. Some studies have shown that upwelling sites can contain high abundances of fauna normally found at much greater depths in the hyporheic zone (see Dole-Olivier *et al.*, 1997, for additional information); however, no controlled experimental manipulations have been conducted to show that the direction and magnitude of hyporheic flow is the causative factor (Palmer, 1993). Sediment grain size or water chemistry, or the interplay between these factors, must be considered when considering the distribution and abundance of hyporheic meiofauna (Ward and Voelz, 1990; Ruse, 1994; Strayer, 1994b; Ward and Palmer, 1994).

The most important physicochemical factor influenced by flow conditions and sediment granulometry is probably dissolved oxygen concentration (Strayer, 1994b; Findlay, 1995), which may in turn control the abundance and distribution of hyporheic meiofauna especially when oxygen concentration is low (Whitman and Clark, 1984; Williams, 1993). Experimental manipulation of sediment grain size has been found to influence the abundances of various stream meiofaunal taxa (Schwoerbel, 1967; but see Hakenkamp, 1997), perhaps by influencing oxygen concentration and interstitial flow rates (Strayer, 1994b; Findlay, 1995). The relationship between water flux, sediment grain size, and oxygen concentration needs to be experimentally examined to determine if the presence of certain meiofaunal taxa could be used to identify areas within the hyporheic zone with specific interstitial flow patterns.

V. TOLERANCE TO ANOXIA, BODY SIZE, AND BIOMASS

Hyporheic meiofauna have been found to exhibit similar tolerance limits to many of the same chemical factors as their marine counterparts. This tolerance includes pH (Rundle and Ormerod, 1991; Rundle, 1993), metals, pesticides (Plenet and Gibert, 1994; Traunspurger and Drews, 1996), and oxygen (Boulton *et al.*, 1992). Recently, Giere (1993) summarized more than 20 years of research in marine, estuarine, and lentic systems on meiofauna that have the ability to live in low or anoxic conditions (see also Pennak, 1985). A wide variety of taxa (including protists, gnathostomulids, nematodes, gastrotrichs, turbellarians, crustaceans, and oligochaetes) contain species that can withstand and actually thrive in low or no oxygen conditions for long periods of time. In fact, facultative anaerobiosis is not uncommon among marine interstitial meiofauna (Moodley *et al.*, 1997). Considerable controversy exists concerning whether these meiofauna simply tolerate low oxygen conditions or actively feed on the rich stocks of sulfide-supported bacteria in such environments (Vopel *et al.*, 1996). Regardless, many examples exist in which meiofauna (including freshwater taxa) survived under completely anoxia for months (Giere, 1993).

The ability of meiofauna to survive under anoxia has important implications for hyporheic research. Stream ecologists will have to reexamine the assumption that biotic activity in low oxygen areas is restricted to microbially associated processes. Further, nonphotosynthetic primary production may support some part of hyporheic secondary production. Obviously, further study is needed; however, worthy of reiteration is that low oxygen areas in the hyporheic zone can have considerable abundances of active meiofauna. Although densities of meiofauna in these habitats may be lower than those in more oxygenated streambed sediments, given that a large part of the streambed can have low oxygen conditions (Baker *et al.*, Chapter 11 in this volume), biomass of invertebrates in low oxygen sediments might be a considerable portion of the total biomass within a stream.

Physiological tolerance and total biomass of hyporheic meiofauna are intimately tied to body size and metabolism. Organismal body size is a topic of wide interest to ecologists because comparisons can be made for entire ecosystems (terrestrial and aquatic, planktonic and benthic) with less taxonomic precision needed (Hanson *et al.*, 1989). Further, size spectra can provide clues as to how different faunal groups interact (e.g., predator–prey interactions; see Strayer, 1991) and about the role of environmental factors on community dynamics (Bourassa and Morin, 1995). For the marine benthos, clear biomass maxima are evident within meiofaunal and macrofaunal size gradients (Warwick, 1984), strongly suggesting clear separation, and even perhaps competitive exclusion, between organisms that live interstitially (meiofauna) and those that must live on top of or create burrows into the sediment (macrofauna). As discussed by Strayer (1991) and Cattaneo

(1993), examining the spectrum of body sizes in a system is important because the occurrence of biomass peaks will influence how organisms (such as meiofauna) feed, find resources, and respond to predation.

Interestingly, presence and clear separation of macrofaunal and meiofaunal biomass peaks have not been found in studies on the size spectra of freshwater fauna (both lentic and lotic; Strayer, 1986; Poff *et al.*, 1993), even though these studies were conducted in habitats with sediment grain sizes (and thus microhabitats) similar to the marine studies. In the freshwater studies, large abundances of insects and bivalves dominated the size spectra which points to potentially fundamental differences (evolutionary and ecological) between marine and freshwater ecosystems. Obviously more data, especially for hyporheic systems, are needed to verify whether the lack of distinct meiofaunal and macrofaunal body size maxima is a general finding for all freshwater systems. Further, more work is needed to explore the implications of size spectra differences for energy and material transfer within these systems (Strayer, 1991). The ecological interpretation of size spectra analysis is still under debate (Morin and Nadon, 1991; Poff *et al.*, 1993; Bourassa and Morin, 1995), in part because it is unknown whether biomass can be meaningfully estimated at fairly coarse taxonomic levels and compared between different habitats.

VI. TROPHIC ROLES, DISPERSAL DYNAMICS, AND RESPONSE TO SPATIAL–TEMPORAL HETEROGENEITY

Controlled experimentation in stream ecology has been traditionally championed by ecologists working on fish and macroinvertebrate communities. Similar work on stream meiofauna is less common, in part because such studies can be methodologically difficult in the hyporheic zone (Palmer, 1993; Palmer and Strayer, 1996). Nonetheless, stream meiofauna have been used to test basic ecological questions and address the role of meiofauna in hyporheic systems.

The importance of hyporheic meiofauna as food for higher trophic levels has been of interest for some time because they can be found in the guts of a wide variety of predators. Amores-Serrano (1991) found copepods, ostracods, and cladocerans in the gut contents of approximately 95% of the fish in an Arkansas stream. Hyporheic meiofauna have also been confirmed as important in the diet of larval fish. In field experiments conducted during summer baseflow in a sandy stream, abundances of meiofauna were reduced significantly inside cages with fish (at densities commonly found in the stream) versus fishless cages or cage controls (Shofner, 1999). Field experiments in an Arkansas stream also suggested that fish predation on stream meiofauna may be significant (Pope and Brown, 1996).

The effect of stream macroinvertebrates (some of which are known meiofauna predators) on hyporheic meiofauna has received little study, though crustaceans have been found in macroinvertebrate guts (Sherberger and Wallace, 1971; Benke and Wallace, 1980). Chironomids in a Austrian stream were found to feed extensively on rotifers, although prey choice in general was found to be strongly related to prey availability (Schmid-Arraya and Schmid, 1995; Schmid and Schmid-Arraya, 1997). In a North Carolina stream, experimental removal of macroinvertebrates from a stream influenced meiofaunal abundances, but the impact was variable through time and space (O'Doherty, 1988). Although the nature of the interaction was indeterminate, this study clearly showed that macroinvertebrates can have an impact on meiofauna.

Macroinvertebrate and fish predators may influence the dispersal dynamics of hyporheic meiofauna. Fish predators increased the frequency and magnitude of meiofaunal movements causing an increase in emigration (Shofner and Palmer, 1997). Changes in flow are also well known to influence meiofaunal movement and distribution. Meiofauna aggregate in areas of the streambed characterized by low shear stress (Robertson *et al.,* 1995, 1997). They also make small-scale vertical migrations within the hyporheic zone in response to small changes in surface velocities but are unable to migrate deep into the streambed in order to avoid high flows during spates (Palmer *et al.,* 1992). During high flows, meiofauna are often eroded just like grains of sand and may be deposited in areas of reduced flows such as backwater regions (Richardson, 1991) or near debris dams (Palmer *et al.,* 1995, 1996a; see also Boulton, Chapter 14 in this volume). Thus, the notion that the hyporheic zone is a refuge for all invertebrates during spates (see review in Palmer *et al.,* 1992) may not be true for meiofauna, particularly in sandy streams. Larger macrofauna, or some species of meiofauna in gravel/cobble streams, may indeed be able to quickly migrate into the hyporheic zone prior to or during spates (Dole-Olivier *et al.,* 1997).

Meiofauna may also enter the water column actively in response to changes in flow or to habitat (food) quality (Palmer, 1992; Tokeshi, 1994). Hyporheic meiofauna may move up to the surface of the streambed and/ or onto the surfaces of wood, leaves, and vegetation when flows are low (less than critical erosion velocity for the sediments) or when they find themselves in unsuitable habitats following spates (Palmer, 1990b, 1992; Palmer *et al.,* 1995, 1996a). Indeed, sand-dwelling meiofauna appear to track the presence and quality of leaf matter on and in hyporheic sediments (Swan, 1997; Palmer *et al.,* 1999). When patch quality was manipulated (amount of microbial and fungal biomass and production) in the hyporheic zone, meiofaunal abundances (especially chironomids) increased dramatically in high quality areas (Swan, 1997). Similarly, fitness (measured as egg output and growth rate) of chironomids on these patches was higher than on patches of lower quality (Ward and Cummins, 1979; Stanko-mischic *et al.,* 1999).

The responses of meiofauna to spatial heterogeneity in food resources and to temporal heterogeneity in flow has been long recognized by marine ecologists (Decho and Fleeger, 1988). Recent work by Swan (1997), Stankomishic *et al.* (1999), Palmer *et al.* (1999) and Tokeshi (1994) suggests that stream meiofauna may be good experimental organisms for testing resource and patch dynamics theory. Manipulations of the magnitude, direction (upwelling vs. downwelling), and variability in hydraulic gradients and measurement of the meiofaunal response may be a fruitful avenue for understanding the importance of environmental heterogeneity (over time or space) to stream ecosystems (Palmer *et al.*, 1997).

VII. MEIOFAUNAL–MICROBIAL INTERACTIONS

Very little is known about the relationship between meiofauna and microbial communities in streams despite their obvious trophic interactions. Meiofauna feed on bacterial cells associated with detritus (Perlmutter and Meyer, 1991) and on small algae and bacteria associated with inorganic sediments (Borchardt and Bott, 1995). Chironomids and copepods have been shown to preferentially assimilate bacterial carbon compared to detrital particulate organic matter (Hall, 1995).

A more comprehensive understanding of the relationship between meiofauna and bacteria exists in marine systems where microbial food web dynamics have been more extensively studied (Table III). Giere (1993) reviewed the marine literature and emphasized six major points. First, although meiofauna feed on bacteria, they rarely reduce bacterial biomass (in part because bacterial productivity is so rapid). Meiofauna are important in removing dead cells from the microbial community and in so doing, enhance bacterial growth rates (Meyer-Reil and Faubel, 1980; Montagna, 1984, 1995). Second, meiofauna may increase microbial growth rates by increasing the supply of limiting resources, such as essential nutrients and providing organic matter (mucus secretions, Gerlach, 1978; Riemann and Schrage, 1978; Alkemade *et al.*, 1992). Third, meiofauna may increase fluxes of oxygen and nutrients to bacteria in deeper sediment layers through bioturbation (creating burrows and tubes that increase water movement through sediments), although the spatial scale of bioturbation effects may be rather small (i.e., a few centimeters; Aller and Aller, 1992; Green and Chandler, 1994). Fourth, meiofauna may make detritus more accessible to bacteria by mechanically breaking it down into smaller pieces (Tenore *et al.*, 1977; Chauvet *et al.*, 1993). Fifth, meiofauna may serve as food for bacteria following death. Sixth, meiofauna may be symbiotic partners for specialized bacteria (e.g., sulfate reducers, Giere and Langheld, 1987; Fenchel and Finlay, 1989).

Examining the role of small invertebrates in microbial dynamics has greatly increased our understanding of ecosystem processes in both plank-tonic (lakes and oceans) and benthic (marine) habitats. Almost nothing is known about these same dynamics in stream hyporheic systems, and this should clearly be an area of exciting future research. Given that meiofauna influence microbial growth and abundances in other benthic systems, meio-fauna likely play a central role in influencing many processes in stream hy-porheic systems including decomposition, primary production, and trophic exchanges.

VIII. MEIOFAUNA AND ECOSYSTEM ECOLOGY

We conclude this chapter with a hypothesis concerning how hyporheic meiofauna may have an impact on stream ecosystem functioning (e.g., oxy-gen consumption, secondary production, decomposition). The importance of meiofauna to hyporheic functioning is virtually unstudied, and we wish to present ideas for generating discussion and future research. Here we fo-cus specifically on meiofauna in the hyporheic zone; the potential importance of all freshwater invertebrates to ecosystem processes in sediments has also been recently reviewed (Palmer *et al.*, 1997).

We begin by reiterating several key points. First, hyporheic meiofauna are abundant, taxonomically diverse, and have rapid generation times. Sec-ond, meiofauna are mobile, both interstitially throughout a wide range of sediment grain sizes, and between different habitat patches within the same stream. Third, at least some species, from a wide variety of taxa, are toler-ant of low oxygen and perhaps anoxia. Fourth, meiofauna are trophic links between the macroinvertebrate/fish community (as prey) and the microbial community (via predation or by stimulating microbial metabolism). Fifth, meiofauna may influence the rates of various microbial processes by en-hancing the availability of limiting resources.

Given the strong evidence that hyporheic microbial processes are ex-tremely important to stream ecosystem functioning (Fischer *et al.*, 1996; see also Findlay and Sobczak, Chapter 12 in this volume; Baker *et al.*, Chap-ter 11 in this volume), we ask the question: under what conditions might meiofaunal processes influence rates of hyporheic microbial processes? We suggest that meiofauna may be most important to ecosystem functioning when bulk interstitial flow rates are very low (i.e., low hydraulic conduc-tivity). We outline our logic through the presentation of three scenarios cho-sen from points on the continuum of low to high interstitial water flow rates. Interstitial water flow rates should be important to a wide variety of both meiofaunal and microbial processes (Fig. 2) especially by controlling how quickly new resources enter the hyporheic zone (Strayer, 1994b; Find-lay, 1995; Kaplan and Newbold, Chapter 10 in this volume; Findlay and Sobczak, Chapter 12 in this volume).

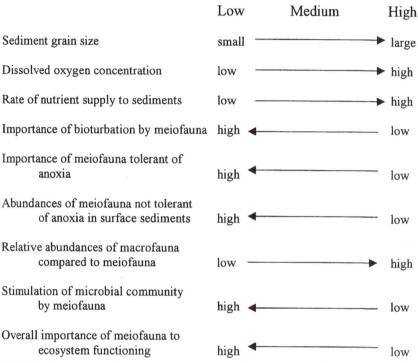

FIGURE 2   Proposed relationship between abiotic and biotic parameters in streams of varying interstitial flow rates.

A. Scenario A: High Interstitial Flow Rates

High interstitial flow rates (>10 cm sec^{-1}; e.g., Flathead River; Ward and Stanford, 1993) may cause selection for species that can withstand erosion and high shear stresses, and the supply of nutrients and oxygen is probably large enough to meet the metabolic demands of the microbial community. Rapid interstitial flow rates are typical in systems with coarse sediments (e.g., cobble, gravel) because there is often a positive relationship between grain size and interstitial flow rates (Findlay, 1995; Harvey and Wagner, Chapter 1 in this volume). Large grain sizes would allow colonization of the hyporheic zone by macroinvertebrates. Whereas meiofauna can be abundant in such systems (Ward and Voelz, 1990), their impact on overall ecosystem processes is probably overwhelmed by large hydrologic flux and may be small relative to the effects of microbes and macroinvertebrates (see Boulton, Chapter 14 in this volume). Thus, we predict that meiofauna will have the least impact on hyporheic ecosystem processes when interstitial water flow rates are very large.

B. Scenario B: Intermediate Interstitial Flow Rates

At intermediate interstitial flow rates (2–7 cm sec^{-1}; e.g., Sycamore Creek; Valett *et al.*, 1990), we suggest that meiofauna should be more important to ecosystem processes than at higher interstitial flow rates. This should be true especially when decreased interstitial flow rates are associated with decreased sediment grain size and smaller interstitial pore spaces, and thus fewer macrofauna relative to meiofauna (see Fig. 1; Boulton, Chapter 14 in this volume). Dissolved oxygen, nutrients, and organic matter should be supplied at sufficient rates from surface waters to maintain oxic conditions. In these systems, meiofauna should be important in the breakdown of larger-sized organic matter, increasing microbial colonization and mineralization, as well as food resources to larger filter-feeding meiofauna such as rotifers, ciliates, and crustaceans. By grazing on attached microbial communities, meiofauna such as nematodes, crustaceans, and oligochaetes should stimulate microbial growth and production. Thus, rates of total meiofaunal community secondary production and organic matter mineralization should be large. Upwelling sites may become "hotspots" of predation on meiofauna by macroinvertebrates and fish given the large secondary production by meiofauna. In such systems, we predict that meiofauna contribute greatly to total hyporheic secondary production both directly via their production and indirectly by increasing rates of microbial production.

C. Scenario C: Low Interstitial Flow Rates

At low interstitial flow rates (<2 cm sec^{-1}; e.g., Goose Creek; Hakenkamp, 1997), microzones should start to form within the hyporheic zone as either dissolved oxygen, organic matter, or nutrients become locally depleted (Paerl and Pinckney, 1996). In this scenario, bioturbation by meiofauna may become important in providing limiting resources to individual microhabitats. Meiofauna will have different roles in this type of system depending on their tolerance for low oxygen conditions. Taxa that are tolerant of low oxygen conditions will be able to move among microhabitats, graze the microbial biofilms, and perhaps provide the only sources of organic matter or nutrients in the form of secretions and fecal material. Taxa that are sensitive to low oxygen conditions will become concentrated in the upper few centimeters of the streambed (e.g., as in silty or muddy sediments). In marine sediments of this type, bioturbation by meiofauna has been found to be an important determinant of solute transport across the sediment–water interface (Aller and Aller, 1992; Green and Chandler, 1994), and predation by fish on meiofauna is often large because meiofaunal biomass is concentrated near the sediment–water interface (Coull, 1990). In streams, this concentration of meiofauna near the streambed surface may increase the number of passively eroded meiofauna that enter the drift (Shiozawa, 1986). In low flow interstitial systems, we predict that meiofauna will cause increased microbial growth and production. Meiofaunal bioturbation will foster solute transport across the sediment–water interface, and predation on meiofauna by fish and macrofauna may be large.

In summary, we suggest that meiofauna may be extremely important to stream ecosystem functioning when interstitial flow rates are intermediate to low (Fig. 2). In low flow sediment systems, meiofauna may have very different roles depending on their depth in the hyporheic zone and sensitivity to anoxic conditions. At the surface of the sediments, where biomass may be greatest, meiofauna would be involved in solute transport through bioturbation and as food for fish and macroinvertebrates. In deeper hyporheic sediments, meiofauna tolerant of low dissolved oxygen conditions may be important in stimulating microbial processes. Given that we know almost nothing about the role of hyporheic meiofauna in stream ecosystem functioning across hyporheic systems, these hypotheses remain to be tested.

IX. CONCLUSIONS AND FUTURE RESEARCH NEEDS

In the last few decades, research on hyporheic meiofauna has greatly increased, aided in part by what was already known about the important role

TABLE VII   Research Needed on Meiofauna That Would Contribute Most to the Advancement of Hyporheic Ecology

1. Broader spatial scale of examination of surface versus hyporheic invertebrate abundance, biomass, and biodiversity to facilitate generalizations about meiofaunal communities in and streams.
2. Greater taxonomic training of hyporheic researchers to expand what little is known about meiofaunal species in streams (especially in North America)
3. Explicit comparison of the usefulness and bias of different sampling techniques, across a variety of sediment grain sizes.
4. Experimental manipulation of critical factors (e.g., oxygen level and food availability) that are thought to control the distribution of meiofauna.
5. Greater study of size and biomass spectra of hyporheic fauna (including protists) to allow comparisons between streams and to facilitate consideration of factors restricting growth and reproduction.
6. Experiments to study the importance of meiofauna to rates of ecological processes in the hyporheic zone, especially nutrient, carbon, and oxygen dynamics.
7. Examination of the nature of interactions (direct and indirect) between macrofauna and meiofauna.
8. Experiments to determine whether predation on meiofauna is enhanced or diminished by meiofaunal movements into and out of the hyporheic zone.
9. Study of the relationship between meiofauna and microbes (bacteria, small protists, fungi) determining whether meiofauna influence rates of microbial processes in hyporheic sediments.

of meiofauna in marine systems. Many issues require further study, and we present a list of some we believe are most important and immediate in Table VII. It is clear that hyporheic meiofauna are likely important players in both community and ecosystem processes in streams. They are often a dominant component of the hyporheos in terms of abundances, species diversity, and total biomass, and they have the ability to move rapidly between different sediment patches within the stream and rapidly recolonize those patches after floods. Meiofauna are also likely to be important trophic links between the macroinvertebrate–fish community and the microbial community and may influence the rates of microbial processes.

REFERENCES

Alkemade, R., A. Wielemaker, S. A. deJong, and A. J. J. Dandee. 1992. Experimental evidence for the role of bioturbation by the marine nematode *Diplolaimelloides dievengatensis* in stimulating the mineralization of *Spartina anglica* detritus. *Marine Ecology Progress Series* 90:149–155.

Aller, R. C., and J. Y. Aller. 1992. Meiofauna and solute transport in marine muds. *Limnology and Oceangraphy* 37:1018–1033.

Amores-Serrano, R. R. 1991. Stream dwelling meiofauna: Temporal abundance, distribution and utilization by larval fish. Ph.D. Dissertation, University of Arkansas, Little Rock.

Benke, A. C., and J. B. Wallace. 1980. The trophic basis of production among net-spinning caddisflies in a southern Applachian stream. *Ecology* 61:108–118.

Borchardt, M. A., and T. L. Bott. 1995. Meiofaunal grazing of bacteria and algae in a Piedmont stream. *Journal of the North American Benthological Society* 14:278–298.

Botosaneanu, L. 1986. "Stygofauna Mundi." Brill, Leiden, The Netherlands.

Boulton, A. J., H. M. Valett, and S. G. Fisher. 1992. Spatial distribution and taxonomic composition of the hyporheos of several Sonoran Desert streams. *Archiv für Hydrobiologie* 125:37–61.

Bourassa, N., and A. Morin. 1995. Relationships between size structure of invertebrate assemblages and trophy and substrate composition in streams. *Journal of the North American Benthological Society* 14:393–403.

Braioni, M. G., and M. Gottardi. 1979. I Rotiferi dell'Adige: Confronto tra il popolamento interstiziale quello bentico-perifitico. *Bolletino de Museo Civico di Storia Naturale di Verona* 6:187–219.

Bretschko, G. 1990. The effect of escape reactions on the quantitative sampling of gravel stream fauna. *Archiv für Hydrobiologie* 120:41–49.

Bretschko, G. 1992. The sedimentfauna in the uppermost parts of the impoundment "Altenworth" (Danube, stream km 2005 and 2007). *Archiv für Hydrobiologie, Supplementband* 84:131–168.

Bretschko, G., and E. Christian. 1989. Collembola in the bed sediments of an alpine gravel stream (RITRODAT-Lunz Study Area, Austria). *Internationalen Revue der gesamten Hydrobiologie* 74:491–498.

Cattaneo, A. 1993. Size spectra of benthic communities in Laurentian streams. *Canadian Journal of Fisheries and Aquatic Sciences* 50:2659–2666.

Chauvet, E., N. Giani, and M. O. Gessner. 1993. Breakdown and invertebrate colonization of leaf litter in two contrasting streams: significance of oligochaetes in a large river. *Canadian Journal of Fisheries and Aquatic Sciences* 50:488–495.

Coireau, N. 1971. Les isopodes interstitiels: Documents sur leur écologie et leur biologie. *Mémoires des Museum National d'Histoire Naturelle, Series A (Paris)* **64**:1–170.

Coull, B. C. 1988. Ecology of marine meiofauna. *In* "Introduction to the Study of Meiofauna" (R. P. Higgins and H. Thiel, eds.), pp. 18–38. Smithsonian Press, Washington, DC.

Coull, B. C. 1990. Are members of the meiofauna food for higher trophic levels? *Transactions of the American Microscopical Society* **109**:233–246.

Coull, B. C., and G. T. Chandler. 1992. Pollution and meiofauna: Field, laboratory, and mesocosm studies. *Oceanography and Marine Biology* **30**:191–271.

Coull, B. C., and O. Giere. 1988. The history of meiofaunal research. *In* "Introduction to the Study of Meiofauna" (R. P. Higgins and H. Thiel, eds.), pp. 14–17. Smithsonian Press, Washington, DC.

Coull, B. C., and M. A. Palmer. 1984. Field experimentation in meiofaunal ecology. *Hydrobiologia* **118**:1–19.

Creuzé des Châtelliers, M., and M. J. Dole-Olivier. 1991. Tests of the Bou-Rouch sampler used for interstitial fauna capture. (I). Chemical tracing with NaCl. *Comptes Rendus des Seances de l'Academie des Sciences* **312**:671–676.

Danielopol, D. L. 1980. Sur la biologie de quelques Ostracodes Canoninae épigés ets hypogés d'Europe. *Bulletin du Museum National d'Histoire Naturelle* **2**:471–506.

Danielopol, D. L. 1982. Phreatobiology reconsidered. *Polskie Archiwum Hydrobiologii* **29**:375–386.

Danielopol, D. L. 1984. Ecological investigations on the alluvial sediments of the Danube in the Vienna area—a phreatobiological project. *Verhandlungen Internationale Vereinigung für Theoretische und Angewandte Limnologie* **22**:1755–1761.

Decho, A. W., and J. W. Fleeger. 1988. Microscale dispersion of meiobenthic copepods in response to food-resource patchiness. *Journal of Experimental Marine Biology and Ecology* **118**:229–244.

Dole-Olivier, M. J., P. Marmonier, and J.-L. Beffy. 1997. Response of invertebrates to lotic disturbance: Is the hyporheic zone a patchy refugium? *Freshwater Biology* **37**:257–276.

Dumont, H. J., and S. Negrea. 1996. Conspectus of the Cladocera of the subterranean waters of the world. *Hydrobiologia* **325**:1–30.

Eder, R. 1983. Nematoden aus dem interstitiial der Donau bei Fischamend (Niederösterreich). *Archiv für Hydrobiologie, Supplementband* **68**:100–113.

Fenchel, T. 1978. The ecology of micro- and meiobenthos. *Annual Review of Ecology and Systematics* **9**:99–121.

Fenchel, T. M., and B. J. Finlay. 1989. *Kentrophoros*: A mouthless ciliate with a symbiotic kitchen garden. *Ophelia* **30**:75–93.

Findlay, S. 1995. Importance of surface-subsurface exchange in stream ecosystems: The hyporheic zone. *Limnology and Oceanography* **40**:159–164.

Fischer, H., M. Pusch, and J. Schwoerbel. 1996. Spatial distribution and respiration of bacteria in stream-bed sediments. *Archiv für Hydrobiologie* **137**:281–300.

Fleeger, J. W., and A. W. Decho. 1987. Spatial variability of interstitial meiofauna: A review. *Stygologia* **3**:35–54.

Fraser, B. G. and D. D. Williams. 1997. Accuracy and precision in sampling hyporheic fauna. *Canadian Journal of Fisheries and Aquatic Sciences* **54**:1135–1141.

Gee, J. M. 1989. An ecological and economic review of meiofauna as food for fish. *Journal of the Linnean Society of London, Zoology* **93**:243–261,

Gerlach, S. A. 1971. On the importance of marine meiofauna for benthos communities. *Oecologia* **6**:176–190.

Gerlach, S. A. 1978. Food-chain relationships in subtidal silty-sand marine sediments and the role of meiofauna in stimulating bacterial growth. *Oecologia* **33**:55–69.

Gibert, J., D. L. Danielopol, and J. A. Stanford. 1994. "Groundwater Ecology." Academic Press, San Diego, CA.

Giere, O. 1993. "Meiobenthology." Springer-Verlag, Berlin.

Giere, O., and C. Langheld. 1987. Structural organization, transfer and biological fate of endosymbiotic bacteria in gutless oligochaetes. *Marine Biology* **93**:641–650.

Gledhill, L. 1982. Water-mites (Hydrachnellae, Limnohalacaridae, Acari) from the interstitial habitat of riverine deposits in Scotland. *Polskie Archiwum Hydrobiologii* **29**:439–451.

Golladay, S. W., and C. L. Hax. 1995. Effects of an engineered flow disturbance on meiofauna in a north Texas prairie stream. *Journal of the North American Benthological Society* **14**:404–413.

Green, A. S., and G. T. Chandler. 1994. Meiofaunal bioturbation effects on the partitioning of sediment-associated cadmium. *Journal of Experimental Marine Biology and Ecology* **180**:59–70.

Hakenkamp, C. C. 1997. Oxygen consumption in streambeds: Examining the impacts of environmental factors and hyporheic fauna. Ph.D. Dissertation, University of Maryland, College Park.

Hall, R. O. 1995. Use of a stable isotope addition to trace bacterial carbon through a stream food web. *Journal of the North American Benthological Society* **14**:269–277.

Hanson, J. M., E. E. Prepas, and W. C. Mackay. 1989. Size distribution of the macroinvertebrate community in a freshwater lake. *Canadian Journal of Fisheries and Aquatic Sciences* **46**:1510–1519.

Higgins, R. P., and H. Thiel, eds. 1988. "Introduction to the Study of Meiofauna." Smithsonian Press, Washington, DC.

Hummon, W. D. 1981. Extraction by sieving: A biased procedure in studies of stream meiobenthos. *Transactions of the American Microscopical Society* **100**:278–284.

Hummon, W. D. 1987. Meiobenthos of the Mississippi headwaters. *In* "Biology of Tardigrades" (R. Bertolani, ed.), pp. 125–140. Mucchi, Modena.

Hummon, W. D., W. A. Evans, M. R. Hummon, F. G. Doherty, R. H. Waiberg, and S. Stanley. 1978. Meiofaunal abundance in sandbars of acid mine polluted, reclaimed, and unpolluted streams in southeastern Ohio. *In* "Energy and Environmental Stress in Aquatic Ecosystems" (J. H. Thorp and J. W. Gibbons, eds.), DOE Symposium Series, pp. 188–203. National Technical Information Services, Springfield, IL.

Kemp, P. F. 1990. The fate of benthic bacterial production. *Reviews in Aquatic Sciences* **2**:109–123.

Klemens, W. E. 1991. Quantitative sampling of bed sediments (Ritrodat-Lunz study area, Austria). *Verhandlungen Internationale Vereinung für Theoretische und Angewandte Limnologie* **24**:1926–1929.

Kolasa, J. 1982. On the origin of stream interstitial Microturbellarians. *Polskie Archiwum Hydrobiologii* **29**:405–413.

Kolasa, J. 1983. Formation of the turbellarian fauna in a submontane stream in Italy. *Acta Zoologica Cracoviensia* **26**:297–354.

Kolasa, J., D. Strayer, and E. Bannon-O'Donnell. 1987. Microturbellarians from interstitial waters, streams, and springs in southeastern New York. *Journal of the North American Benthological Society* **6**:125–132.

Kowarc, A. V. 1990. Production biology of *Limnocamptus echinatus* (Harpacticoida) in a second order gravel stream. *Stygologia* **5**:25–32.

Kowarc, A. V. 1992. Depth distribution and mobility of a harpacticoid copepod within the bed sediment of an alpine brook. *Regulated Rivers* **7**:57–63.

Lafont, M., A. Durbec, and C. Ille. 1992. Oligochaete worms as biological describers of the interactions between surface and groundwaters: A first synthesis. *Regulated Rivers* **7**:65–73.

Lescher-Moutoné, F. 1973. Sur la biologie et l'ecologie des Copépodes Cyclopides hypogés (Crustacés). *Annales de Spéléologie* **28**:429–502, 581–674.

Mare, M. F. 1942. A study of a marine benthic community with special reference to the microorganisms. *Journal of Marine Biology* **25**:517–554.

Marmonier, P., and M. Creuzé des Châtelliers. 1992. Biogeography of the benthic and interstitial living ostracods (Crustacea) of the Rhone River (France). *Journal of Biogeography* 19:693–704.

Marmonier, P., and J. Ward. 1990. Superficial and interstitial Ostracoda of the South Platte River (Colorado, U.S.A.)—systematics and biogeography. *Stygologia* 5:225–239.

Marmonier, P., P. Vervier, J. Gibert, and M.-J. Dole-Olivier. 1993. Biodiversity in ground waters. *Trends in Ecology and Evolution* 8:392–395.

Mauclaire, L., P. Marmonier, and J. Gibert. 1998. Sampling water and sediment in interstitial habitats: A comparison of coring and pumping techniques. *Archiv für Hydrobiologie* 142:111–123.

McElravy, E. P., and V. H. Resh. 1991. Distribution and seasonal occurrence of the hyporheic fauna in a northern California stream. *Hydrobiologia* 220:233–246.

Meyer, J. L. 1990. A blackwater perspective on riverine ecosystems. *BioScience* 40:643–651.

Meyer-Reil, L.-A., and A. Faubel. 1980. Uptake of organic matter by meiofauna organisms and interrelationships with bacteria. *Marine Ecology: Progress Series* 3:251–256.

Montagna, P. A. 1984. In situ measurement of meiobenthic grazing rates on sediment bacteria and edaphic diatoms. *Marine Ecology: Progress Series* 18:119–130.

Montagna, P. A. 1995. Rates of metazoan meiofaunal microbivory: A review. *Vie Milieu* 45: 1–9.

Moodley, L., G. J. van der Zwaan, P. M. J. Herman, L. Kempers, and P. van Breugel. 1997. Differential response of benthic meiofauna to anoxia with special reference to Foraminifera (Protista: Sarcodina). *Marine Ecology: Progress Series* 158:151–163.

Morin, A., and D. Nadon. 1991. Size distribution of epilithic lotic invertebrates and implications for community metabolism. *Journal of the North American Benthological Society* 10:300–308.

O'Doherty, E. C. 1988. The ecology of meiofauna in an Appalachian headwater stream. Ph.D. Dissertation, University of Georgia, Athens.

Paerl, H. W., and J. L. Pinckney. 1996. A mini-review of microbial consortia: their roles in aquatic production and biogeochemical cycling. *Microbial Ecology* 31:225–247.

Palmer, M. A. 1990a. Temporal and spatial dynamics of meiofauna within the hyporheic zone of Goose Creek, Virginia. *Journal of the North American Benthological Society* 9:17–25.

Palmer, M. A. 1990b. Understanding the movement dynamics of a stream-dwelling meiofauna community using marine analogs. *Stygologia* 5:67–74.

Palmer, M. A. 1992. Incorporating lotic meiofauna into our understanding faunal transport processes. *Limnology and Oceanography* 37:329–341.

Palmer, M. A. 1993. Experimentation in the hyporheic zone: challenges and prospectus. *Journal of the North American Benthological Society* 12:84–93.

Palmer, M. A., and D. L. Strayer. 1996. Meiofauna. In "Methods in Stream Ecology" (F. R. Hauer and G. A. Lamberti, eds.), pp. 315–337. Academic Press, San Diego, CA.

Palmer, M. A., A. E. Bely, and K. E. Berg. 1992. Response of invertebrates to lotic disturbance: A test of the hyporheic refuge hypothesis. *Oecologia* 89:182–194.

Palmer, M. A., P. Arensburger, P. S. Botts, C. C. Hakenkamp, and J. W. Reid. 1995. Disturbance and the community structure of stream invertebrates: patch-specific effects and the role of refugia. *Freshwater Biology* 34:343–356.

Palmer, M. A., P. Arensburger, A. P. Martin, and D. W. Denman. 1996a. Disturbance and patch-specific responses: The interactive effects of woody debris and floods on lotic invertebrates. *Oecologia* 105:247–257.

Palmer, M. A., C. C. Hakenkamp, and K. Nelson-Baker. 1996b. Ecological heterogeneity in streams: Why variance matters. *Journal of the North American Benthological Society* 16:189–202.

Palmer, M. A., A. P. Covich, B. Finlay, J. Gibert, K. D. Hyde, R. K. Johnson, T. Kairesalo, S. Lake, C. R. Lovell, R. J. Naiman, C. Ricci, F. Sabater, and D. Strayer. 1997. Biodiversity and ecosystem function in freshwater sediments. *Ambio* 26:571–577.

Palmer, M. A., C. M. Swan, K. Nelson, P. Silver, and R. Alvestad. 1999. Streambed landscapes:

Evidence that stream invertebrates respond to the type and spatial arrangement of patches. *Landscape Ecology,* In Press.

Pennak, R. W. 1951. Comparative ecology of the interstitial fauna of fresh-water and marine beaches. *Année Biologique* 27:217–249.

Pennak, R. W. 1985. The fresh-water invertebrate fauna: Problems and solutions for evolutionary success. *American Zoologist* 25:671–687.

Pennak, R. W. 1988. Ecology of freshwater meiofauna. *In* "Introduction to the Study of Meiofauna" (R. P. Higgins and H. Thiel, eds.), pp. 39–60. Smithsonian Press, Washington, DC.

Pennak, R. W., and J. V. Ward. 1986. Interstitial fauna communities of the hyporheic and adjacent groundwater biotopes of a Colorado mountain stream. *Archiv für Hydrobiologie, Supplementband* 74:356–396.

Perlmutter, D. G., and J. L. Meyer. 1991. The impact of a stream-dwelling harpacticoid copepod upon detritially associated bacteria. *Ecology* 72:2170–2180.

Plenet, S., and J. Gibert. 1994. Invertebrate community responses to physical and chemical factors at the river/aquifer interaction zone I. Upstream from the city of Lyon. *Archiv für Hydrobiologie* 132:165–189.

Poff, N.L., M. A. Palmer, P. L. Angermeier, R. L. Vadas, Jr., C. C. Hakenkamp, A. Bely, P. Arensburger, and A. P. Martin. 1993. The size structure of a metazoan community in a Piedmont stream. *Oecologia* 95:202–209.

Pope, M. L., and A. V. Brown. 1996. Effects of flow rate, predation and diel periodicity on drifting and benthic meiofauna densities in artificial streams. *Bulletin of the North American Benthological Society* 13:135.

Reid, J. W., and D. L. Strayer. 1994. *Diacyclops dimorphus*, a new species of copepod from Florida, with comments on morphology of interstitial cyclopine cyclopoids. *Journal of the North American Benthology Society* 13:250–265.

Richardson, W. B. 1991. Seasonal dynamics, benthic habitat use, and drift of zooplankton in a small stream in southern Oklahoma, U.S.A. *Canadian Journal of Zoology* 69:748–756.

Richardson, W. B. 1992. Microcrustacea in flowing water: Experimental analysis of washout times and a field test. *Freshwater Biology* 28:217–230.

Riemann, F., and M. Schrage. 1978. The mucus-trap hypothesis on feeding of aquatic nematodes and implications for biodegradation and sediment texture. *Oecologia* 34:75–88.

Robertson, A.L., J. Lancaster, and A. G. Hildrew. 1995. Stream hydraulics and the distribution of microcrustacea: A role for refugia? *Freshwater Biology* 33:469–484.

Robertson, A. L., J. Lancaster, L. R. Belyea, and A. G. Hildrew. 1997. Hydraulic habitat and the community structure of stream benthic microcrustacea. *Journal of the North American Benthological Society* 16:562–575.

Rouch, R. 1968. Contribution à la connaissance des Harpacticides hypogés (Crustacés Copépodes). *Annales de Spéléologie* 23:5–167.

Rouch, R. 1988. Sur la répartition spatiale des Crustacés dans le sous-écoulement d'un ruisseau des Pyrénées. *Annales de Limnologie* 24:213–234.

Rouch, R. 1991. Structure du peuplement des Harpacticides dans le milieu hyporheique d'un ruisseau des Pyrénées. *Annales de Limnologie* 27:227–241.

Rouch, R. 1995. Peuplement des Crustacés dans la zone hyporheique d'un ruisseau des Pyrénées. *Annales de Limnologie* 31:9–28.

Rouch, R., and D. L. Danielopol. 1987. L'origine de la faune aquatique souterraine; entre le paradigme du refuge et le modele de la colonisation active. *Stygologia* 3:345–372.

Rundle, S. D. 1993. Temporal and demographic patterns of microcrustacean populations in upland Welsh streams of contrasting pH. *Archiv für Hydrobiologie* 128:91–106.

Rundle, S. D., and M. J. Attrill. 1995. Comparison of meiobenthic crustacean community structure across freshwater acidification gradients. *Archiv für Hydrobiologie* 133:441–456.

Rundle, S. D., and S. J. Ormerod. 1991. The influence of chemistry and habitat features on the microcrustacea of some upland Welsh streams. *Freshwater Biology* 26:439–451.

Ruse, L. P. 1994. Chironomid microdistribution in gravel of an English chalk river. *Freshwater Biology* **32**:533–551.

Schmid, P. E. 1992. Community structure of larval Chironomidae (Diptera) in a backwater area of the River Danube. *Freshwater Biology* **27**:151–167.

Schmid, P. E. 1993. Random patch dynamics of larval Chironomidae (Diptera) in the bed sediments of a gravel stream. *Freshwater Biology* **30**:239–255.

Schmid, P. E., and J. M. Schmid-Araya. 1997. Predation on meiobenthic assemblages: Resource use of a tanypod guild (Chironomidae, Diptera) in a gravel stream. *Freshwater Biology* **38**:67–91.

Schmid-Araya, J. M. 1993. Benthic rotifera inhabiting the bed sediments of a mountain gravel stream. *Journal of the Biological Station Lunz* **14**:75–101.

Schmid-Araya, J. M. 1994a. Spatial and temporal distribution of micro-meiofaunal groups in an alpine stream. *Verhandlungen Internationale Vereingung für Theoretische und Angewandte Limnologie* **25**:2649–1655.

Schmid-Araya, J. M. 1994b. Temporal and spatial distribution of benthic microfauna in sediments of a gravel streambed. *Limnology and Oceanography* **39**:1813–1821.

Schmid-Araya, J. M. 1995. Disturbance and population dynamics of rotifers in bed sediments. *Hydrobiologia* **313/314**:279–290.

Schmid-Araya, J. M., and P. E. Schmid. 1995. Preliminary results on diet of stream invertebrate species: The meiofaunal assemblages. *Journal of the Biological Station Lunz* **15**:23–31.

Schwank, P. 1981–1982. Turbellarien, Oligochaeten und Archianneliden des Breitenbachs und anderer oberhessischer Mittelgebirgsbache. I-IV. *Archiv für Hydrobiologie, Supplementband* **62**:1–85, 86–147, 191–253, 254–290.

Schwoerbel, J. 1961. Subterrane wassermilben (Acari: Hydrachnellae, Porohalacaridae und Stygothrombiidae), ihre Ökologie und Bedeutung für Abgrenzung eines aquatischen Lebensraumes zwischen Oberfläche und Grundwasser. *Archiv für Hydrobiologie, Supplementband* **25**:242–306.

Schwoerbel, J. 1967. Das hyporheische interstitial als Grenzbiotop zwischen oberirdischem und subterranem Ökosystem und seine Bedeutung für die Primär-Evolution vo Kleinsthöhlenbeworhnern. *Archiv für Hydrobiologie, Supplementband* **33**:1–62.

Sherberger, F. F., and J. B. Wallace. 1971. Larvae of the southerastern species of *Mollana*. *Journal of the Kansas Entomological Society* **44**:217–224.

Shiozawa, D. K. 1986. The seasonal community structure and drift of microcrustaceans in Valley Creek, Minnesota. *Canadian Journal of Zoology* **64**:1655–1664.

Shiozawa, D. K. 1991. Microcrustacea from the benthos of nine Minnesota streams. *Journal of the North American Benthological Society* **10**:286–299.

Shofner, M. A. 1999. Predation, habitat patchiness, and prey exchange: Interactions between stream meiofauna and juvenile fish. Ph.D. Dissertation, University of Maryland, College Park.

Shofner, M. A., and M. A. Palmer. 1997. Predation risk and prey exchange across patch types: Interactions between juvenile fish and meiofauna. *Bulletin of the North American Benthological Society* **14**:136.

Stanko-mishic, S., J. K. Cooper, and P. Silver. 1999. Manipulation of habitat quality and effects on chironomid life history traits. *Freshwater Biology* **41**:1–9.

Strayer, D. 1985. The benthic micrometazoans of Mirror lake, New Hampshire. *Archiv für Hydrobiologie, Supplementband* **72**:287–426.

Strayer, D. 1986. The size structure of a lacustrine zoobenthic community. *Oecologia* **69**:513–516.

Strayer, D. L. 1988a. Crustaceans and mites (Acari) from hyporheic and other underground waters in southeastern New York. *Stygologia* **4**:192–207.

Strayer, D.L. 1988b. New and rare copepods (Cyclopoida and Harpacticoida) from freshwater interstitial habitats in southeastern New York. *Stygologia* **4**:279–291.

Strayer, D. L. 1991. Perspectives on the size structure of lacustrine zoobenthos, its causes, and its consequences. *Journal of the North American Benthological Society* 10:210–221.

Strayer, D.L. 1994a. Body size and abundance of benthic animals in Mirror Lake, New Hampshire. *Freshwater Biology* 32:83–90.

Strayer, D. L. 1994b. Limits to biological distributions in groundwater. *In* "Groundwater Ecology" (J. Gibert, D. L. Danielopol, and J. A. Stanford, eds.), pp. 287–310. Academic Press, San Diego, CA.

Strayer, D. L., and E. B. Bannon-O'Donnell. 1988. Aquatic microannelids (Oligochaeta and Aphanoneura) of underground waters of southeastern New York. *American Midland Naturalist* 119:327–335.

Strayer, D. L., and E. B. O'Donnell. 1992. The hyporheic nematode community of some streams in southeastern New York State, U.S.A. *Stygologia* 7:143–148.

Strayer, D. L., D. R. Nelson, and E. B. O'Donnell. 1994. Tardigrades from shallow groundwaters in southeastern New York, with the first record of *Thulinia* from North America. *Transactions of the American Microscopical Society* 113:325–332.

Strommer, J. L., and L. A. Smock. 1989. Vertical distribution and abundances of invertebrates within the sandy substrate of a low-gradient headwater stream. *Freshwater Biology* 22:263–274.

Swan, C. M. 1997. Heterogeneity in patch quality: Microbial—invertebrate dynamics in a sandy bottom stream. Master's Thesis, University of Maryland, College Park.

Tenore, K. R., J. H. Tietjen, and J. H. Lee. 1977. Effect of meiofauna on incorporation of aged eelgrass, *Zostera marina*, detritus by the polychaete *Nephthys incisa*. *Journal of the Fisheries Research Board of Canada* 34:563–567.

Tietjen, J. H. 1980. Microbial-meiofaunal interrelationships: A review. *In* "Aquatic Microbial Ecology" (R. R. Colwell and J. Foster, eds.), pp. 130–140. University of Maryland Press, College Park.

Tokeshi, M. 1994. Community ecology and patchy freshwater habitats. *In* "Aquatic Ecology: Scale, Pattern and Process" (P. S. Giller, A. G. Hildrew and D. G Raffaelli, eds.), pp. 63–90. Blackwell, Oxford.

Traunspurger, W., and C. Drews. 1996. Toxicity analysis of freshwater and marine sediments with meio- and macrobenthic organisms: A review. *Hydrobiologia* 328:215–261.

Turner, P.N., and M. A. Palmer. 1996. Notes on the species composition of the rotifer community inhabiting the interstitial sands of Goose Creek, Virginia with comments on habitat preferences. *Quekett Journal of Microscopy* 37:552–565.

Valett, H. M., S. G. Fisher, and E. H. Stanley. 1990. Physical and chemical characteristics of the hyporheic zone of a Sonoran Desert stream. *Journal of the North American Benthological Society* 9:201–215.

Vila, P. B. 1989. The occurrence and significance of Cladocera (Crustacea) in some streams of central Indiana. *Hydrobiologia* 171:201–214.

Vopel, K., J. Dehmlow, and G. Arlt. 1996. Vertical distribution of *Cletocamptus confluencs* (Copepoda, Harpacticoida) in relation to oxygen and sulphide microprofiles of a brackish water sulphuretum. *Marine Ecology: Progress Series* 141:129–137.

Wagner, H. P. 1994. A monographic review of the Thermosbaenacea. *Zoologische Verhandelingen* 291:1–338.

Ward, G. M., and K. W. Cummins. 1979. Effects of food quality on growth of a stream detritivore. *Paratendipes albimanus* (Meigen) (Diptera: Chironomidae). *Ecology* 60:57–64.

Ward, J. A., and J. V. Stanford. 1993. An ecosystem perspective of alluvial rivers: Connectivity and the hyporheic corridor. *Journal of the North American Benthological Society* 12:48–60.

Ward, J. V., and M. A. Palmer. 1994. Distribution patterns of interstitial freshwater meiofauna over a range of spatial scales, with emphasis on alluvial river-aquifer systems. *Hydrobiologia* 287:147–156.

Ward, J. V., and N. J. Voelz. 1990. Gradient analysis of interstitial meiofauna along a longitudinal stream profile. *Stygologia* 5:93–100.

Ward, J.V., and N. J. Voelz. 1994. Groundwater fauna of the South Platte System, Colorado. *In* "Groundwater Ecology" (J. Gibert, D. L. Danielopol, and J. A. Stanford, eds.), pp. 391–421. Academic Press, San Diego, CA.

Warwick, R. M. 1984. Species size distributions in marine benthic communities. *Oecologia* 61:32–41.

Warwick, R. M. 1989. The role of meiofauna in the marine ecosystem: evolutionary considerations. *Journal of the Linnean Society of London, Zoology* 96:229–241.

Warwick, R. M. 1993. Environmental impact studies on marine communities: Pragmatical considerations. *Australian Journal of Ecology* 18:63–80.

Whitman, R. L., and W. J. Clark. 1984. Ecological studies of the sand-dwelling community of an east Texas stream. *Freshwater Invertebrate Biology* 3:59–79.

Williams, D. D. 1984. The hyporheic zone as habitat for aquatic insects and associated arthropods. *In* "The Ecology of Aquatic Insects" (V. H. Resh and D. M. Rosenberg, eds.), pp. 430–455. Praeger, New York.

Williams, D. D. 1989. Towards a biological and chemical definition of the hyporheic zone in two Canadian rivers. *Freshwater Biology* 22:189–208.

Williams, D. D. 1993. Nutrient and flow vector dynamics at the hyporheic/groundwater interface and their effects on the interstitial fauna. *Hydrobiologia* 251:185–198.

Williams, D. D., and H. B. N. Hynes. 1974. The occurrence of benthos deep in the substratum of a stream. *Freshwater Biology* 4:233–256.

Wolz, E. R., and D. K. Shiozawa. 1995. Soft sediment benthic macroinvertebrate communities of the Green River at the Ouray National Wildlife Refuge, Uintah County, Utah. *Great Basin Naturalist* 55:213–224.

14

The Subsurface Macrofauna

Andrew Boulton

Division of Ecosystem Management
University of New England
Armidale, 2350 New South Wales
Australia

I. Introduction   337
II. Functional Classification of Subsurface Macrofauna   338
III. Factors Influencing the Distribution of Subsurface
Macrofauna   345
IV. The Functional Role of Subsurface Macrofauna   349
V. Potential Role as Biomonitors of Deteriorating Groundwater
Quality   354
VI. Conclusions   355
References   356

I. INTRODUCTION

The existence of invertebrate macrofauna (and meiofauna; Hakenkamp and Palmer, Chapter 13 in this volume) in subsurface waters below rivers and their banks has been known since early this century (Karaman, 1935), but it has only been in the last 15 years that the *functional* significance of this fauna to river ecosystems and their ecological processes has attracted interest. Although the subsurface habitat might be considered physically homogenous compared to the epigean (surface) habitat, it can harbor a diverse fauna that varies in body size, activity, phenology, and abundance. What factors influence the distribution of this fauna at fine and broad scales? How do subsurface macrofauna contribute to ecological processes occurring in this zone, and how does this tie in with their distribution? Does the significance of subsurface macrofauna vary among streams, and if so, why? Might subsurface macrofauna (e.g., biomonitoring groundwater pollution) potentially play roles in management of river ecosystems and their subsurface zones?

Conventionally, macrofauna are defined as invertebrate fauna retained by a 500-µm mesh even though the early life stages of many of these pass through mesh openings of this size (Hauer and Resh, 1996). To distinguish between definitions of macrofauna and meiofauna in the present text, I have generally followed the usage by Palmer and Strayer (1996) of a mesh-size of 40–500 µm to define the size-range of meiofauna, and greater than 500 µm to define macrofauna. Further, in this book we have distinguished taxonomic groups such as microcrustaceans, rotifers, tardigrades, nematodes and small oligochaetes that typify the meiofauna from peracarids, stonefly and mayfly nymphs, and other insects that are commonly recognized as macrofauna. In most subsurface environments where interstitial spaces are small, the invertebrate faunal diversity and production are dominated by meiofauna (Palmer and Strayer, 1996; Hakenkamp and Palmer, Chapter 13 in this volume). However, by virtue of their body size and activity, macrofauna may play an important role in some subsurface ecosystem processes. They are also likely to be constrained in their distribution by some features that are not as relevant to meiofauna (e.g., physical size of interstitial spaces, interstitial current velocity).

In this chapter, I discuss the present functional classification of the subsurface macrofauna, summarize the literature on physical and chemical parameters that influence their distribution in subsurface sediments at fine and broad scales, and review the scant data on the functional role of subsurface macrofauna, emphasizing the scope for experimental research to elucidate this further. I conclude with a discussion of their potential value as biomonitors of the effects of land use and pollution, especially on groundwater quality. I shall use the term "hyporheic" interchangeably with "subsurface," and reserve the term "hyporheos" to refer to the *fauna* of the hyporheic zone. The focus of this chapter is on fluvial systems, but in places, I shall draw parallels with other groundwater habitats.

II. FUNCTIONAL CLASSIFICATION OF SUBSURFACE MACROFAUNA

Macrofauna collected from subsurface zones are taxonomically diverse (Table I) representing most phyla that occur in surface aquatic habitats. However, few macrofauna are obligate "permanent hyporheos" (*sensu* Williams and Hynes, 1974) spending their entire life cycles below the bed of the stream; the majority are occasional occupants either entering the hyporheic zone as juvenile stages or seeking refuge from flooding or drying of the surface streams (Clifford, 1966; Williams and Hynes, 1977; Marchant, 1988; Boulton, 1989; Cooling and Boulton, 1993; Marchant, 1995). Nonetheless, these occasional visitors influence subsurface processes and food webs, for example, acting as predators or prey (Gibert *et al.*, 1994).

TABLE I Broad Taxonomic List of Macrofauna Reported from Subsurface Zones of Some Streams and Rivers[a]

Taxon	Ecological category	Reference
Coelenterata		
Hydrozoa	OH	Ward et al. (1992)
		Ward and Voelz (1994)
Nematomorpha		Marchant (1988)
Gordiidae	OH	Marchant (1995)
Mollusca		
Gastropoda		
Ancylidae	OH	Maridet et al. (1996)
Bythinellidae	OH	Maridet et al. (1992)
		Maridet et al. (1996)
Hydrobiidae	PH	Marchant (1988)
		Cooling and Boulton (1993)
		Dole-Olivier et al. (1994)
Lymnaeidae	OH	Creuzé des Châtelliers et al. (1992)
Bivalvia		
Sphaeriidae	OH	Williams and Hynes (1974)
		Creuzé des Châtelliers et al. (1992)
Annelida		
Oligochaeta	OH, PH	Williams and Hynes (1974)
		Pennak and Ward (1986)
		Marchant (1988)
		Boulton et al. (1992)
		Creuzé des Châtelliers et al. (1992)
		Maridet et al. (1992)
		Sterba et al. (1992)
		Cooling and Boulton (1993)
		Dole-Olivier et al. (1994)
		Stanford et al. (1994)
		Ward and Voelz (1994)
Hirudinea	OH	Sterba et al. (1992)
		Maridet et al. (1996)
Arthropoda		
Crustacea		
Isopoda		
Asellidae	OH, PH	Henry (1976)
		Lattinger-Penko (1976)
		Creuzé des Châtelliers et al. (1992)
		Danielopol et al. (1994)
		Dole-Olivier et al. (1994)
Microparasellidae	PH	Boulton et al. (1992)
		Dole-Olivier et al. (1994)
		Stanford et al. (1994)
Amphipoda		
Crangonyctidae	PH	Pennak and Ward (1986)
		Boulton et al. (1992)

(continued)

TABLE I (continued)

Taxon	Ecological category	Reference
		Ward *et al.* (1992)
		Cooling and Boulton (1993)
		Dole-Olivier *et al.* (1994)
		Stanford *et al.* (1994)
		Ward and Voelz (1994)
Gammaridae	OH	Creuzé des Châtelliers *et al.* (1992)
		Maridet *et al.* (1996)
Talitridae	PH	Boulton and Stanley (1992)
Niphargidae	PH	Dole-Olivier *et al.* (1994)
		Maridet *et al.* (1996)
Salentinellidae	PH	Dole-Olivier *et al.* (1994)
Syncarida		
Bathynellidae	PH	Pennak and Ward (1986)
		Ward *et al.* (1992)
		Cooling and Boulton (1993)
		Dole-Olivier *et al.* (1994)
		Stanford *et al.* (1994)
		Ward and Voelz (1994)
Parabathynellidae	PH	Boulton *et al.* (1992)
Insecta		
Ephemeroptera		
Ameletopsidae	OH, PH	Pennak and Ward (1986)
		Stanford *et al.* (1994)
		Ward and Voelz (1994)
Baetidae	OH	Pennak and Ward (1986)
		Creuzé des Châtelliers *et al.* (1992)
		Maridet *et al.* (1992)
		Cooling and Boulton (1993)
		Ward and Voelz (1994)
		Maridet *et al.* (1996)
Brachycentridae	OH	Maridet *et al.* (1996)
Caenidae	OH	Williams and Hynes (1974)
		Marchant (1988)
		Creuzé des Châtelliers *et al.* (1992)
		Maridet *et al.* (1992)
		Cooling and Boulton (1993)
		Gibert *et al.* (1994)
		Maridet *et al.* (1996)
Coloburiscidae	OH	Marchant (1988)
Ephemeridae	OH	Maridet *et al.* (1992)
		Maridet *et al.* (1996)
Ephemerellidae	OH	Pennak and Ward (1986)
		Maridet *et al.* (1992)
		Maridet *et al.* (1996)
Heptageneiidae	OH, PH	Maridet *et al.* (1992)
		Maridet *et al.* (1996)

(continued)

TABLE I (*continued*)

Taxon	Ecological category	Reference
Leptophlebiidae	OH	Williams and Hynes (1974)
		Pennak and Ward (1986)
		Stanford *et al.* (1994)
		Ward and Voelz (1994)
Tricorythidae	OH	Ward *et al.* (1992)
Plecoptera		
Capniidae	OH, AM	Pennak and Ward (1986)
		Stanford *et al.* (1994)
		Ward and Voelz (1994)
		Maridet *et al.* (1996)
Chloroperlidae	OH, AM	Stanford and Ward (1993)
		Pennak and Ward (1986)
		Stanford *et al.* (1994)
		Ward and Voelz (1994)
		Maridet *et al.* (1996)
Leuctridae	OH, AM	Pennak and Ward (1986)
		Maridet *et al.* (1992)
		Stanford and Ward (1993)
		Stanford *et al.* (1994)
		Maridet *et al.* (1996)
Nemouridae	OH	Maridet *et al.* (1996)
Perlidae	OH	Ward and Voelz (1994)
Perlodidae	OH	Ward and Voelz (1994)
Taeniopterygidae	OH	Ward and Voelz (1994)
Odonata		
Coenagrionidae	OH	Cooling and Boulton (1993)
Libellulidae	OH	Cooling and Boulton (1993)
Diptera		
Athericidae	OH	Maridet *et al.* (1992)
		Sterba *et al.* (1992)
		Ward and Voelz (1994)
		Maridet *et al.* (1996)
Ceratopogonidae	OH	Gray and Fisher (1981)
		Pennak and Ward (1986)
		Boulton *et al.* (1991)
		Boulton *et al.* (1992)
		Creuzé des Châtelliers *et al.* (1992)
		Maridet *et al.* (1992)
		Sterba *et al.* (1992)
		Ward *et al.* (1992)
		Cooling and Boulton (1993)
		Dole-Olivier *et al.* (1994)
		Ward and Voelz (1994)
		Boulton and Stanley (1995)
		Marchant (1995)
		Maridet *et al.* (1996)

(*continued*)

TABLE I (continued)

Taxon	Ecological category	Reference
Chironomidae	OH, AM	Gray and Fisher (1981)
		Pennak and Ward (1986)
		Marchant (1988)
		Boulton *et al.* (1992)
		Creuzé des Châtelliers *et al.* (1992)
		Sterba *et al.* (1992)
		Ward *et al.* (1992)
		Cooling and Boulton (1993)
		Dole-Olivier *et al.* (1994)
		Stanford *et al.* (1994)
		Ward and Voelz (1994)
		Marchant (1995)
		Maridet *et al.* (1996)
Corethrellidae	OH	Boulton *et al.* (1992)
Empididae	OH	Pennak and Ward (1986)
		Marchant (1995)
Limoniidae	OH	Maridet *et al.* (1992)
		Maridet *et al.* (1996)
Muscidae	OH	Ward and Voelz (1994)
Psychodidae	OH	Maridet *et al.* (1992)
		Sterba *et al.* (1992)
Simuliidae	OH	Boulton *et al.* (1992)
		Maridet *et al.* (1992)
		Ward *et al.* (1992)
		Cooling and Boulton (1993)
		Ward and Voelz (1994)
		Maridet *et al.* (1996)
Tabanidae	OH	Gray and Fisher (1981)
		Boulton *et al.* (1992)
		Maridet *et al.* (1996)
Tanyderidae	OH	Ward and Voelz (1994)
Tipulidae	OH	Pennak and Ward (1986)
		Boulton *et al.* (1992)
		Ward *et al.* (1992)
		Dole-Olivier *et al.* (1994)
		Ward and Voelz (1994)
		Marchant (1995)
Megaloptera		
Corydalidae	OH	Boulton *et al.* (1992)
Coleoptera		
Dytiscidae	OH, AM[b]	Dole-Olivier *et al.* (1994)
Elmidae	OH	Stanford and Gaufin (1974)
		Williams and Hynes (1974)
		Pennak and Ward (1986)
		Marchant (1988)

(continued)

TABLE I (*continued*)

Taxon	Ecological category	Reference
		Maridet *et al.* (1992)
		Marchant (1995)
		Maridet *et al.* (1996)
Hydraenidae	OH	Maridet *et al.* (1996)
Psephenidae	OH	Williams and Hynes (1974)
		Marchant (1995)
Ptilodactylidae	OH	Marchant (1995)
Trichoptera		
Conoesucidae	OH	Marchant (1995)
Glossosomatidae	OH	Marchant (1988)
		Maridet *et al.* (1992)
		Maridet *et al.* (1996)
Helicopsychidae	OH	Williams and Hynes (1974)
		Marchant (1995)
Hydrobiosidae	OH	Cooling and Boulton (1993)
		Marchant (1995)
Hydropsychidae	OH	Williams and Hynes (1974)
		Marchant (1988)
		Maridet *et al.* (1992)
		Dole-Olivier *et al.* (1994)
		Marchant (1995)
		Maridet *et al.* (1996)
Hydroptilidae	OH	Lattinger-Penko (1976)
		Maridet *et al.* (1996)
Lepidostomatidae	OH	Maridet *et al.* (1992)
Leptoceridae	OH	Williams and Hynes (1974)
		Dole-Olivier *et al.* (1994)
		Marchant (1995)
		Maridet *et al.* (1996)
Limnephilidae	OH	Maridet *et al.* (1992)
		Maridet *et al.* (1996)
Philopotamidae	OH	Maridet *et al.* (1992)
Philorheithridae	OH	Marchant (1988)
Rhyacophilidae	OH	Maridet *et al.* (1992)
		Maridet *et al.* (1996)
Sericostomatidae	OH	Maridet *et al.* (1996)

[a]For more details, see References. Classification into the ecological categories of occasional hyporheos (OH), amphibite (AM), or permanent hyporheos (PH) (including some ubiquitous stygobites) follows that described in the text. Note that some broad taxonomic groupings span several ecological categories.

[b]*Siettitia avenionensis* is a phreatic water beetle from the Rhône (Dole-Olivier *et al.*, 1994).

Of the many categorizations of the hyporheos based on their phenology, the most versatile appears to be that proposed by Gibert *et al.* (1994) because it includes groundwater and cave/karst fauna as well, acknowledging the similarities in selection pressures. There are three main groups:

1. Stygoxenes are taxa with no affinities for ground water, occurring accidentally in caves or stream sediments. Macrofaunal examples include megalopteran and odonatan larvae.

2. Stygophiles actively exploit subsurface resources, seeking protection from unfavorable surface conditions. They are further divided into occasional hyporheos, amphibites, and permanent hyporheos. Occasional hyporheos are primarily larval forms of aquatic benthic insects that can develop totally in surface conditions but occasionally (as small instars) enter the hyporheic zone. The life cycle of amphibites necessitates use of both subsurface and surface waters, and often aerial habitats as well. The best examples are some members of the stonefly (Plecoptera) genera *Isocapnia, Paraperla,* and *Kathroperla* whose larval stages occupy the hyporheic zone and lateral alluvial sediments of the Flathead River in Montana before returning to the river to emerge as adults (Stanford and Ward, 1988, 1993). Midge larvae of *Krenosmittia* (Diptera: Chironomidae) also exemplify amphibites (Ferrington, 1984). The permanent hyporheos does not need to enter the surface water (but can live in the surface stream; P. Marmonier, personal communication) and is composed of numerous meiofaunal taxa (naidid oligochaetes, mites, nematodes, tardigrades, ostracods, copepods and cladocerans) as well as macrofauna such as amphipods and isopods (Table I).

3. Stygobites are specialized subterranean taxa. The distinction between permanent hyporheos and stygobites appears to be one of degree of adaptation to a subsurface existence (i.e., lack of pigmentation, regression of eyes and photosensitive organs, hypertrophy of sensory organs). Stygobites are further categorized into ubiquitous stygobites common in a range of subsurface habitats including cave streams and phreatobite species restricted to deep ground water of alluvial aquifers (phreatic zones).

This classification is useful when we consider functional roles of macrofauna in subsurface sediments by partly reflecting the physical size and residence time of fauna in a more useful manner than a taxonomic approach. The classification is also relevant when generalizing about life-history strategies that typify these groups. Most occasional hyporheos, such as the majority of surface stream benthos, are relatively fecund, short-lived, mobile, and exhibit other traits associated with physically variable environments (Townsend, 1989). Conversely, the more stable environmental conditions in subsurface habitats would be expected to favor longevity, low rates of development and growth, and low fecundity which are features found in many stygobites and permanent hyporheos (Gibert *et al.,* 1994; Rouch, 1986; Danielopol, 1989; Strayer, 1994). These different life-history attributes may directly influence the energetics and trophic dynamics of subsurface macrofauna. For equivalent densities, the contribution of respiration to the total stream ecosystem may be lower (see later), and reproductive recruitment after disturbances may be slow, implying the fauna is at risk from human activities (e.g., alteration of the exchanges with surface stream water, toxic or organic contamination; Danielopol, 1989; Brunke and Gonser, 1997; Boulton *et al.*, 1998).

III. FACTORS INFLUENCING THE DISTRIBUTION OF SUBSURFACE MACROFAUNA

Although subsurface processes can be examined at a range of scales (Gibert *et al.*, 1994; Boulton *et al.*, 1998), the two most relevant scales for examination of the factors influencing the distribution (and therefore potential roles) of macrofauna are the sediment (microhabitat) and reach (mesohabitat) scales (Boulton *et al.*, 1998). The microhabitat scale is represented by the local conditions occurring at the scale of the interstitial pore or fissure (Fig. 1). The mesoscale ranges from the sequences of upwelling and downwelling zones that occur along a river channel in response to geomorphological features to the interactions occurring with the riparian zone and parafluvial sediments at a catchment scale (Fig. 2; Boulton *et al.*, 1998). These spatial scales correspond to matching temporal scales ranging up to decades in response to substantial fluvial reworking by major floods that may lead to local extinction of some macrofaunal taxa. In this ecological hierarchy, the interactions of processes whose effects span scales contributes to the complexity of variables influencing the macrofauna (Gibert *et al.*, 1994; Boulton *et al.*, 1998).

Numerous authors (Pennak and Ward, 1986; Danielopol, 1989; Boulton *et al.*, 1992, 1997; Dole-Olivier and Marmonier, 1992a; Maridet *et al.*, 1992, 1996; Creuzé des Châtelliers *et al.*, 1994; Culver, 1994; Danielopol *et al.*, 1994; Gibert *et al.*, 1994; Stanford *et al.*, 1994; Strayer, 1994; Boulton and Stanley, 1995; de Bovee *et al.*, 1995; Brunke and Gonser, 1997) suggest that the primary microscale determinants of subsurface macrofauna (Fig. 1) appear to be physical pore size (which is a function of grain size, lithology, and sedimentation rates), interstitial water velocity, dissolved oxygen and water temperature, and available energy resources (e.g., biofilms, prey). These physical factors are likely to similarly influence biotic variables (e.g., intra- and interspecific competition, predation, prey availability) that also must influence the subsurface macrofauna but have received far less attention (Henry, 1976; Schmid, 1993; Culver, 1994). At finer scales, variables affecting the microbial loop and the consumption of bacteria and fungi by meiofauna are also relevant (Hakenkamp and Palmer, Chapter 13; Karaman, 1935; Hauer and Resh, 1996; Hendricks, 1996; Palmer *et al.*, 1997; Boulton *et al.*, 1998).

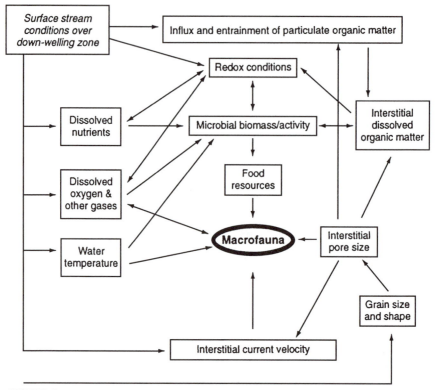

FIGURE 1 Microscale variables influencing the distribution of subsurface macrofauna in streams. Biotic interactions such as predation and competition are not included. Double-headed arrows signify that the interactions may occur in both directions. Some interactions have been omitted for clarity of the diagram.

The complexity of the interactions among fine-scale factors and their temporal dynamism caused by hydrological and chemical fluctuations lead to considerable patchiness (Pennak and Ward, 1986; Danielopol, 1989; Boulton *et al.*, 1991; Dole-Olivier and Marmonier, 1992a; Schmid, 1993; Ward and Voelz, 1994; Brunke and Gonser, 1997; Boulton *et al.*, 1998). Such variability has complicated sampling efficiency of this rather intractable habitat (Hakenkamp and Palmer, Chapter 13 in this volume; Hakenkamp and Palmer, 1992; Palmer, 1993; Boulton *et al.*, 1998), but some studies have suggested that many of these microhabitat-scale factors influencing macrofaunal distribution are reasonably well predicted by a few mesoscale variables. For example, Strayer (1994) suggests that as few as two integrative variables (hydraulic conductivity and the age of the water) may suffice to develop simple models to describe biological distributions within and across habitats. Hydraulic conductivity is a measure of the ease with which water flows through sediments, determining renewal rates of water, dissolved gases, and nutrients. The age of the water (hyporheic residence time) is related to the time for weathering and consumption of materials by biogeochemical processes (Findlay, 1995; Boulton *et al.*, 1998).

Marmonier and Dole (1986) demonstrated the primary influences of mesoscale hydrological exchange between surface and subsurface water, hydraulic conductivity, and hyporheic residence time on the distribution of macrofauna in the sediments of a by-passed channel of the River Rhône. Stygobites and permanent hyporheos were more common where ground water upwelled into the sediments, whereas epigean (surface) taxa appeared to be carried passively into downwelling areas, more so during high flows. Some epigean taxa also actively colonized the hyporheic zone during flooding (P. Marmonier, personal communication). Similar responses have been reported for hyporheic invertebrates in arid zone streams in the Sonoran Desert and in South Australia where much of the variance in community composition can be explained by proximity to upwelling and downwelling zones where water is exchanged between the surface and subsurface zones, residence time of the hyporheic water, and possibly interstitial pore size (Boulton and Stanley, 1995, 1996; Cooling and Boulton, 1993).

Increasingly, longer-term studies of the ecology of subsurface macrofauna have been performed at several sites, and the role of mesoscale events such as droughts and floods are becoming evident (Dole-Olivier *et al.*, 1994; Boulton and Stanley, 1995). In mesic streams, flooding and increased discharge often result in higher numbers of occasional hyporheos in the sediments (Williams and Hynes, 1974; Marchant, 1988, 1995), but it is not clear whether the movement is active or the fauna passively "drift," carried by strong local downwelling (Marmonier, 1988; Creuzé des Châtelliers and

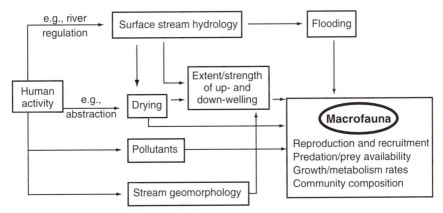

FIGURE 2 Mesoscale variables influencing the distribution of subsurface macrofauna in streams. See text for detailed explanation of impacts of pollution, flooding, and drying.

Marmonier, 1990; Schmid, 1993). However, flooding does not always alter the vertical distribution of the hyporheos (Whitman and Clark, 1984; Giberson and Hall, 1988; McElravy and Resh, 1991) and further work is needed to understand the processes that occur in the subsurface zones of streams during floods (Boulton *et al.*, 1998).

In downwelling zones of the Rhône River, surface-water infiltration increased during spates, causing a "washout" effect resulting in reduced abundance and diversity of interstitial fauna. Postspate, there was invasion of the subsurface zone by epigean taxa (Creuzé des Châtelliers and Marmonier, 1990). Upwelling zones were also subject to surface-water infiltration during spates, acting as a stream storage zone for trapped material (Creuzé des Châtelliers and Marmonier, 1990). Recolonization to preflood distributions appeared complete after 19 days (Dole-Olivier *et al.*, 1994). Rearrangement of sediments by floods in the Miribel Canal, France, altered subsurface water chemistry and the composition of the hyporheos (Marmonier, 1991). Sedimentation reduced surface–subsurface water exchange and the fauna became dominated by stygobites. Local hydrogeological features of the channel influenced the extent of this change (Marmonier, 1991; Dole-Olivier and Marmonier, 1992a), illustrating the interaction between mesoscale and microscale features. In general, postflood recovery appears to reflect disturbance regime (flood frequency) and season (Dole-Oliver and Marmonier, 1992b; Dole-Oliver *et al.*, 1994; Marchant, 1995) whereas the significance of the hyporheic zone to postflood recovery by surface taxa may be related to the severity of the flood and the removal of fauna from other refuges (Brooks and Boulton, 1991; Marchant, 1995).

In Sycamore Creek, an arid zone stream in the Sonoran Desert, spates had variable effects on invertebrate assemblage composition (Stanley and Boulton, 1993; Boulton and Stanley, 1995). Phreatic fauna occupying deeper sediments seemed least affected. Floods caused changes in the composition of shallow hyporheic assemblages, but recovery to preflood densities by most taxa was complete after 5 days (Boulton and Stanley, 1995), reflecting the high mobilities of the hyporheos (Boulton *et al.*, 1991). Another mesoscale event that occurs in streams is drying. This "press" disturbance caused major and long-lasting changes in the composition of the hyporheos of Sycamore Creek (Stanley and Boulton, 1993; Boulton and Stanley, 1995), presumably because of the complete removal of input of nutrients and energy from the surface stream and the physical absence of water. Many macrofaunal stygobites are intolerant of desiccation (Schminke, 1974; Williams, 1987), and evidence indicates that the hyporheos migrates into deeper, saturated sediments during drying of the surface stream (Griffith and Perry, 1993; Clinton *et al.*, 1996). Drying may also have an impact on microbial and meiofaunal assemblages, depriving subsurface macrofauna of food.

Both flooding and drying are hydrodynamic processes, but their effects upon the fluxes of organic matter and energy, and hence macrofaunal productivity, in the subsurface are not well known. Standing waves that form during flash floods in streams in the arid Flinders Ranges, South Australia, pound particulate organic matter into the sediments, creating local pockets of physically comminuted leaf litter and sticks (A. J. Boulton, unpublished data). The removal by flash floods of benthic algae and other surface organic matter (Grimm and Fisher, 1989) interrupts the potential supply of materials and energy to the subsurface zone via downwelling hydrological pathways, but data are lacking on the impacts of this on macrofaunal distribution. Redistribution of silt and inorganic fines also occurs during floods (Creuzé des Châtelliers and Marmonier, 1990), potentially influencing pore size and interstitial current velocity and hence the distribution of macrofauna (de Bovee *et al.*, 1995; Maridet *et al.*, 1996). Thus, the influence of flooding upon the hyporheic zone (i.e., fluxes of nutrients and fauna, modification of organic matter and sediment structure, energy fluxes) is analogous to the effects of floods on floodplains and riparian zones of rivers, and predictions of the Flood Pulse Concept (Junk *et al.*, 1989) may apply equally well.

Human activities such as river regulation alter the hydrological regime of rivers from their normal pattern of flooding and drying, usually with negative impacts upon surface processes and biota (reviews in Petts and Calow, 1996). Impacts of river regulation upon subsurface macrofauna are less studied (perhaps because they are not as apparent) but appear no less deleterious (Creuzé des Châtelliers *et al.*, 1994; Dole-Oliver *et al.*, 1994; Stanford *et al.*, 1994; Brunke and Gonser, 1997; Boulton *et al.*, 1998). Ground water provides a key source of potable water for many cities and towns, but little legislation exists concerning the potential impacts of groundwater abstraction upon subsurface invertebrates. Organic pollution has been shown to alter subsurface communities (see later), whereas the effects of gravel extraction and anthropogenic sedimentation upon the distribution of the macrofauna have been rarely studied. Overall, the hyporheic zone is considered to contribute to resistance by stream communities to many types of disturbance (Townsend and Hildrew, 1994; Brunke and Gonser, 1997), but this generalization may not always hold true (Boulton *et al.*, 1998).

IV. THE FUNCTIONAL ROLE OF SUBSURFACE MACROFAUNA

Although the hyporheic zone initially attracted the scientific attention because it sometimes contained a diverse array of invertebrates (Danielopool, 1989; Gibert *et al.*, 1994; Jones and Holmes, 1996a), many of which proved new to science, the contribution of these invertebrates to the total metabolism occurring in the sediments may be trivial. In a third-order stream in the Black Forest, Germany, hyporheic macrofauna (>100 μm) consumed <5% of the oxygen respired in the stream sediments (Pusch,

1996; Fischer *et al.*, 1996; Pusch and Schwoerbel, 1994). On this basis, it could be considered that macrofauna play no important role in the functioning of subsurface ecosystem processes (Jones and Holmes, 1996b), but an argument based entirely on contribution to total metabolism would ignore the importance of top predators in all food webs and would be unreasonably narrow. There have been few comparisons of the relative contributions of various stream compartments (e.g., channel, floodplain, hyporheic zone) to invertebrate production. Smock *et al.* (1992) demonstrated that the hyporheic zone accounted for 65% of invertebrate production in the channel of a Virginian sand-bottomed stream, suggesting the figure would be higher in streams with coarser sediments. These authors further suggested that the high contribution to total productivity made by the hyporheic zone in many streams is due to the magnitude of the hyporheic zone rather than to the density of macrofauna. Several other authors report relatively high abundances of hyporheic macrofauna (Williams and Hynes, 1974; Pugsley and Hynes, 1983) and it would be interesting to know what proportion of total invertebrate production occurred in the subsurface zones of these streams.

Although numerous descriptive studies have examined the factors correlated with the distribution of subsurface macrofauna, experimental evidence to demonstrate their role in subsurface ecosystem processes is scarce. This scarcity probably represents the technical difficulties of sampling and conducting manipulative experiments in this habitat (Hakenkamp and Palmer, 1992; Palmer, 1993) as well as our limited knowledge of the ecology of most hyporheic invertebrates. The following discussion is therefore largely speculative and will hopefully encourage the collection of data to test whether macrofauna play any key roles in the functioning of subsurface ecosystems (Boulton and Stanley, 1996; Palmer *et al.*, 1997; Table II). For example, if macrofauna are removed from the hyporheic zone, are rates of subsurface decomposition of organic matter or nutrient transformation altered? Do seasonal variations in macrofaunal abundance correlate with the rates of some of these processes?

Whereas some macrofauna are vermiform (worm-shaped) and able to negotiate interstitial spaces with little rearrangement of particles, others such as peracarid and syncarid crustaceans are relatively robust and may actively burrow in fine sediments, increasing bacterial dispersal (Edler and Dodds, 1992). This physical alteration of pore size has ramifications for local hydrological exchange (Torreiter *et al.*, 1994) and associated transport of nutrients and dissolved gases, which, in turn, may be important for the existence of other fauna and provide suitable conditions for incubation of the eggs of some fish (Pollard, 1955; Johnson, 1980; Boulton *et al.*, 1998).

Similarly, feeding and egestion serve to alter fine-sediment structure. Husmann (1974) proposed that through their feeding, macrofauna such as oligochaetes could regulate bacterial growth and prevent the premature clog-

TABLE II   Potential Functional Roles of Subsurface Macrofauna

Activity	Effect	Relevant reference
Burrowing	Fine-scale alteration of pore size and water velocity, changing redox gradients and supply of matter/energy to microbes and meiofauna, aeration of sediments, dispersal of bacteria	Edler and Dodds (1992) Edler and Dodds (1996) Palmer *et al.* (1997)
Egestion of coarse fecal pellets	Fine-scale alteration of pore size (as above), compaction of fine sediments, localized production of particulate organic matter	Danielopol (1984) Danielopol (1989) Palmer *et al.* (1997)
Excretion	Increase labile dissolved organic carbon, enhancing release of nutrients, providing favorable habitat for microbes	Gallepp (1979) Hendricks (1996)
Grazing on microbial biofilms	Promotion of biofilm activity, role in natural gravel-bar filters	Gibert *et al.* (1994) Torreiter *et al.* (1994) Palmer *et al.* (1997)
Feeding on particulate organic matter (e.g., leaves, sticks, bark, finer fragments)	Comminution of coarse particulate organic matter buried in the sediments (acting as "shredders" and "collectors" *sensu* Cummins and Klug, 1979)	Smith and Lake (1993) Danielopol *et al.* (1994) Palmer *et al.* (1997)
Predation on meiofauna and other macrofauna	Alteration of population sizes and structure due to "top-down" effects	Hakenkamp and Palmer, Chapter 13 in this volume Stanford and Gaufin (1974) Pennak and Ward (1986)
Prey for macrofauna and surface fauna, especially near upwelling zones	As above ("bottom-up"), supply of energy to surface predators	Gray and Fisher (1981) Smock *et al.* (1992) Strayer (1994)
Movement and migration	Transfer of matter and energy among subsurface zones	Sterba *et al.* (1992) Stanford and Ward (1993) Danielopol *et al.* (1994) Stanford *et al.* (1994)
Emergence of amphibites	Removal of matter and energy from subsurface zone to terrestrial and riverine habitats, food for riparian predators	Perry and Perry (1986) Stanford *et al.* (1994)

ging of sand filters. However, some studies have shown that subsurface oligochaetes may not occur in sufficient densities or be active enough to markedly improve the filtration efficiency of sand filters (Torreiter *et al.,* 1994). Subsurface isopods and amphipods egest large fecal pellets that accumulate in large numbers in the interstitial spaces, and such pelletization of silty material plays a role in maintaining "open" interstitial systems (Palmer *et al.,* 1997; Danielopol, 1989; Husmann, 1974).

Most subsurface macrofauna eat a range of foods (polyphagous;

Danielopol, 1989; Culver, 1994), and their distribution often correlates with that of particulate organic matter and high-protein microbial biofilms that coat the large interstitial surface area (Williams and Hynes, 1974; Bärlocher and Murdoch, 1989; Pusch, 1996). Grazing activity apparently promotes biofilm activity (Danielopol, 1989; Torreiter *et al.*, 1994; Palmer *et al.*, 1997) and may alter microbial composition (Edler and Dodds, 1992) with consequent effects on rates of microbially mediated nutrient transformation. In the Flathead River alluvial aquifers, the energy base of the food web appears to be the diverse microbial biofilm of fungi and bacteria, and associated microconsumers. The productivity of this microbial loop in the oligotrophic Flathead River alluvial aquifer is probably limited by the availability of short-chain dissolved organic matter. Top consumers in this relatively simple food web appear to be the macrofauna *Paraperla* (Plecoptera) and *Stygobromus* (Amphipoda) and perhaps salamanders (Stanford *et al.*, 1994).

The importance of the hyporheic zone as a transitional ecotone is discussed elsewhere in this book (see Holmes, Chapter 5 in this volume; Hill, Chapter 3 in this volume; Baker *et al.*, Chapter 11 in this volume). These ecotones serve to control subsurface metabolism because they may function as temporary or permanent sinks of inorganic and organic nutrients and organic matter from the catchment (Pinay *et al.*, 1990). Such ecotones also act as natural filtering and buffering systems, thus improving water quality (Danielopol, 1989; Gibert *et al.*, 1994). Macrofauna diversity and abundance is usually greatest in these ecotones (Palmer *et al.*, 1997), and their feeding and excretion may play a key role in the functions of the ecotone (Mestrov and Lattinger-Penko, 1981; Danielopol, 1989; Torreiter *et al.*, 1994; Brunke and Gonser, 1997; Palmer *et al.*, 1997). For example, at a fine scale, the excretion by high densities of macrofauna (oligochaetes, chironomid larvae) in sediments may increase labile dissolved organic carbon, enhancing release of inorganic nitrogen and phosphorus (Gallepp, 1979; Fukuhara and Sakamoto, 1987). Thus, dense assemblages of hyporheos and high levels of dissolved oxygen in downwelling zones may provide a more favorable habitat for aerobic microbes (Palmer *et al.*, 1997).

Although subsurface habitats usually have less detritus than epigean ones (Culver, 1994), dissolved and fine particulate organic matter may be quite high in the hyporheic zone (Marchant, 1988; Stanford and Ward, 1988; Metzler and Smock, 1990; Leichtfried, 1991). Fragments of particulate organic matter have been found in the guts of subsurface peracarids (isopods and amphipods), bathynellaceans, and chironomids (Cooling and Boulton, 1993; Boulton *et al.*, 1992; Culver, 1994), implying that some macrofauna may occupy functional feeding groups analogous to the "shredders" and "collectors" (*sensu* Cummins and Klug, 1979) in surface streams. Smith and Lake (1993) reported accelerated breakdown of buried eucalypt leaf packs relative to surface-placed leaves in an Australian upland stream and speculated that the hyporheos of Australian streams may significantly contribute to decomposition of particulate organic matter. This was not the case for buried leaves in a sand-bottomed stream in Virginia but the large quantity of subsurface detritus and the presence of hyporheic invertebrates implies that the subsurface zone is important for trophic dynamics and energetics in this stream ecosystem as well (Metzler and Smock, 1990; Strommer and Smock, 1989).

Some subsurface taxa are predatory, including members of the crustacean groups Syncarida, Isopoda, and Amphipoda; dipteran families Ceratopogonidae, Tipulidae, Tabanidae, and Chironomidae; plecopteran families Eustheniidae and some Gripopterygidae and Chloroperlidae; and megalopterans (Stanford and Gaufin, 1974; Henry, 1976; Pennak and Ward, 1986; Danielopol, 1989; Boulton *et al.*, 1992; Stanford and Ward, 1993; Dainelopol *et al.*, 1994). The role of predators in the trophic webs of the hyporheic zone is poorly known but presumably equivalent to predation in other ecosystems where top-down controls occur. Strayer (1994) suggests that the rarity of subsurface macrofauna from near-surface environments might be due to excessive mortality from epigean predators but could be equally due to mortality from spates. Predation pressure has also been suggested to control the distribution of other subsurface fauna (Pennak and Ward, 1986). Subsurface macrofauna sometimes occur in benthic stream samples, especially near upwelling zones, and provide a food resource for surface invertebrates. In spatially intermittent Brachina Creek (South Australia), gut contents of several aeshnid dragonfly nymphs collected from a pool below an upwelling zone contained fragments of interstitial amphipods (A. J. Boulton, unpublished data). Other studies have also suggested that hyporheic production is available to epigean consumers as individuals move to the surface (Smock *et al.*, 1992; Brunke and Gonser, 1997).

At the micro- and mesoscale, some subsurface macrofauna are capable of substantial migrations vertically, laterally, and downstream (Marchant, 1988; Boulton *et al.*, 1991; Sterba *et al.*, 1992; Danielopol *et al.*, 1994; Hendricks, 1996; the "hyporheic corridor concept," Stanford and Ward, 1993) and, as such, represent transfers of energy in the subsurface environment. For example, Sterba *et al.* (1992) suggest that migration of Trichoptera, Chironomidae, and Hirudinea to depths of 50–70 cm in the bed of the River Morava (Czechoslovakia) occurred in response to organic pollution and declining water temperatures. The hyporheic zone has also been hypothesized as a refuge from the toxic effects of pesticides (Jeffrey *et al.*, 1986). Emerging amphibites (e.g., Table I) represent a loss of energy and nutrients from the subsurface system and presumably provide food resources for epigean invertebrate and vertebrate predators, riparian birds, etc. The amphibite stoneflies of the Flathead River are large (2–3.5 cm in length as mature nymphs (Stanford *et al.*, 1994)) and move from the alluvium to rivers or spring brooks to emerge during spring (Stanford and Ward, 1988). Areal rates of productivity are high due to the large volume of ground water, and the amphibite stoneflies dominate drift during emergence and are an important food source for riverine fish and other top consumers (Perry and Perry, 1986). Similar amphibitic assemblages probably exist in other well-oxygenated riverine aquifers in North America and elsewhere (Stanford and Ward, 1993; Stanford *et al.*, 1994).

Many of these roles of subsurface macrofauna parallel those of their epigean counterparts (Palmer *et al.*, 1997). However, their lower diversity and abundance, apparently in response to lower habitat heterogeneity and energy influx (Gibert *et al.*, 1994), imply they are of less importance in this habitat (especially if their metabolic rates are similarly low) than at the surface. Experimental manipulations of macrofaunal density are required to test this hypothesis. If subsurface macrofauna are removed from a stream reach, does infiltration of fine sediments occur more rapidly? Do more localized pockets of anoxia that surround entrapped particulate organic matter exist? Does this consequently alter microbial activities in a type of subsurface cascade mechanism? Do changes in the nature or rates of surface–subsurface exchanges of materials occur (see Dent *et al.*, Chapter 16 in this volume)? When subsurface fauna are removed, do rates of production of surface fauna (or flora) change? Do anthropogenic changes to groundwater quality influence macrofaunal distribution? Does this alteration of macrofaunal distribution further alter groundwater quality?

V. POTENTIAL ROLE AS BIOMONITORS OF DETERIORATING GROUNDWATER QUALITY

By virtue of their size and ubiquity, subsurface macrofauna have often been suggested as potential biomonitors of groundwater quality (Danielopol, 1989; Ward *et al.*, 1992; Notenboom *et al.*, 1994; Plénet *et al.*, 1996; Brunke and Gonser, 1997; Palmer *et al.*, 1997; Boulton *et al.*, 1998). For example, groundwater wells in a limited area of intensive housing development and individual sewage drain fields in the Kalispell Valley (Montana) had lower dissolved oxygen, higher coliform bacteria counts, higher concentrations of nitrate and dissolved organic matter, and fewer groundwater biota than wells elsewhere in the valley (Stanford *et al.*, 1994). Groundwater contamination (organic pollution) changed community structure in a French karst system from a composition of essentially stygobites to a community dominated by epigean, ubiquitous species (Malard *et al.*, 1994). Invertebrate community composition had not recovered a year later after a 3-day pumping test in another karst system (Rouch *et al.*, 1993). Another key point is that macrofauna are most diverse in shallow groundwater zones and alluvial aquifers. Pollutants enter these zones before contaminating deeper layers (Notenboom *et al.*, 1994) so changes in the macrofaunal community structure of these ecotones may provide early warning of expanding contamination. The higher concentrations of heavy metals in interstitial water, and hence more toxic effects on the hyporheos, led Nelson *et al.*, (1993) to suggest this group as a useful bioindicator for assessing mine drainage impacts in the upper Arkansas River.

Despite these promising attributes, we are still hampered by a lack of ecological information on subsurface macrofaunal distribution and abundance, life histories, roles in nutrient cycling and organic matter processing, food web dynamics, and ecotoxicology. The conclusion from a recent review (Notenboom *et al.*, 1994) was that groundwater invertebrates appear to be good biomonitors where pollution is severe. But in the absence of information on chronic responses to low level contamination or validation of laboratory experiments to field situations, evidence for the macrofauna being useful biomonitors of the more subtle effects of groundwater pollution is equivocal. In the few riverine systems that have been studied, most authors have reported responses by hyporheic invertebrates to pollution (Mestrov and Lattinger-Penko, 1981; Creuzé des Châtelliers *et al.*, 1992; Nelson *et al.*, 1993; Notenboom *et al.*, 1994; Torreiter *et al.*, 1994) but this does not seem universal (Ward *et al.*, 1992).

Interestingly, groundwater macrofauna themselves may be perceived as a problem, and knowledge of their biology is needed for their control. For example, invasion of the water mains of a large Austrian city by the groundwater crustacean isopod *Proasellus strouhali* caused concern to the public and posed a problem for water managers wishing to control the isopod using chlorine (Ward *et al.*, 1992). A similar situation led to the discovery of the now-famous amphibitic stonefly nymphs of the Flathead River, Montana (Stanford *et al.*, 1994).

VII. CONCLUSIONS

A diverse taxonomic array of macrofauna occupies subsurface zones of most rivers, especially where hydraulic conductivities and exchange with the surface water are high. Their relatively large body size and activity (i.e., feeding, excretion, migration) may mean that they have an effect on the functioning of subsurface ecosystems that is disproportionate to their metabolic activity compared to meiofaunal and microbial respiration. However, in subsurface zones where pore size is small and hydraulic conductivity is low, macrofauna are likely to be uncommon and not particularly important. More information on life histories, food web dynamics, and the roles that macrofauna play in nutrient transformations and biofilm composition is required; so far, the focus has been on simply describing their distribution and identifying correlations with physical and chemical variables. As we begin to recognize the environmental features that influence the distribution of macrofauna at micro- and mesoscales, there is more opportunity for identifying their potential as biomonitors of changes in groundwater conditions due to human activities.

Aside from the utilitarian reasons concerning their likely role in subsurface habitats, there are strong scientific and cultural grounds for protecting hyporheic macrofaunal diversity (Danielopol *et al.*, 1997; Rouch and Danielopol, 1997). Present interest in the links between biodiversity and ecosystem function in sediments (Boulton *et al.*, 1998) and concerns about the preservation of aquatic biodiversity in general (Frissell and Bayles, 1996) should prompt scientists to transfer information about the hyporheic zone to water resource managers (Gibert *et al.*, 1997). Although some elements of protection of the hyporheic zone are readily communicated because they are well known [e.g., siltation of trout spawning gravels (Pollard, 1955; Brunke and Gonser, 1997)], the more poorly understood aspects such as the functional role of macrofauna can only be hypothesized because we have few reliable data (Palmer *et al.*, 1997), and therefore a strong case for their protection is harder to make to river managers. Nonetheless, we understand enough about the hyporheic ecotone and its susceptibility to human activities (Brunke and Gonser, 1997; Gibert *et al.*, 1997; Boulton *et al.*, 1998) that explicit recognition and assessment of this ecosystem compartment should be mandatory in all river management plans.

REFERENCES

Bärlocher, F., and J. H. Murdoch. 1989. Hyporheic biofilms—a potential food source for interstitial animals. *Hydrobiologia* **184**:61–67.

Boulton, A. J. 1989. Over-summering refuges of aquatic macroinvertebrates in two intermittent streams in central Victoria. *Transactions of the Royal Society of South Australia* **113**:23–34.

Boulton, A. J., and E. H. Stanley. 1995. Hyporheic processes during flooding and drying in a Sonoran Desert stream. II. Faunal dynamics. *Archiv für Hydrobiologie* **134**:27–52.

Boulton, A. J., and E. H. Stanley. 1996. But the story gets better: Subsurface invertebrates in stream ecosystems. *Trends in Ecology and Evolution* **11**:430.

Boulton, A. J., S. E. Stibbe, N. B. Grimm, and S. G. Fisher. 1991. Invertebrate recolonization of small patches of defaunated hyporheic sediments in a Sonoran Desert stream. *Freshwater Biology* **26**:267–277.

Boulton, A. J., H. M. Valett, and S. G. Fisher. 1992. Spatial distribution and taxonomic composition of the hyporheos of several Sonoran Desert streams. *Archiv für Hydrobiologie* **125**:37–61.

Boulton, A. J., M. R. Scarsbrook, J. M. Quinn, and G. P. Burrell. 1997. Land-use effects on the hyporheic ecology of five small streams near Hamilton, New Zealand. *New Zealand Journal of Marine and Freshwater Research* **31**:609–622.

Boulton, A. J., S. Findlay, P. Marmonier, E. H. Stanley, and H. M. Valett. 1998. The functional significance of the hyporheic zone in streams and rivers. *Annual Review of Ecology and Systematics* **29**:59–81.

Brooks, S. S., and A. J. Boulton. 1991. Recolonization dynamics of benthic macroinvertebrates after artificial and natural disturbances in an Australian temporary stream. *Australian Journal of Marine and Freshwater Research* **42**:295–308.

Brunke, M., and T. Gonser. 1997. The ecological significance of exchange processes between rivers and groundwaters. *Freshwater Biology* 37:1–33.

Clifford, H. G. 1966. The ecology of invertebrates in an intermittent stream. *Investigations of Indiana Lakes and Streams* 7:57–98.

Clinton, S. M., N. B. Grimm, and S. G. Fisher. 1996. Response of a hyporheic invertebrate assemblage to drying disturbance in a desert stream. *Journal of the North American Benthological Society* 15:700–712.

Cooling, M. P., and A. J. Boulton. 1993. Aspects of the hyporheic zone below the terminus of a South Australian arid-zone stream. *Australian Journal of Marine and Freshwater Research* 44:411–26.

Creuzé des Châtelliers, M., and P. Marmonier. 1990. Macrodistribution of Ostracoda and Cladocera in a by-passed channel: Exchange between superficial and interstitial layers. *Stygologia* 5:17–24.

Creuzé des Châtelliers, M., P. Marmonier, M. -J. Dole-Olivier, and E. Castella. 1992. Structure of interstitial assemblages in a regulated channel of the River Rhine (France). *Regulated Rivers* 7:23–30.

Creuzé des Châtelliers, M., D. Poinsart, and J. P. Bravard. 1994. Geomorphology of alluvial groundwater ecosystems. *In* "Groundwater Ecology" (J. Gibert, D. L. Danielopol and, J. A. Stanford, eds.), pp. 271–285. Academic Press, San Diego, CA.

Culver, D. C. 1994. Species interactions. *In* "Groundwater Ecology" (J. Gibert, D. L. Danielopol, and J. A. Stanford, eds.), pp. 271–285. Academic Press, San Diego, CA.

Cummins, K. W., and M. J. Klug. 1979. Feeding ecology of stream invertebrates. *Annual Review of Ecology and Systemics* 10:147–172.

Danielopol, D. L. 1984. Ecological investigations on the alluvial sediments of the Danube in the Vienna area—a phreatobiological project. *Verhandlungen Internationale Vereinigung für Theoretische ünd Angewandte Limnologie* 22:1755–1761.

Danielopol, D. L. 1989. Groundwater fauna associated with riverine aquifers. *Journal of the North American Benthological Society* 8:18–35.

Danielopol, D. L., M. Creuzé des Châtelliers, F. Moeszlacher, P. Pospisil, and R. Popa. 1994. Adaptation of Crustacea to interstitial habitats: A practical agenda for ecological studies. *In* "Groundwater Ecology" (J. Gibert, D. L. Danielopol, and J. A. Stanford, eds.), pp. 217–243. Academic Press, San Diego, CA.

Danielopol, D. L., R. Rouch, P. Pospisil, P. Torreiter, and F. Moeszlacher. 1997. Ecotonal animal assemblages; their interest for groundwater studies. *In* "Groundwater/Surface Water Ecotones: Biological and Hydrological Interactions and Management Options" (J. Gibert, J. Mathieu, and F. Fournier, eds.), pp. 11–20. UNESCO, Cambridge, UK.

de Bovee, F., M. Yacoubi-Khebiza, N. Coineau, and C. Boutin. 1995. Influence du substrat sur la répartition des Crustacés stygobies interstitiels du Haut-Atlas occidental. *Internationale Revue der Gesamten Hydrobiologie* 80:453–468.

Dole-Olivier, M.-J., and P. Marmonier. 1992a. Patch distribution of interstitial communities: Prevailing factors. *Freshwater Biology* 27:177–191.

Dole-Olivier, M.-J., and P. Marmonier. 1992b. Effects of spates on the vertical distribution of the interstitial community. *Hydrobiologia* 230:49–61.

Dole-Olivier, M.-J., P. Marmonier, M. Creuzé des Châtelliers, and D. Martin. 1994. Interstitial fauna associated with the alluvial floodplains of the Rhône River (France) *In* "Groundwater Ecology" (J. Gibert, D. L. Danielopol, and J. A. Stanford, eds.), pp. 313–346. Academic Press, San Diego, CA.

Edler, C., and W. K. Dodds. 1992. Characterization of a groundwater community dominated by *Caecidotea tridentata* (Isopoda). *In* "Proceedings of the First International Conference on Ground Water Ecology" (J. A. Stanford and J. J. Simons, eds.), pp. 91–99. American Water Resources Association, Bethesda, MD.

Edler, C., and W. K. Dodds. 1996. The ecology of a subterranean isopod, *Caecidotea tridentata*. *Freshwater Biology* 35:249–259.

Ferrington, L. C. 1984. Evidence for the hyporheic zone as a microhabitat of *Krenosmittia* spp. larvae (Diptera: Chironomidae). *Journal of Freshwater Ecology* 2:353–358.

Findlay, S. 1995. Importance of surface-subsurface exchange in stream ecosystems: The hyporheic zone. *Limnology and Oceanography* 40:159–164.

Fischer, H., M. Pusch, and J. Schwoerbel. 1996. Spatial distribution and respiration of bacteria in stream-bed sediments. *Archiv für Hydrobiologie* 137:281–300.

Frissell, C. A., and D. Bayles. 1996. Ecosystem management and the conservation of aquatic biodiversity and ecological integrity. *Water Resources Bulletin* 32:229–240.

Fukuhara, H., and M. Sakamoto. 1987. Enhancement of inorganic nitrogen and phosphate release from lake sediment by tubificid worms and chironomid larvae. *Oikos* 48:312–320.

Gallepp, G. W. 1979. Chironomid influence on phosphorus release in sediment-water microcosms. *Ecology* 60:547–556.

Giberson, D. J., and R. J. Hall. 1988. Seasonal variation in faunal distribution within the sediments of a Canadian Shield stream, with emphasis on responses to spring floods. *Canadian Journal of Fisheries and Aquatic Sciences* 45:1994–2002.

Gibert, J., J. A. Stanford, M.-J. Dole-Olivier, and J. V. Ward. 1994. Basic attributes of groundwater systems and prospects for research. *In* "Groundwater Ecology" (J. Gibert, D. L. Danielopol, and J. A. Stanford, eds.), pp. 7–40. Academic Press, San Diego, CA.

Gibert, J., F. Fournier, and J. Mathieu. 1997. The groundwater/surface water ecotone perspective: state of the art. *In* "Groundwater/Surface Water Ecotones: Biological and Hydrological Interactions and Management Options" (J. Gibert, J. Mathieu, and F. Fournier, eds.), pp. 3–8. UNESCO, Cambridge, UK.

Gray, L. J., and S. G. Fisher. 1981. Postflood recolonization pathways of macroinvertebrates in a lowland Sonoran desert stream. *American Midland Naturalist* 106:249–257.

Griffith, M. B., and S. A. Perry. 1993. The distribution of macroinvertebrates in the hyporheic zone of two small Appalachian headwater streams. *Archiv für Hydrobiologie* 126:373–384.

Grimm, N. B., and S. G. Fisher. 1989. Stability of periphyton and macroinvertebrates to disturbance by flash floods in a desert stream. *Journal of the North American Benthological Society* 8:293–307.

Hakenkamp, C. C., and M. A. Palmer. 1992. Problems associated with quantitative sampling of shallow groundwater invertebrates. *In* "Proceedings of the First International Conference on Ground Water Ecology" (J. A. Stanford and J. J. Simons, eds.), pp. 101–110. American Water Resources Association, Bethesda, MD.

Hauer, F. R., and V. H. Resh. 1996. Benthic macroinvertebrates. *In* "Methods in Stream Ecology" (F. R. Hauer and G. A. Lamberti, eds.), pp. 339–369. Academic, Press, San Diego, CA.

Hendricks, S. P. 1996. Bacterial biomass, activity, and production within the hyporheic zone of a north-temperate stream. *Archiv für Hydrobiologie* 136:467–487.

Henry, J. P. 1976. Recherches sur les Asellidae hypogées de la lignée *Cavaticus*. Ph.D. Thesis, University of Dijon (cited by Creuzé des Châtelliers *et al.*, 1994).

Husmann, S. 1974. Die ökologische Bedeutung der Mehrzellfauna bei der natürlichen und künstlichen Sandfiltration. Künstliche Grundwasseranreicherung am Rhein. *Wissenschaftliche Berihte der Untersuchungen und Planungen der Stadtwerke Wiesbaden* AG. 2:173–183 (cited by Torreiter *et al.*, 1994).

Jeffrey, K. A., F. W. Beamish, S. C. Ferguson, R. J. Kolton, and P. D. MacMahon. 1986. Effects of the lampricide, 3-trifluoromethyl-4-nitrophenol (TFM) on the macroinvertebrates within the hyporheic region of a small stream. *Hydrobiologia* 134:43–51.

Johnson, R. A. 1980. Oxygen transport in salmon spawning gravels. *Canadian Journal of Fisheries and Aquatic Sciences* 37:155–162.

Jones, J. B., and R. M. Holmes. 1996a. Surface-subsurface interactions in stream ecosystems. *Trends in Ecology and Evolution* 11:239–242.

Jones, J. B., and R. M. Holmes. 1996b. Reply from J. B. Jones and R. M. Holmes. *Trends in Ecology and Evolution* 11:430.

Junk, W. J., P. B. Bayley, and R. E. Sparks. 1989. The flood pulse concept in river-floodplain systems. *Canadian Journal of Fisheries and Aquatic Sciences, Special Publication* **106**: 110–127.

Karaman, S. L. 1935. Die Fauna unterirdischer Gewässer Jugoslawiens. *Verhandlungen Internationale Vereinigung für Theoretische und Angewandte Limnologie* 7:46–53.

Lattinger-Penko, R. 1976. Quelques données sur la population de *Proasellus slavus* ssp. n. Sket (Crustacea, Isopoda) dans l'hyporhéique de la rivière Drave près de Legrad. *International Journal of Speleology* 8:83–97.

Leichtfried, M. 1991. POM in bed sediments of a gravel stream (Ritrodat-Lunz study area, Austria). *Verhandlungen Internationale Vereinigung für Theoretische und Angewandte Limnologie* 24:1921–1925.

Malard, F., J. L. Reygrobellet, J. Mathieu, and M. Lafont. 1994. The use of invertebrate communities to describe groundwater flow and contaminant transport in a fractured rock aquifer. *Archiv für Hydrobiologie* 131:93–110.

Marchant, R. 1988. Vertical distribution of benthic invertebrates in the bed of the Thomson River, Victoria. *Australian Journal of Marine and Freshwater Research* **39**:775–784.

Marchant, R. 1995. Seasonal variation in the vertical distribution of hyporheic invertebrates in an Australian upland river. *Archiv für Hydrobiologie* 134:441–457.

Maridet, L., J. G. Wasson, and M. Philippe. 1992. Vertical distribution of fauna in the bed sediment of three running water sites: Influence of physical and trophic factors. *Regulated Rivers* 7:45–55.

Maridet, L., M. Philippe, J. G. Wasson, and J. Mathieu. 1996. Spatial and temporal distribution of macroinvertebrates and trophic variables within the bed sediment of three streams differing by their morphology and riparian vegetation. *Archiv für Hydrobiologie* 136:41–64.

Marmonier, P. 1988. Biocénoses interstitielles et circulation des eaux dans le sous-écoulement d'un chenal aménagé du Haut Rhône français. Ph.D. Thesis; Université Claude Bernard, Lyon.

Marmonier, P. 1991. Effect of alluvial shift on the spatial distribution of interstitial fauna. *Verhandlungen Internationale Vereinigung für Theoretische und Angewandte Limnologie* 24:1613–1616.

Marmonier, P., and M.-J. Dole. 1986. Les amphipodes des sédiments d'un bras court-circuité du Rhône. Logique de répartition et réaction aux crues. *Sciences de l'eau* 5:461–486.

McElravy, E. P., and V. H. Resh. 1991. Distribution and seasonal occurrence of the hyporheic fauna in a northern California stream. *Hydrobiologia* 220:233–246.

Mestrov, M., and R. Lattinger-Penko. 1981. Investigation of the mutual influence between a polluted river and its hyporheic. *International Journal of Speleology* 11:159–171.

Metzler, G. M., and L. A. Smock. 1990. Storage and dynamics of subsurface detritus in a sand-bottomed stream. *Canadian Journal of Fisheries and Aquatic Sciences* 47:588–594.

Nelson, S. M., R. A. Roline, and A. M. Montano. 1993. Use of hyporheic samplers in assessing mine drainage impacts. *Journal of Freshwater Ecology* 8:103–109.

Notenboom, J., S. Plénet, and M. J. Turquin. 1994. Groundwater contamination and its impact on groundwater animals and ecosystems. *In* "Groundwater Ecology" (J. Gibert, D. L. Danielopol, and J. A. Stanford, eds.), pp. 477–504. Academic Press, San Diego, CA.

Palmer, M. A. 1993. Experimentation in the hyporheic zone: Challenges and prospectus. *Journal of the North American Benthological Society* 12:84–93.

Palmer, M. A., and D. L. Strayer. 1996. Meiofauna. *In* "Methods in Stream Ecology" (F. R. Hauer and G. A. Lamberti, eds.), pp. 315–337. Academic Press, San Diego, CA.

Palmer, M. A., A. P. Covich, B. J. Findlay, J. Gibert, K. D. Hyde, R. K. Johnson, T. Kairesalo, S. Lahe, C. R. Lovell, R. J. Naiman, C. Ricci, F. Sabater, and D. Strayer. 1997. Biodiversity and ecosystem processes in freshwater sediments. *Ambio* 26:571–577.

Pennak, R. W., and J. V. Ward. 1986. Interstitial faunal communities of the hyporheic and adjacent groundwater biotopes of a Colorado mountain stream. *Archiv für Hydrobiologie, Supplementband* 74:356–396.

Perry, S. A., and W. B. Perry. 1986. Effects of experimental flow regulation on invertebrate drift and stranding in the Flathead and Kootenai Rivers, Montana, USA. *Hydrobiologia* **134**:171–182.

Petts, G., and P. Calow, eds. 1996. "River Restoration." Blackwell, Oxford.

Pinay, G., H. Décamps, E. Chauvet, and E. Fustec. 1990. Functions of ecotones in fluvial systems. *In* "The Ecology and Management of Aquatic-Terrestrial Ecotones" (R. J. Naiman and H. Décamps, eds.), pp. 141–169. UNESCO and Parthenon, Paris and London.

Plénet, S., H. Hugueny, and J. Gibert. 1996. Invertebrate community responses to physical and chemical factors at the river/aquifer interaction zone II. Downstream from the city of Lyon. *Archiv für Hydrobiologie* **136**:65–88.

Pollard, R. A. 1955. Measuring seepage through salmon spawning gravel. *Journal of the Fisheries Research Board of Canada* **12**:706–741.

Pugsley, C. W., and H. B. N. Hynes. 1983. A modified freeze-core technique to quantify the depth distribution of fauna in stony streambeds. *Canadian Journal of Fisheries and Aquatic Sciences* **40**:637–643.

Pusch, M. 1996. The metabolism of organic matter in the hyporheic zone of a mountain stream, and its spatial distribution. *Hydrobiologia* **323**:107–118.

Pusch, M., and J. Schwoerbel. 1994. Community respiration in hyporheic sediments of a mountain stream (Steina, Black Forest). *Archiv für Hydrobiologie* **130**:35–52.

Rouch, R. 1986. Sur l'écologie des eaux souterraines dans le karst. *Stygologia* **2**:352–398.

Rouch, R., and D. L. Danielopol. 1997. Species richness of microcrustacea in subterranean freshwater habitats. Comparative analysis and approximate evaluation. *Internationale Revue der Gesamten Hydrobiologie* **82**:121–145.

Rouch, R., A. Pitzalis, and A. Descouens. 1993. Effets d'un pompage à gros débit sur le peuplement des Crustacés d'un aquifère karstique. *Annales de Limnologie* **29**:15–29.

Schmid, P. E. 1993. Random patch dynamics of larval Chironomidae (Diptera) in the bed sediments of a gravel stream. *Freshwater Biology* **30**:239–255.

Schminke, H. K. 1974. Mesozoic intercontinental relationships as evidenced by bathynellid Crustacea (Syncarida: Malacostraca). *Systematic Zoology* **23**:157–164.

Smith, J. J., and P. S. Lake. 1993. The breakdown of buried and surface-placed leaf litter in an upland stream. *Hydrobiologia* **271**:141–148.

Smock, L. A., J. E. Gladden, J. L. Riekenberg, L. C. Smith, and C. R. Black. 1992. Lotic macroinvertebrate production in three dimensions: Channel surface, hyporheic, and floodplain environments. *Ecology* **73**:876–886.

Stanford, J. A., and A. R. Gaufin. 1974. Hyporheic communities of two Montana rivers. *Science* **185**:700–702.

Stanford, J. A., and J. V. Ward. 1988. The hyporheic habitat of river ecosystems. *Nature (London)* **335**:64–66.

Stanford, J. A., and J. V. Ward. 1993. An ecosystem perspective of alluvial rivers: Connectivity and the hyporheic corridor. *Journal of the North American Benthological Society* **12**:48–60.

Stanford, J. A., J. V. Ward, and B. K. Ellis. 1994. Ecology of the alluvial aquifers of the Flathead River, Montana. *In* "Groundwater Ecology" (J. Gibert, D. L. Danielopol, and J. A. Stanford, eds.), pp. 367–390. Academic Press, San Diego, CA.

Stanley, E. H., and A. J. Boulton. 1993. Hydrology and the distribution of hyporheos: Perspectives from a mesic river and a desert stream. *Journal of the North American Benthological Society* **12**:79–83.

Sterba, O., V. Uvira, P. Mathur, and M. Rulik. 1992. Variations of the hyporheic zone through a riffle in the R. Morava, Czechoslovakia. *Regulated Rivers* **7**:31–43.

Strayer, D. L. 1994. Limits to biological distributions in groundwater. *In* "Groundwater Ecology" (J. Gibert, D. L. Danielopol, and J. A. Stanford, eds.), pp. 287–310. Academic Press, San Diego, CA.

Strommer, J. L., and L. A. Smock. 1989. Vertical distribution and abundance of invertebrates within the sandy substrate of a low-gradient headwater stream. *Freshwater Biology* **22**:263–274.

Torreiter, P., P. Pitaksintorn-Watanamahart, and D. L. Danielopol. 1994. The activity of oligochaetes in relation to their ecological role in slow filtration columns. *In* "Proceedings of the Second International Conference on Ground Water Ecology" (J. A. Stanford and J. J. Simons, eds.), pp. 86–94. American Water Resources Association, Herndon, VA.

Townsend, C. R. 1989. The patch dynamics concept of stream community ecology. *Journal of the North American Benthological Society* **8**:36–50.

Townsend, C. R., and A. G. Hildrew. 1994. Species traits in relation to a habitat templet for river systems. *Freshwater Biology* **31**:265–275.

Ward, J. V., and N. J. Voelz. 1994. Groundwater fauna of the South Platte River system, Colorado. *In* "Groundwater Ecology" (J. Gibert, D. L. Danielopol, and J. A. Stanford, eds.), pp. 391–423. Academic Press, San Diego, CA.

Ward, J. V., N. J. Voelz, and P. Marmonier. 1992. Groundwater faunas at riverine sites receiving treated sewage effluent. *In* "Proceedings of the First International Conference on Ground Water Ecology" (J. A. Stanford and J. J. Simons, eds.), pp. 351–364. American Water Resources Association, Bethesda, MD.

Whitman, R. L., and W. J. Clark. 1984. Ecological studies of the sand-dwelling community of an east Texas stream. *Freshwater Invertebrate Biology* **3**:59–79.

Williams, D. D. 1987. "The Ecology of Temporary Waters." Croom Helm, London.

Williams, D. D., and H. B. N. Hynes. 1974. The occurrence of benthos deep in the substratum of a stream. *Freshwater Biology* **4**:233–256.

Williams, D. D., and H. B. N. Hynes. 1977. The ecology of temporary streams. II. General remarks on temporary streams. *Internationale Revue der Gesamten Hydrobiologie* **62**:189–208.

15

Lotic Macrophytes and Surface–Subsurface Exchange Processes

David S. White and Susan P. Hendricks

Hancock Biological Station and Center for Reservoir Research
Murray State University
Murray, Kentucky

I. Introduction   363
II. Macrophytes and Aquatic Habitats   366
III. Macrophytes as Indicators of Surface and Subsurface
    Conditions   367
IV. Macrophytes and Surface- and Subsurface-Water Flow
    Patterns   369
V. Nutrient Uptake from Sediments   371
VI. Processes at the Rhizosphere   373
VII. Summary and Avenues for Further Study   375
    References   376

I. INTRODUCTION

The roles of aquatic macrophytes as sources of organic matter and as habitat for periphyton, aquatic invertebrates, and fish are well documented in the stream ecology literature (Allan, 1995). The role of lotic macrophytes in biogeochemical cycles, particularly with respect to surface–subsurface exchange processes, is less well known but is expected to be of importance in streams with well-developed plant beds. Lotic macrophytes include the macro-algae, liverworts and mosses, and vascular plants attached to or rooted in the substrata of streams and rivers (as defined by Sculthorpe, 1967). Liverworts and mosses (bryophytes), most macroalgae with the exception of

Charophyta (*Chara, Nitella*), and a few vascular species (e.g., *Podostemum*) live anchored to the surface of rocks and other solid surfaces. Some anchored forms may influence surface-water flow and velocity patterns; however, the effects on surface–subsurface exchange processes probably are minimal in most streams. The charophytes, with rootlike rhizoids, and the true rooted vascular plants are unique among stream organisms in that a single plant may exist simultaneously within both interstitial and surface-water environments. Emergent species function in the aerial environment as well.

The capability to link surface and subsurface environments underscores the potential role of macrophytes in exchange processes. Rooted macrophytes obtain nutrients and ions from both the water column and the sediments, acting as conduits between the two environments. Macrophytes may actively alter the availability of nutrients through both biological and physical processes (Sculthorpe, 1967; Jaynes and Carpenter, 1986; Wetzel, 1990; Barko *et al.,* 1991). In maintaining position within a lotic environment, vascular plants and charophytes alter sediment deposition patterns, influencing both surface- and subsurface-water flow. The potential for elemental cycling combined with the potential for altering surface- and subsurface-water flow implies a complex set of interactions. In this chapter, we review the literature on lotic macrophytes with emphasis on macrophyte roles in surface–subsurface exchange, particularly processes at the rhizosphere. We primarily consider species of the open water that are submersed or emergent (Table I). Riparian taxa (*Typha,* sedges, reeds) and their relationships to sediment biogeochemistry are well known in the wetland literature (see reviews in Mitsch

TABLE I   A Systematic Listing of Some of the Lotic Macrophyte Taxa Mentioned in the Text or Commonly Associated with Streambed Modifications and Surface–Subsurface Processes

Family	Genus/species	Common name	General form
CHAROPHYTA			
Characeae			
	Chara hispida	Stonewort	Submersed
	Chara minor	Stonewort	Submersed
	Chara vulgaris	Stonewort	Submersed
	Nitella confervacea	Stonewort	Submersed
	Nitella flexilis	Stonewort	Submersed
RHODOPHYTA			
	Batrachospermum sp.	Red algae	Submersed
MUSCI			
	Fontinalis antipyretica	Moss	Submersed
SPERMATOPHYTA			
Typhaceae			
	Typha spp.	Cat tail	Emergent

(*continued*)

TABLE I (continued)

Family	Genus/species	Common name	General form
Sparganiaceae			
	Sparganium chlorocarpum	Bur reed	Emergent
	Sparganium emersum	Bur reed	Emergent
Potamogetonaceae			
	Potamogeton berchtoldii	Pondweed	(Genus primarily
	Potamogeton coloratus	Pondweed	submersed with
	Potamogeton crispus	Pondweed	some emergent
	Potamogeton filiformis	Pondweed	leaves and
	Potamogeton foliosus	Pondweed	flowers)
	Potamogeton gramineus	Pondweed	
	Potamogeton pectinatus	Pondweed	
	Potamogeton richardsonii	Pondweed	
Zannichelliaceae			
	Zannichellia palustris	Common poolmat	Emergent
Alismataceae			
	Sagittaria latifolia	Arrowhead	Emergent
Hydrocharitaceae			
	Elodea canadensis	Waterweed	Submersed
	Elodea cophocarba	Waterweed	Submersed
	Elodea nuttallii	Waterweed	Submersed
	Vallisneria americana	Eelgrass	Submersed
Cruciferae			
	Nasturtium officinale (also see *Rorippa*)	Watercress	Emergent
Podostemaceae			
	Podostemon ceratophyllum	Thread-foot	Rock surfaces
Ranunculaceae			
	Caltha palustris	Marsh marigold	Emergent
	Ranunculus aquatilis	Buttercup	(Submersed and
	Ranunculus fluitans	Buttercup	emergent)
	Ranunculus peltatus	Buttercup	
	Ranunculus septentrionalis	Buttercup	
	Ranunculus trichophyllus	Buttercup	
Callitrichaceae			
	Callitriche cophocarpa	Water starwort	Submersed
	Callitriche hermaphroditica	Water starwort	Submersed
	Callitriche obtusangula	Water starwort	Submersed
	Callitriche platycarpa	Water starwort	Submersed
Haloragaceae			
	Myriophyllum heterophyllum	Water milfoil	Submersed
	Myriophyllum spicatum	Water milfoil	Submersed
	Myriophyllum exalbescens	Water milfoil	Submersed
Scrophulariaceae			
	Veronica catenata	Speedwell	Emergent
Acanthaceae			
	Justicia americana	Water willow	Emergent

Primarily after Correll and Correll, 1972, and Voss, 1985.

and Gosselink, 1993) and are not considered here. Taxa that are strictly littoral or are phreatophytes also are not discussed here, but many taxa of trees and shrubs (alders, willows, cottonwoods, etc.) are known to influence both the hydrology and geochemistry of streams and rivers.

II. MACROPHYTES AND AQUATIC HABITATS

Although published more that 30 years ago, Sculthorpe's *The Biology of Aquatic Vascular Plants* (1967) remains the single-most valuable resource for basic information on the physiology and ecology of aquatic plants. Reviews of the distributions and general ecology of aquatic macrophytes with emphasis on lentic and wetland habitats can be found in several texts, for example Fassett (1969), Correll and Correll (1972), Hutchinson (1975), Beal (1977), Wetzel (1983), and Mitsch and Gosselink (1993). Hynes (1970), Westlake (1975), Haslam (1978), Allan (1995) and others have reviewed the biology and ecology of stream and riverine taxa. Much of what has been published on lotic macrophytes has described plant distributions and "community succession." Biogeochemical processes occurring at the rhizosphere have been well described in lentic habitats over the past 30 years (Barko *et al.*, 1991), but few studies have considered similar processes in lotic habitats. Few vascular aquatic macrophytes are restricted to or specifically adapted to flowing waters (Hynes, 1970), and it seems reasonable to assume that plant functions, particularly processes at the rhizosphere, will be similar regardless of the extent of surface-water movement. It should be noted that aquatic botany is a growing field of study, and there is an extensive literature on root–sediment relationships. Therefore, we have been selective in summarizing the lotic and lentic literature to assess the potential role of macrophytes in lotic surface–subsurface exchange.

Factors influencing the distribution and growth of lotic macrophytes are known to include turbulence and water velocity, seasonal temperature patterns, surface and interstitial water chemistry, sediment composition, light, and other biota (Spence, 1967; Peltier and Welch, 1968; Hutchinson, 1975; Haslam, 1978; Fortner and White, 1988; Barko *et al.* 1991; Thiébaut and Muller, 1998). Establishment of well-developed lotic macrophyte populations and "communities," however, appears to be related primarily to three essential factors: constant or predictable discharge (i.e., lower dynamic discharge patterns), turbidity, and canopy cover (Biggs, 1996). Obviously, not all streams and rivers have a well-developed macrophyte flora. Large rivers that carry high loads of suspended solids usually are unfavorable for plant growth because of light limitation. However, where suspended solids are low, even very large rivers may contain extensive macrophyte beds (e.g., Amazon River, Detroit River, also see examples in Hutchinson, 1975, and Haslam,

1978). Similarly, very small streams often lack aquatic plants because of variable discharge regimes or if the canopy prevents much light from reaching the stream surface. First-order streams, that begin from permanent springs, can be rich in macrophytes, particularly bryophytes. High gradient streams contain primarily well-anchored mosses and liverworts and macroalgae along with a limited number of anchored vascular taxa (Hynes, 1970; Biggs, 1996).

In North America, the streams and rivers presently appearing to have the greatest diversity of aquatic macrophytes are located in once-glaciated regions and in some areas of the southwest. Not surprisingly, much of our knowledge of aquatic macrophytes comes from these regions and similar areas in Europe (see reviews in Fassett, 1969; Correll and Correll, 1972; Haslam, 1978). Streams with well-developed macrophyte communities or assemblages tend to have low levels of disturbance or seasonally predictable disturbance patterns (reviewed in Bornette *et al.*, 1994), the most important of which appears to be surface-water velocity (Biggs, 1996). Higher velocities in spring can remove organic matter and silt accumulations promoting vegetation growth, whereas high discharges and accompanying high velocities during the growing season may eliminate macrophytes through scour. Prior to European settlement and extensive agricultural development, many parts of North America may have contained streams with well-developed macrophyte populations. For example, Upper Three Runs Creek, located mostly on the protected forest lands of the Savannah River Plant in South Carolina, may be typical of southeastern streams prior to European habitation. This low gradient stream, which is not particularly subject to flooding and scour following rain events, is very rich in macrophytes (David Lenat, North Carolina Department of Environmental Conservation, personal communication) and supports an aquatic invertebrate fauna of more than 500 species. Well-developed macrophyte populations often occur in northern deranged systems (i.e., streams between lakes) where predictably high velocities occur only after spring snowmelt (Fortner and White, 1988; Hendricks and White, 1995).

III. MACROPHYTES AS INDICATORS OF SURFACE AND SUBSURFACE CONDITIONS

Patterns of long-term macrophyte community succession in lake ontogeny are well known (Hutchinson, 1975; Wetzel, 1983). The concept of lotic plants occurring as communities where niche or habitat partitioning occurs (sensu Tilman, 1994) is still debated because of the erosive nature of streams and rivers. Few macrophytes are known to have specific adaptations to flowing waters, and most species will grow where favorable combinations of substrate type, water depth, surface velocity, nutrient availability, etc., occur. Once established, continued production throughout the growing season often produces extensive monospecific patches which can significantly alter channel topography and hydraulics (Gessner, 1955; Minckley, 1963). Patches may reoccur roughly in the same location over several years, and specific distributions within a stream appear to be related to flood events, sediment structure, and modes of reproduction (Haslam, 1978). Larger patches, or plant beds of one species, may alter sedimentation and surface flow conditions, promoting the establishment of other species and producing community-like successional responses (French and Chambers, 1996; Wigand *et al.*, 1997).

Species and communities of lotic aquatic macrophytes have been correlated with both surface-water and groundwater (water table) physicochemical conditions (reviewed in Carbiener *et al.*, 1995). As bioindicators, aquatic macrophytes have advantages of being attached within the bed sediments, perennial, and relatively easy to identify (Haslam, 1982; Carbiener *et al.*, 1990; Muller, 1990). Many species accumulate metal ions and will respond to increased nitrogen and phosphorus levels in both hard and soft waters (Carbiener *et al.*, 1990; Muller, 1990; Schütz, 1995; Small *et al.*, 1996; Elgin *et al.*, 1997; Carr and Chambers, 1998) or are tolerant of eutrophication (e.g., *Ranunculus fluitans* and *Potamogeton crispus;* Tremp and Kohler, 1995). Several taxa including *Batrachospermum* spp., *Chara* spp., *Fontinalis antipyretica, Caltha palustris, Nasturtium officinale,* and *Ranunculus septentrionalis* have been used as indicators of areas of seepage or direct hydrologic discharge from ground water to surface water (Fortner and White, 1988; Carbiener *et al.*, 1990; Trémolières *et al.*, 1993; Bornette *et al.*, 1996; Elgin *et al.*, 1997). Ground water may present a constant-temperature environment as well as a source of free (unbound) carbon dioxide and nutrients (Sculthorpe, 1967).

Although Westlake (1975) suggested that surface flow may be more important than substrate type, lotic charophyte and aquatic vascular plant growth and diversity appear to be highest in medium-textured sediments, particularly medium sands, where dissolved nitrogen, phosphorus, and ions are more mobile (Haslam, 1978; Barko *et al.*, 1991). Macrophytes are less common where there are surface accumulations of particulate organic matter or clays that may impede subsurface–surface exchange or strongly adsorb nutrients such as ammonium (NH_4^+) or phosphate (PO_4^{3-}). Organic matter accumulation can create localized anoxic conditions that release nitrite (NO_2^-), sulfide (S^{2-}), ferrous iron (Fe^{2+}), and organic by-products that limit plant growth. Microbial metabolism also may out-compete macrophytes for nutrients where there are large accumulations of organic matter; however, plants have been shown to respond positively to lower amounts of organic matter where microbial metabolism may act as a source of nutrients (Barko *et al.*, 1991; also see discussions in Mitsch and Gosselink, 1993).

IV. MACROPHYTES AND SURFACE- AND SUBSURFACE-WATER FLOW PATTERNS

Figures 1–3 summarize discussions in the following sections. The figures are idealized in that most of what is known is descriptive as very few experimental studies have been conducted on lotic macrophytes. Surface-water velocities are reduced by lotic macrophyte growth with the amount of reduction being related to plant morphology and the size and density of patches or beds (Haslam, 1978; Sand-Jensen and Mebus, 1996). Decreased surface velocities promote bed load and organic matter deposition within a plant bed (Fig. 1). For many species, the increased sedimentation may create unfavorable surface flow and chemical conditions for optimal growth in the middle and downstream ends of bed, thus beds tend to grow toward the upstream and lateral edges.

Sediment deposition for some species results in the formation of wedge-shaped hummocks (Fig. 2). Taxa known to form hummocks include species of *Chara, Elodea, Myriophylum, Nitella,* and *Potamogeton*. Reduced water velocity at the downstream end of the hummock causes deposition of sand and other fine material. Plants producing hummocks are rooted near the tops of the deposits, and vegetative masses continue to grow in a downstream direction as long as the hummocks continue to elongate (Butcher, 1933; Hynes, 1970; Gessner, 1955; Minckley, 1963; Haslam, 1978; Hendricks and White, 1988). The vegetative mass itself tends to flatten in the current, increasing hummock density and decreasing roughness, which further stabilizes the hummock. Vortexes created by increased water column depth at the downstream end of hummocks may force water upstream through the plant mass, maintaining favorable metabolic conditions for the entire colony (Minckley, 1963). Many of the hummock-forming taxa are annual species, and the resulting hummocks are often removed during annual flooding cycles.

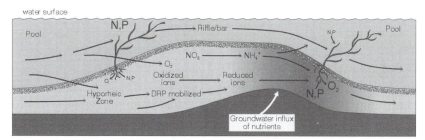

FIGURE 1   Hypothetical structure of a lotic macrophyte bed showing movement of water and nutrients. Finer sediments within the bed are composed of organic and bed load deposits where decomposition may provide additional sources of nitrogen and phosphorus. Surface flow is diverted over and around the bed. Decreased velocity over the bed coupled with increased sediment density deflects subsurface (hyporheic) flow upward into the surficial along with ions and nutrients.

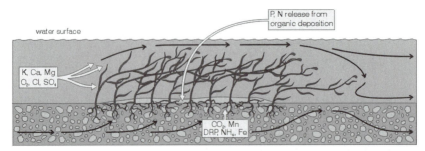

FIGURE 2 Hypothetical structure of a lotic macrophyte hummock (e.g., *Chara*) showing movement of water and nutrients. The hummock grows in a downstream direction as bed load sediments are deposited at the foot. Flocculent organic matter also may be deposited along with sands providing an additional source of nutrients. Unlike macrophyte beds where sediments are deposited over the roots (Fig. 1), hummock plant roots (rhizoids in charophytes) remain near the surface of deposits. Decreased velocity over the hummock coupled with increased sediment density deflects subsurface (hyporheic) flow upward into the hummock along with reduced ions and nutrients. Surface-water backflow distributes surface and subsurface nutrients throughout the plant mass.

For both bed and hummock-forming species, water velocity is decreased within the plant mass but increased at the sides (Gessner, 1950, 1955). Further, deposited sediments are often finer and denser than surrounding areas of the streambed not containing macrophytes, impeding water downwelling into the streambed. The changes in surface velocity, bed surface density (plant and/or sediment), and streambed topography are known to alter subsurface flowpaths and, consequently, surface–subsurface exchange patterns. As bed surface density increases and surface velocity decreases, interstitial water tends to upwell (Vaux, 1968; Hendricks and White, 1988; White, 1990). The result may be delivery of nutrient-rich interstitial water to the root mass. Depending on hummock location within the streambed topography, increased surface velocity along the sides of plant beds and hummocks is expected to further alter subsurface upwelling–downwelling patterns (White *et al.*, 1992).

Most lotic macrophytes are not very deeply rooted (generally <15 cm) nor do they have extensive root systems (Sculthorpe, 1967), but notable exceptions include the emergent *Justicia americana* where the root masses may constitute a substantial portion of the streambed. Thus, the effects of root masses on sediment hydraulic conductivity and permeability may not be as significant as they are in terrestrial systems. This topic remains essentially unstudied in stream ecology, although it could be quite important to the movement of water and nutrients through bed sediments. Similarly unstudied are root–invertebrate associations. Because a majority of plants are expected to have roots in anaerobic sediments (see Section V), a highly developed hyporheos would not be expected. The larvae (and pupae) of some species of ephydrid flies, however, are known to tap the oxygen-rich roots while feeding on accumulated organic matter and associated bacteria (Stuart Neff, Temple University, personal communication).

V. NUTRIENT UPTAKE FROM SEDIMENTS

Nitrogen may prove to be more limiting than phosphorus to aquatic macrophyte growth in some lotic and lentic systems (Grimm and Fisher, 1984; Barko et al., 1991). Much more is known, however, of the limiting role of phosphorus (Wetzel, 1983; Grimm and Fisher, 1984). The origins, cycling, and fates of nutrients and ions in streambed sediments are discussed in detail in other chapters of this volume, as are processes along hyporheic underflow paths; therefore, the following discussion is limited to specific macrophyte influences. Nearly 100 years ago, Pond (1905) recognized the importance of sediment fertility in aquatic macrophyte growth. The knowledge that some taxa (particularly submersed species) could exhibit expansive areal coverage either with little or no root mass or in very infertile sediments has led to a long-standing debate on the importance of root versus shoot as the primary site of nutrient uptake, particularly phosphorus, in both lentic (e.g., Sculthorpe, 1967; Hutchinson, 1975; Carignan and Kalff, 1980; Barko and Smart, 1981; Barko et al., 1991) and lotic (e.g., Chambers et al., 1989; Robach et al., 1995) species. Most aquatic vascular plants, however, appear to obtain at least some portions of nutrient requirements through both root and shoot avenues. Also, some charophytes may absorb and translocate nutrients from sediments through both stalks and rhizoids (especially phosphorus, Littlefield and Forsberg, 1965). Anaerobic sediments provide the richest nutrient environment for rooted taxa (Sculthorpe, 1967; Hutchinson, 1975; Jaynes and Carpenter, 1986; Barko et al., 1991). Barko et al., (1991) provide the listing for the sources of nutrients for macrophytes in Table II.

TABLE II   Potential Primary Sources of Nutrients for Aquatic Macrophytes

Sediment as a primary source	Surface water as a primary source
Nitrogen	Calcium
Phosphorus	Magnesium
Iron	Sodium
Manganese	Potassium
Micronutrients	Sulfate
Dissolved inorganic carbon[a]	Chloride

[a]Primarily for submersed, isoetid plant forms.
After Barko et al. (1991).

Carignan (1982) proposed an empirical model to estimate the importance of roots versus shoots in phosphorus uptake. If sediment soluble reactive phosphorus (SRP) concentration was at least four times greater than overlying water SRP concentration, 50% or more of phosphorus requirements might be met through root uptake. Except under highly eutrophic conditions, sediment SRP in lakes is usually much more than four times the concentration in surface water, particularly in fine-grained, anaerobic sediments (Carignan and Flett, 1981; Wetzel, 1981). Similarly high proportions of sediment SRP might be expected in highly reduced streambed sediments associated with long-residence times (underflow paths) of hyporheic water or where ground water is a significant source of phosphorus to the stream (Hendricks and White, Chapter 9 in this volume).

The vertical hydraulic gradients and underflow paths created by streambed features such as riffles and bars (Vaux, 1968; White *et al.*, 1987; Harvey and Bencala, 1993; Hendricks and White, 1991, 1995) may play a strong role in the distribution and nutrition of lotic macrophytes (Fig. 3). Macrophytes growing in hyporheic downwelling areas where sediment and water column phosphorus concentrations are similar would be expected to obtain more phosphorus from the water column. For example (following the Carignan, 1982, model), White *et al.* (1992) observed that *Potamogeton gramineus,* which is known to use primarily water column phosphorus (Denny, 1972), was more randomly distributed over the streambed, including downwelling areas than species known to use sediment phosphorus. Conversely, species known to rely on sediment phosphorus were found growing most often in reduced sediments associated with hyporheic upwelling areas

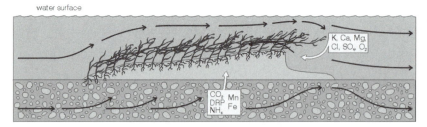

FIGURE 3 Hypothetical subsurface flowpaths beneath a topographical streambed feature (e.g., a riffle or gravel/sand bar) in relation to nutrient availability for plants. Downwelling at the head of the feature carries oxygenated surface water into the sediments. Microbial and chemical processes along underflow paths create reducing conditions, mobilizing soluble reactive phosphorus (SRP) and ammonium (NH_4^+). Macrophytes growing at the head of the feature would be expected to rely primarily on surface nitrogen and phosphorus and release little oxygen to the sediments as surface and subsurface phosphorus concentrations would be similar. Macrophytes growing at the downstream end of feature (end of a hyporheic underflow paths) might be expected to utilize nitrogen and phosphorus from the sediments because of higher sediment concentrations. These plants would be expected to release more oxygen at the rhizosphere.

containing higher concentrations of phosphorus and ammonium (Fortner and White, 1988; White *et al.*, 1992; Hendricks and White, 1995; Idestam-Almquist and Kautsky, 1995). Apparently, however, not all species of lotic macrophytes rely primarily on phosphorus from sediments, and many taxa are opportunistic in being able to take in phosphorus from the source where it is most available (Carignan and Kalff, 1980; Carignan, 1982; White *et al.*, 1992; Robach *et al.*, 1995). Idestam-Almquist and Kautsky (1995) note that in transplant experiments, *Potamogeton pectinatus* (found in both lotic and lentic habitats) is able to adapt to both still and flowing water and to differences in sediment nutrient concentrations. Dudley and Valett (1991) transplanted three macrophyte species into both hyporheic upwelling and downwelling areas of an Arizona desert stream, observing somewhat more growth in upwelling areas which was attributed to greater nutrient supplies. Unfortunately, few experimental studies have been attempted because of the difficulties in transplanting lotic aquatic species. Because phosphorus readily complexes with various metals under aerobic conditions (Hendricks and White, Chapter 9 in this volume), macrophytes that rely on phosphorus from sediments essentially must have roots in reduced sediments (Fig. 3; Mantai and Newton, 1982; Wigand *et al.*, 1997).

VI. PROCESSES AT THE RHIZOSPHERE

Vascular macrophytes release oxygen from their roots with the amount dependent on species, time of day, season, and condition of the plant (Sculthorpe, 1967; Sand-Jensen *et al.*, 1982; Wigand *et al.*, 1997). Downward transport of oxygen is essential for normal root and root hair respiration, and to overcome potential anaerobiosis in reduced sediments (Sculthorpe, 1967; Sand-Jensen *et al.*, 1982). Of the few species examined, lotic taxa appear to release less oxygen than do lake or wetland taxa (Sand-Jensen *et al.*, 1982). The role of plant types and morphology present in streams may be important. Isoetid forms (greater root/shoot ratios, e.g., *Vallisneria*) appear to transmit more oxygen to sediments than elodeid forms (lower root/shoot ratios, e.g., some *Potamogeton* species). Plants known to derive most phosphorus requirements from sediments, such as *Vallisneria americana*, which is common in both lotic and lentic waters, tend to release more oxygen (see Fig. 3) than do taxa that are expected to meet most phosphorus requirements from the surface waters (Wigand *et al.*, 1997). Released oxygen affects sediment oxidation–reduction potential, pH, and the form and retention of nutrients and metals (Wigand *et al.*, 1997). Oxygen release into anaerobic sediments promotes precipitation of iron and manganese, sulfate oxidation, coprecipitation of phosphorus, nitrification, and denitrification and may aid in the aerobic microbial metabolism of particulate and dissolved organic matter. Organic matter oxidation further increases the amount of dissolved phosphorus and nitrogen available for plant uptake (van Wijck *et al.*, 1992). The extent of oxygen release on interstitial chemistry, however, seems to be dependent on the fertility of the sediments and may be minimal to undetectable except for microzones surrounding the roots in highly reduced sediments (Carpenter *et al.*, 1983; Carpenter and Lodge, 1986; Barko *et al.*, 1991).

Oxygen release also may facilitate carbon dioxide (CO_2) uptake through the roots; however, few macrophytes (mostly isoetid forms) are expected to exhibit net uptake of carbon dioxide from sediments. Submersed species may use interstitial carbon dioxide more than emergent taxa due to differences in vascular systems. Many aquatic macrophytes have the ability to use bicarbonate (HCO_3^-) as well as carbon dioixde, but uptake of bicarbonate through root systems is not well known (for in-depth discussions on inorganic carbon uptake, see Sand-Jensen, 1983; Madsen, 1993). The distributions of several macrophytes have been related to surface-water alkalinity (e.g., *Justicia americana*, Hill, 1981), but the same species may occur in low alkalinity systems if there is a subsurface source of calcium carbonate ($CaCO_3$; Lee and White, 1995).

Many aquatic macrophytes are known to generate enough upward water transport through their root systems to meet nutrient requirements if nutrients and ions are sequestered from the sediments (Pedersen and Sand-Jensen, 1993). Most aquatic macrophytes preferentially take up phosphorus as SRP and nitrogen as ammonium (Barko *et al.*, 1991; Short and McRoy, 1983). Although oxygen release can reduce the availability of SRP, ammonium, and some ions, oxygen also can facilitate nutrient uptake through roots. For example, *Vallisneria americana* roots have been shown to harbor aerobic symbiotic micorrhizal fungi that enhance uptake of phosphate and aid in plant growth (Wigand *et al.*, 1997). Root–bacteria–micorrhizal interactions are well known for terrestrial ecosystems (Atlas and Bartha, 1993) and for wetland plants (Wetzel, 1990; Mitsch and Gosselink, 1993) but rarely have been studied in lotic ecosystems. Aquatic macrophyte uptake can significantly reduce phosphorus and nitrogen concentrations in lake and wetland sediments (Barko *et al.*, 1991; Mitsch and Gosselink, 1993; Wigand *et al.*, 1997), releasing them to surface waters through leaching and plant senescence (Carpenter, 1980; Barko *et al.*, 1991).

Historically, organic matter deposition and nutrient diffusion (sometimes aided by bioturbation) have been thought sufficient to replace phosphorus, nitrogen, and ions lost from the sediments by macrophyte uptake (Barko and Smart, 1986; Barko *et al.*, 1991). Lodge *et al.* (1989) and Lillie and Barko (1990) noted, however, that greater macrophyte production occurs in areas of either advective surface-water or groundwater movement through lake bottom sediments, particularly where surface-water nutrient concentrations may be limited. The effects of nutrient flux to the rhizosphere of lotic macrophytes via subsurface flow have yet to be examined in any depth; from earlier discussions, however, the potential for relatively fast water renewal probably could quickly overcome losses through uptake.

VII. SUMMARY AND AVENUES FOR FURTHER STUDY

Macrophytes appear to be unique in the sense that growth forms can alter streambed topography, hydrology, and nutrient flux paths to their advantage. Macrophytes may greatly enhance surface–subsurface biogeochemistry and exchange processes by increasing surface–subsurface complexity, by decreasing nutrient spiraling length through nutrient uptake, and through plant–microbial interactions. Perhaps of all the major stream organisms (with the possible exception of the sediment–microbial community), the functions of macrophytes in lotic ecosystems, however, are least understood. The fate of macrophyte decomposition is well studied (Webster and Benfield, 1986; Allan, 1995), but the roles as living organisms are not. The reason for this simply may be that macrophytes are not (or no longer) a common component of all stream systems to the extent of aquatic insects, algae, and fish. Compared with any of the other major lotic taxa, fewer researchers focus on the ecology of aquatic macrophytes, and much of what is published comes from attempts to control plant growth rather than to understand functions. Except as a source for organic matter and as a habitat for invertebrates and algae, macrophyte importance to biogeochemical processes in the normal functioning of a stream ecosystem has been largely overlooked. Once depleted from streams and rivers through human alteration of discharge, channel form, and water quality patterns, macrophytes do not easily reestablish themselves (Bornette *et al.,* 1994; Barrat-Segretain and Amoros, 1996; Schütz, 1995). Indeed, those species that do become established or reestablished can find themselves in direct opposition to human uses of waterways (Dunderdale and Morris, 1996).

Much of the subsurface processing of dissolved organic carbon (DOC), phosphorus, nitrogen, metals, etc., is linked to microbial activity, and apparently plant–microbial interactions at the rhizosphere drive many of the subsurface geochemical processes involving vascular aquatic plants. Most vascular plant species also have mycorrhizal fungi associated with the roots that are capable of altering the forms and mobility of DOC, nitrogen, phosphorus, iron, manganese, and sulfate (Harley, 1989). Plant–mycorrhizal fungi–bacterial associations have been demonstrated for wetlands (Mitsch and Gosselink, 1993) but are little studied in lotic systems. Besides promoting chemical reactions, mycorrhizal fungi increase the effective surface areas of roots creating more extensive microzones where chemical reactions can occur. The microzones (or perhaps plant root biofilms might be a more appropriate term) associated with large lotic plant beds could play a very significant role in nutrient dynamics of stream ecosystems. This appears to be a promising area for further research, particularly when one considers how important it may be in stream restoration efforts.

Although, lotic macrophytes do not seem to establish interacting communities in surface-water environments (as do algae, insects, and fish) or the successional patterns of lentic species, macrophyte beds can become substantial features of streams and rivers. Macrophytes appear to be sensitive indicators of eutrophication and are generally intolerant of highly variable hydraulic regimes. Human activities that increase runoff to streams (agriculture, urbanization, deforestation, etc.) probably are responsible in many cases for the loss of macrophyte populations. Although studies of plant–sediment interactions have steadily increased over the past two decades, research on lotic taxa still lags behind our knowledge of lentic taxa. Based on this very specific review, some promising areas for laboratory and field research in conjunction with rhizosphere processes might include examination of the (1) ability of species to physiologically adapt to changing nutrient levels in water and sediments, (2) plant community interactions (competition) at the rhizosphere level, (3) potential role of lotic macrophytes in bioremediation, and 4) greater documentation of the effects of macrophyte loss (or gain) on whole system processes.

REFERENCES

Allan, J. D. 1995. "Stream Ecology: Structure and Function of Running Waters." Chapman & Hall, New York.

Atlas, R. M., and B. Bartha. 1993. "Microbial Ecology," 3rd ed. Benjamin/Cummins, Menlo Park, CA.

Barko, J. W., and R. M. Smart. 1981. Sediment-based nutrition of submersed macrophytes. *Aquatic Botany* 5:109–117.

Barko, J. W., and R. M. Smart. 1986. Sediment-related mechanisms of growth limitation in submersed macrophytes. *Ecology* 67:1328–1340.

Barko, J. W., D. Gunnison, and S. R. Carpenter. 1991. Sediment interactions with submersed macrophyte growth and community dynamics. *Aquatic Botany* 41:41–65.

Barrat-Segretain, M. H., and C. Amoros. 1996. Recovery of riverine vegetation after experimental disturbance: A field test of the patch dynamics concept. *Hydrobiologia* 321:53–68.

Beal, E. O. 1977. "A Manual of Marsh and Aquatic Vascular Plants of North Carolina, with Habitat Data." North Carolina Agricultural Experiment Station, North Carolina State University, Raleigh.

Biggs, B. J. F. 1996. Hydraulic habitat of plants in streams. *Regulated Rivers:* Restoration and Management 12:131–144.

Bornette, G., C. Amoros, and C. Castella. 1994. Succession and fluctuation in the aquatic vegetation of two former Rhône River channels. *Vegetatio* 110:171–184.

Bornette, G., M. Guerlesquin, and C. P. Henry. 1996. Are the Characeae able to indicate the origin of groundwater in former river channels? *Vegetatio* 125:207–222.

Butcher, R. W. 1933. Studies on the ecology of rivers. I. On the distribution of macrophyte vegetation in the rivers of Britain. *Journal of Ecology* 21:58–89.

Carbiener, R., M. Trémolières, J. L. Mercier, and A. Ortscheit. 1990. Aquatic macrophyte communities as bioindicators of eutrophication in calcareous oligosaprobe stream waters (Upper Rhine plain, Alsace). *Vegetatio* **86**:71–88.

Carbiener, R., M. Trémolières, and S. Muller. 1995. Végétation des eaux courantes et qualité des eaus: Une thèse, des débats, des perspectives. *Acta Botanica Gallica* **142**:489–531.

Carignan, R. 1982. An empirical model to estimate the relative importance of roots in phosphorus uptake by aquatic macrophytes. *Canadian Journal of Fisheries and Aquatic Sciences* **39**:243–247.

Carignan, R., and R. F. Flett. 1981. Postdepositional mobility of phosphorus in lake sediments. *Limnology and Oceanography* **29**:667–670.

Carignan, R., and J. Kalff. 1980. Phosphorus sources for aquatic weeds: Water or sediments? *Science* **207**:987–989.

Carpenter, S. R. 1980. Enrichment of Lake Wingra, Wisconsin, by submersed macrophyte decay. *Ecology* **61**:1145–1155.

Carpenter, S. R., and D. M. Lodge. 1986. Effects of submersed macrophytes on ecosystem processes. *Aquatic Botany* **26**:341–370.

Carpenter, S. R., J. J. Elser, and K. M. Olsen. 1983. Effects of roots on *Myriophyllum verticillatum* L. on sediment redox conditions. *Aquatic Botany* **17**:243–249.

Carr, G. M., and P. A. Chambers. 1998. Macrophyte growth and sediment phosphorus and nitrogen in a Canadian prairie river. *Freshwater Biology* **39**:525–536.

Chambers, P. A., E. E. Prepas, M. L. Bothwell, and H. R. Hamilton. 1989. Roots versus shoots in nutrient uptake by aquatic macrophytes in flowing waters. *Canadian Journal of Fisheries and Aquatic Sciences* **46**:435–439.

Correll, D. S., and H. B. Correll. 1972. Aquatic and wetland plants of southwestern United States. *Water Pollution Control Research Series* **16030 dsl 01/72**.

Denny, P. 1972. Sites of nutrient absorption in aquatic macrophytes. *Journal of Ecology* **60**:819–829.

Dudley, T., and M. Valett. 1991. Macrophyte responses to hyporheic nutrient fluxes. *Bulletin of the North American Benthological Society* **8**:135.

Dunderdale, J. A. L., and J. Morris. 1996. The economics of aquatic vegetation removal in rivers and land drainage systems. *Hydrobiologia* **340**:157–161.

Elgin, I., U. Roeck, F. Robach, and M. Trémolières. 1997. Macrophyte biological methods used in the study of exchange between the Rhine River and the groundwater. *Water Research* **31**:503–514.

Fassett, N. C. 1969. "A Manual of Aquatic Plants." University of Wisconsin Press, Madison.

Fortner, S. L., and D. S. White. 1988. Interstitial water patterns: A factor influencing the distributions of lotic aquatic vascular macrophytes. *Aquatic Botany* **31**:1–13.

French, T. D., and P. A. Chambers. 1996. Habitat partitioning in riverine macrophyte communities. *Freshwater Biology* **36**:509–520.

Gessner, F. 1950. Die ökologische Bedeutung der Strömungsgeschwindigkeit fliessender Gewässer und ihre Messung auf kleinstem Raum. *Archiv für Hydrobiologie* **43**:159–165.

Gessner, F. 1955. "Hydrobotanik. Die physiologischen Grundlagen der Pflanzenverbreitung im Wasser. I. Energiehaushalt." VEB Dtch. Verlag Wiss., Berlin.

Grimm, N. B., and S. G. Fisher. 1984. Exchange between interstitial and surface water: Implications for stream metabolism and nutrient cycling. *Hydrobiologia* **111**:219228.

Harley, J. L. 1989. The significance of mycorrhiza. *Mycology Research* **92**:129–139.

Harvey, J. W., and K. E. Bencala. 1993. The effect of streambed topography on surface-subsurface water exchange in mountain catchments. *Water Resources Research* **29**:89–98.

Haslam, S. M. 1978. "River Plants: The Macrophytic Vegetation of Watercourses." Cambridge University Press, Cambridge, UK.

Haslam, S. M. 1982. A proposed method for monitoring river pollution using macrophytes. *Environmental Technology Letters* **3**:19–43.

Hendricks, S. P., and D. W. White. 1988. Hummocking by lotic *Cara:* Observations on alterations of hyporheic temperature patterns. *Aquatic Botany* **31**:13–22.

Hendricks, S. P., and D. W. White. 1991. Physicochemical patterns within a hyporheic zone of a northern Michigan river, with comments on surface water patterns. *Canadian Journal of Fisheries and Aquatic Sciences* **48**:1645–1654.

Hendricks, S. P., and D. W. White. 1995. Seasonal biogeochemical patterns in surface water, subsurface hyporheic, and riparian ground water in a temperate stream ecosystem. *Archiv für Hydrobiologie* **134**:459–490.

Hill, B. H. 1981. Distribution and production of *Justicia americana* in the New river, Virginia. *Castenea*, pp. 162–169.

Hutchinson, G. E. 1975. "A Treatise on Limnology. Vol. III. Limnological Botany." Wiley, New York.

Hynes, H. B. N. 1970. "The Ecology of Running Waters." University of Toronto Press, Toronto.

Idestam-Almquist, J., and L. Kautsky. 1995. Plastic responses in morphology of *Potamogeton pectinatus* L. to sediment and above sediment conditions at two sites in the northern Baltic Proper. *Aquatic Botany* **52**:205–216.

Jaynes, M. L., and S. R. Carpenter. 1986. Effects of vascular and nonvascular macrophytes on sediment redox and solute dynamics. *Ecology* **67**:875–882.

Lee, B., and D. White. 1995. Littoral zones in a reservoir: Groundwater flowpaths and *Justicia americana*. *Bulletin of the North American Benthological Society* **13**:98.

Lillie, R. A., and J. W. Barko. 1990. Influence of sediment and groundwater on the distribution and biomass of *Myriophyllum spicatum* L. in Devil's Lake, Wisconsin. *Journal of Freshwater Ecology* **5**:417–426

Littlefield, L., and C. Forsberg. 1965. Absorption and translocation of phosphorus-32 by *Chara globulalris* Thuill. *Physiologia Plantarum* **18**:219–296.

Lodge, D. M., D. P. Krabbenhoff, and R. G. Striegal. 1989. A positive relationship between groundwater velocity and submersed macrohute biomass in Sparkling lake, Wisconsin. *Limnology and Oceanography* **34**:235–239.

Madsen, T. V. 1993. Growth and photosynthetic acclimation by *Ranunculus aquatilis* L. in response to inorganic carbon availability. *New Phytologist* **125**:707–715.

Mantai, K. E., and M. E. Newton. 1982. Root growth in *Myriophyllum:* A specific plant response to nutrient availability. *Aquatic Botany* **13**:45–55.

Minckley, W. L. 1963. The ecology of a spring stream, Doe Run, Meade County, Kentucky. *Wildlife Monographs* **11**:1–124.

Mitsch, W. J., and J. G. Gosselink. 1993. "Wetlands," 2nd ed. Van Nostrand-Reinhold, New York.

Muller, S. 1990. Une séquence de groupments végétaux bioindicateurs d'eutrophisation croissante des cours d'eau faiblement minéralisés des Basses Vosges gréseuses du Nort. *Comptes Rendus des Seances de l'Academie des Sciences* **310**:509–514.

Pedersen, O., and K. Sand-Jensen. 1993. Water transport in submerged macrophytes. *Aquatic Botany* **44**:385–406.

Peltier, W. H., and E. B. Welch. 1968. Factors affecting growth of rooted aquatics in a river. *Weed Science* **17**:412–416.

Pond, R. H. 1905. The relation of aquatic plants to the substratum. *Report of the U.S. Fish and Wildlife Communication* **19**:483–526.

Robach, F., I. Hajnsek, I. Elgin, and M. Trémolières. 1995. Phosphorus sources for aquatic macrophytes in running waters: Water or sediment? *Acta Botanica Gallica* **142**:719–731.

Sand-Jensen, K. 1983. Photosynthetic carbon sources of stream macrophytes. *Journal of Experimental Botany* **34**:198–210.

Sand-Jensen, K., and J. R. Mebus. 1996. Fine-scale patterns of water velocity within macrophyte patches in streams. *Oikos* **76**:169–180.

Sand-Jensen, K., C. Prahl, and H. Stokholm. 1982. Oxygen release from roots of submersed aquatic macrophytes. *Oikos.* **62**:349–354.

Schütz, W. 1995. Vegetation of running waters in Southwestern Germany—pristine conditions and human impact. *Acta Botanica Gallica* **142**:571–584.

Sculthorpe, C. D. 1967. "The Biology of Aquatic Vascular Plants." St. Martin's Press, New York.

Short, R., and M. McRoy. 1983. Nitrogen uptake by leaves and roots of the seagrass *Zostera marina* L. *Botanica Marina* **27**:547–555.

Small, A. M., W. H. Adey, S. M. Lutz, E. G. Reese, and D. L. Roberts. 1996. A macrophyte-based rapid biosurvey of stream water quality: Restoration at the watershed scale. *Restoration Ecology* **4**:124–145.

Spence, N. N. H. 1967. Factors controlling the distributions of freshwater macrophytes with particular reference to the lochs of Scotland. *Journal of Ecology* **55**:147–169.

Thiébaut, G., and S. Muller. 1998. Les communautés de macrophytes aquatiques comme descripteurs de la qualité de l'eau: Exemple de la rivière Moder (Nord-Est France). *Annals of Limnology* **34**:141–153.

Tilman, D. 1994. Competition and biodiversity in spatially structured habitats. *Ecology* **75**: 2–16.

Trémolières, M., I. Elgin, U. Roeck, and R. Carbiener. 1993. The exchange process between river and groundwater on the Central Alsace floodplain (Eastern France). *Hydrobiologia* **254**:133–148.

Tremp, H., and A. Kohler. 1995. The usefulness of macrophyte monitoring-systems, exemplified on eutrophication and acidificaiton of running waters. *Acta Botanica Gallica* **142**:541–550.

van Wijck, C., C. J. de Groot, and P. Grillas. 1992. The effect of anaerobic sediment on the growth of *Potamogeton pectinatus* L.: The role of organic, sulfide and ferrous iron. *Aquatic Botany* **44**:31–49.

Vaux, W. G. 1968. Intergravel flow and interchange of water in a streambed. *Fisheries Bulletin* **66**:479–489.

Voss, E. G. 1985. Michigan flora. Part II. Dicots (Saururaceae—Cornaceae). *Cranbrook Institute of Science Bulletin* **59**:1–724.

Webster, J. R., and E. F. Benfield. 1986. Vascular plant breakdown in freshwater ecosystems. *Annual Review of Ecology and Systematics* **17**:567–594.

Westlake, D. F. 1975. Macrophytes. *In* "River Ecology" (B. A. Whitton, ed.), pp. 106–128. University of California Press, Berkeley.

Wetzel, R. G. 1983. "Limnology," 2nd ed. Saunders College Publishing, New York.

Wetzel, R. G. 1990. Land-water interfaces: metabolic and limnological regulators. *Verlandlungen—Internationale Vereinigung für Theoretische und Angewandte Limnologie* **24**:6–24.

White, D. S. 1990. Biological relationships to convective flow patterns within streambeds. *Hydrobiologia* **196**:148–159.

White, D. S., C. H. Elzinga, and S. P. Hendricks. 1987. Temperature patterns within the hyporheic zone of a northern Michigan River. *Journal of the North American Benthological Society* **6**:85–91.

White, D. S., S. P. Hendricks, and S. L. Fortner. 1992. Groundwater-surface water interactions and the distributions of aquatic macrophytes. *In* "Proceedings of the First International Conference on Ground Water Ecology" (J. A. Stanford and J. J. Simons, eds.), pp. 247–256. American Water Resources Association, Washington, DC.

Wigand, C., J. C. Stevenson, and J. C. Cornwell. 1997. Effects of different submersed macrophytes on sediment biogeochemistry. *Aquatic Botany* **56**:233–244.

16

Subsurface Influences on Surface Biology

C. Lisa Dent, John D. Schade, Nancy B. Grimm, and Stuart G. Fisher

Department of Biology
Arizona State University
Tempe, Arizona

I. Introduction   381
II. Effects on Primary Producers   382
III. Effects on Microorganisms and Microbial Processes   386
IV. Effects on Invertebrates   388
V. Effects on Fish   389
VI. Heterogeneity and Scale   391
    A. Temporal Heterogeneity   391
    B. Spatial Heterogeneity   392
    C. Scale   393
VII. Human Impacts   394
VIII. Summary   396
    References   397

I. INTRODUCTION

Since stream ecologists began to appreciate the four-dimensional nature of lotic ecosystems (e.g., Ward, 1989), recognition of the role of subsurface environments in surface stream structure and functioning has increased. Interactions between surface and subsurface zones occur primarily via the flow of water; however, water carries dissolved and suspended materials as well as organisms, thus linkages are potentially both hydrologic and biologic. In this chapter, we highlight recent research that deals specifically with the effects of subsurface processes on surface biology as mediated through these linkages. Considerations of the reciprocal interaction (surface stream effects on subsurface biology) by others in this book complete the picture, for stream ecosystems are now best understood as heterogeneous landscapes that include both the wetted stream and connected lateral and vertical subsystems (e.g., parafluvial, riparian, and hyporheic zones).

The terms used to describe the spatial components of a stream ecosystem are varied and sometimes conflicting. Our definitions follow Dahm *et al.* (1998). We refer to three surface components of the stream ecosystem: *surface stream,* parts of the active channel covered by surface water; *parafluvial zone,* the region of the active channel without surface water during low discharge; and *riparian zone,* the region bordering the active channel that supports longer-lived, higher stature vegetation (Fig. 1). The *hyporheic zone* is the region of saturated sediments underlying the surface, parafluvial and riparian zones, where subsurface and surface water are actively exchanged. We will consider here influences of processes occurring in hyporheic sediments below the surface stream and below parafluvial zones on surface stream biology. Discrete regions of subsurface–surface exchange where water discharges vertically from hyporheic sediments to the surface are called *upwelling zones;* water re-enters the hyporheic zone at *downwelling zones.* Water also enters the surface from parafluvial subsurface sediments along stream margins at *outwelling zones* and infiltrates parafluvial gravel bars at *inwelling zones* (Fig. 1). In some streams that flow through narrow valleys, a parafluvial zone per se may be difficult to identify. We will not consider here the direct interaction of riparian zones with the surface stream, as this topic is well covered in other chapters of the book.

The currencies of exchange that exert effects on biological patterns and processes are physical (e.g., temperature, current), chemical (nutrients, organic matter, oxygen), and biological (use of refuges, organism movement). Stream discharge dictates much of the nature and magnitude of hydrologic linkages among subsystems (Morrice *et al.,* 1997), and superimposed on this is the effect of flooding as a disturbance to stream biota. An understanding of the effects of subsurface zones on surface biology is incomplete without considering variations in linkage that are incorporated as temporal or spatial scale is increased (Fisher *et al.,* 1998a,b). We will begin by reviewing how subsurface processes affect different groups of surface organisms and then step back to consider landscape-level effects. Finally, we will review the impacts of human activity on subsurface–surface interactions and how these may influence stream biology.

II. EFFECTS ON PRIMARY PRODUCERS

Distribution and abundance of autotrophs in streams has been the subject of numerous treatises pointing to the importance of light, current, temperature, substrate, scouring by floods, grazing, and water chemistry (see

Hynes, 1970; Allan, 1995). Recent work has explicitly examined the role of subsurface processes, showing that exchange with subsurface zones can affect primary production in the surface stream, mainly due to enhanced nutrient concentrations at upwelling and outwelling zones. An early indication of the importance of nutrient input from subsurface zones came from a comparison of algal abundance in two New Mexican streams differing in number of retentive structures and hyporheic processing of organic matter (Coleman and Dahm, 1990). Although periphyton in both streams were nitrogen-limited, higher algal abundance in the beaver-dominated stream was attributed to greater inorganic nitrogen availability at the sediment–water interface, where interstitial ammonium concentration was 25 times that of the incised, poorly retentive stream. In Sycamore Creek, Arizona, nitrogen-limited algae in upwelling zones were more abundant and recovered faster following floods than those at downwelling zones (Fig. 2; Valett *et al.*, 1994). As in New Mexico, concentration of inorganic nitrogen (nitrate, in this case) was much higher at upwelling zones than at downwelling zones. A similar result was found in a back channel of the Queets River, Washington, where standing stocks of benthic algae in upwelling zones were more than double those in downwelling zones (K. Fevold, University of Washington, personal communication).

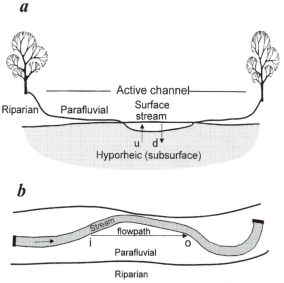

FIGURE 1   Cross-section (a) and overhead view (b) of a stream channel showing the location of the parafluvial and riparian zones relative to the surface stream. Letters indicate hydrologic connections between these surface subsystems and subsurface waters; u, upwelling; d, downwelling; i, inwelling; o, outwelling.

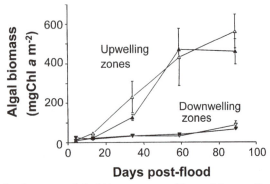

FIGURE 2   Postflood recovery of algal biomass at upwelling and downwelling zones. Reprinted with permission from Valett *et al.*, 1994.

Preliminary evidence indicates that macrophytes are also differentially affected by conditions at upwelling and downwelling zones. In studies of macrophyte distribution patterns in three Michigan streams, most species were more abundant at locations of subsurface water upwelling, whereas one was more likely to be found at downwelling zones (Fortner and White, 1988; White *et al.*, 1992). Although the causes of these relationships were not investigated experimentally, ammonium concentration near macrophyte roots in upwelling zones was again considerably higher than in surface water, pointing to nutrient availability as a probable factor (White *et al.*, 1992). In a similar study, macrophytes transplanted into upwelling zones in Sycamore Creek, a nitrogen-limited stream with elevated nutrient concentrations at upwelling zones, grew faster than those transplanted into downwelling zones (Dudley and Valett, 1991).

Evidence that lateral inputs of subsurface water affect surface-dwelling periphyton is accumulating as well. Many studies suggest that inputs from parafluvial gravel bars are high in nutrients relative to surface water (Holmes *et al.*, 1994a, Boissier *et al.*, 1996; Wondzell and Swanson, 1996; Claret *et al.*, 1997). Surface flow originating in gravel bars that flows into the main channel at some downstream point has been called a spring, source, or parafluvial drainage channel (Fig. 3). Parafluvial drainage channels often support very high standing crops of algae. For example, chlorophyll *a* standing crops were higher at sources than in the Rhone River into which the parafluvial channels drained (Claret and Fontvieille, 1997). These observations could be explained either by differences in physical conditions (temperature, light, current velocity) in the parafluvial drainage channels relative to the main channel or by nutrient subsidies from the subsurface.

Changes in surface stream discharge affect patterns of exchange between surface and subsurface zones (Marti *et al.*, 1997; Morrice *et al.*, 1997). During low flow periods, a greater proportion of surface discharge passes through subsurface sediments, maximizing the potential for subsurface processes to affect surface biology (Stanley and Valett, 1992; Findlay, 1995). In streams such as Sycamore Creek, where low discharge is associated with low surface-water nutrient concentrations (due to high algal uptake), the difference between subsurface and surface concentrations is also at a maximum during this time, enhancing the influence of subsurface nutrient subsidies (Holmes *et al.*, 1994b).

Local changes in hydrology may also alter subsurface–surface connections, particularly at low discharges. For example, there is evidence that primary producers themselves may create local hydrologic conditions that enhance nutrient availability. In a Michigan stream, *Chara* growth altered surface–subsurface flow patterns via sand deposition on the downstream side of the plant, eventually forming hummocks several meters in length (Hendricks and White, 1988; White, 1990). This alteration of the bedform caused hyporheic water to upwell through the base of the hummock, potentially providing elevated nutrient levels to the plant and creating a positive feedback between growth and nutrient availability. Surface-water flow patterns are also altered by periphyton growth (Katznelson, 1989; Mulholland *et al.*, 1994). Extensive growth of periphyton at upwellings and outwellings

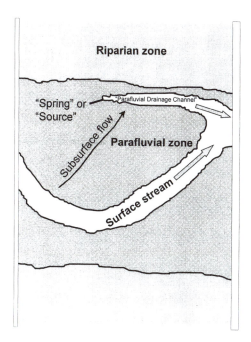

FIGURE 3 Diagram illustrating surface flow originating at a source or spring and flowing down a parafluvial drainage channel into the main channel.

could potentially trap high-nutrient water and prevent dilution from main channel flow.

Subsurface zones are not necessarily a source of nutrients to the surface stream; they may also be nutrient sinks (Pinay *et al.*, 1994; Stanley and Boulton, 1995; Claret *et al.*, 1997). Further work is needed to clarify how variation in subsurface processes affects surface autotrophic production and to examine differences between outwellings and upwellings and/or macrophytes and periphyton.

III. EFFECTS ON MICROORGANISMS AND MICROBIAL PROCESSES

Few studies have directly investigated the effect of subsurface–surface exchange on surface microorganisms and microbial processes; however, several studies provide information from which we can infer likely effects. Potentially significant effects are of two types: (1) changes in the availability of resources, either organic matter or inorganic nutrients, which may alter the activity, density, or composition of microbial communities, and (2) direct transport of microorganisms in water moving from subsurface to the surface stream.

Several studies have suggested that the supply of organic matter at points of hydrological exchange stimulates the activity of microorganisms. In the Rhone River, activity and numbers of bacteria were higher in benthic sediments collected from parafluvial drainage channels (see Fig. 3) than in benthic sediments collected from the main channel (Claret and Fontvieille, 1997). The difference was attributed to an increase in organic matter supply at sources due to increased algal growth. As discussed in the previous section, enhanced algal growth may result from nutrient supplements from subsurface outwelling. Organic matter also may be supplied directly from subsurface water in the form of dissolved organic carbon (DOC). In the Breitenbach, a first-order stream in Germany, bacterial carbon demand far exceeded surface DOC flux (Fiebeg, 1995). Fiebeg suggested that delivery of DOC from upwelling subsurface water (ground water) might balance the sediment carbon budget. However, subsurface zones are not necessarily a source of DOC. As with nutrients, some subsurface zones are DOC sinks. In Sycamore Creek, rates of microbial activity in sediments were higher at downwelling than upwelling zones (Jones *et al.*, 1995a,b), and the activity and numbers of bacteria were greater at inwelling zones than at points farther along parafluvial flowpaths (Holmes *et al.*, 1998). Apparently, input of organic matter from the surface stream to the subsurface stimulates microbial activity, but as water moves through subsurface sediments, labile material is depleted. Thus whether or not bacterial activity is stimulated at points of subsurface–surface exchange (upwellings or outwellings) depends on whether subsurface waters are depleted or enriched in labile DOC (Claret *et al.*, 1997).

These studies suggest that microbial activity is carbon limited, and that hydrologic exchange may provide organic matter (either directly or via enhancement of primary production) that alleviates this limitation, leading to high rates of heterotrophic microbial activity and high microbial densities. Decomposition rates of terrestrially derived organic matter also may be limited by availability of inorganic nutrients, particularly nitrogen or phosphorus. Microbial activity and decomposition rates of allochthonous organic matter substrates (i.e., leaves and wood) are often stimulated by the addition of nitrogen or phosphorus (Howarth and Fisher, 1976; Elwood *et al.*, 1981; Meyer and Johnson, 1983; Webster and Benfield, 1986; Tank and Webster, 1998). Because subsurface–surface exchange can add nutrient-rich water to the surface stream, it follows that breakdown rates and microbial activity of leaves and wood might be higher at upwellings and outwellings (Boulton, 1993; Tank and Webster, 1998).

Water upwelling from the subsurface may also carry microorganisms, particularly bacteria, into the surface stream. Several studies have measured fluxes of bacteria from subsurface zones, both directly (Boissier *et al.*, 1996) and indirectly (Baker and Farr, 1977; McDowell, 1984; Edwards *et al.*, 1990; Wainwright *et al.*, 1992). These inputs often appear to make a significant contribution to bacterial numbers in surface waters. We would expect high influx of bacteria at up- or outwelling zones linked to sites of high bacterial production, such as riparian soils or parafluvial sediments that are rich in organic matter. Inputs of bacteria may be significant because of their potential effect on bacterial biodiversity, rates and types of microbial processes occurring in the surface stream, and movement of genetic information within the stream-riparian corridor (e.g., McArthur and Marzolf, 1986).

The preceding discussion is centered on heterotrophic microorganisms, particularly bacteria. Obviously, this is a subset of the microbial community, which also includes chemo- and photoautotrophs, fungi and protists. The most functionally important chemoautotrophic organisms are the nitrifying bacteria (Jones *et al.*, 1994). In Sycamore Creek, nitrification rates are highest at downwelling and inwelling zones (Holmes *et al.*, 1994b, 1998; Jones *et al.*, 1995a), due to higher levels of ammonium (from mineralization of organic matter) and oxygen, both of which are necessary for nitrification. Although stimulation of nitrification at inwelling zones does not make a large contribution to ecosystem productivity (Jones *et al.*, 1995a), it does have a strong effect on nitrogen cycling (Holmes *et al.*, 1994b; Jones *et al.*, 1995a). Fungi play a major role in nutrient dynamics during organic matter decomposition (Tank and Webster, 1998), and are likely to be affected by nutrient inputs at upwelling zones. Protists are important grazers on bacteria (Carlough and Meyer, 1990; Allan, 1995) and may influence all of the processes discussed earlier. The interaction of fungi and protists with hydrologic exchange is little known and represents an important area for future research.

IV. EFFECTS ON INVERTEBRATES

Hydrologic exchange between surface and subsurface water may strongly influence the distribution and abundance of invertebrates. As mentioned previously, nutrient inputs at upwelling zones stimulate primary production and recovery of algal biomass after floods. A positive association between periphyton abundance and invertebrate density has been well established by both correlation and experiment (Hawkins *et al.*, 1982; Wallace and Gurtz, 1986; Allan, 1995). Although periphyton abundance in these studies was governed largely by light availability, the principle should hold regardless of the mechanism involved. Thus we would expect higher invertebrate density and faster postflood recovery of invertebrate communities at upwelling zones due to stimulation of production by nutrient inputs. Indeed, in the Flathead River, where water seeping from the riparian zone stimulates primary production in a normally unproductive river, the growth period of many zoobenthic species coincides with the development of algal assemblages (Ward, 1989). Invertebrate production and density have also been closely linked with the input and decomposition of leaf litter. Conditioning of leaf material by microbes greatly enhances leaf material as a food source (Cummins, 1974; Arsuffi and Suberkropp, 1984; Lawson *et al.*, 1984; Suberkropp and Arsuffi, 1984; Allan, 1995). Because leaf litter breakdown at upwelling zones may be accelerated (see Section III), these zones should have higher rates of secondary production and different assemblages of detritivores. Despite these seemingly obvious connections, direct studies of the impact of nutrient inputs at up- and outwelling zones on invertebrate assemblages are lacking. A better understanding of hydrologic linkages may influence the interpretation of past and future studies of invertebrate abundance and distribution (Boulton, 1993).

Subsurface zones may also influence benthic invertebrates by providing a refuge from disturbance. In many streams, floods cause severe reductions in the abundance of benthic organisms, which then recolonize once flood waters have receded. The hyporheic zone is a potential source of colonists, either by organisms that moved into sediments to avoid floods, or by organisms originally living deeper in sediments not affected by flood waters. Drought may also be an important form of disturbance in stream ecosystems and can have a profound effect on invertebrate assemblages (Stanley *et al.*, 1997). Under these conditions, subsurface sediments may serve as a refuge into which organisms may move in search of water (Bishop, 1973; Williams, 1977; Sedell *et al.*, 1990; Griffith and Perry, 1993). After rewetting, these organisms then recolonize the surface stream. Desert streams may be an exception, however, as evidenced by studies on Sycamore Creek, which show a lack of overlap between hyporheic and benthic communities (Boulton *et al.*, 1992) and recolonization of benthic invertebrates through oviposition by aerial adults (Gray and Fisher, 1981; Jackson and Fisher, 1986).

In addition to providing refuges and nutrient inputs, hydrologic exchange between surface and subsurface waters change invertebrate assemblages simply through physical effects on invertebrate movement. At upwelling zones, the upward movement of water at upwelling zones may dislodge benthic organisms, increasing invertebrate drift (Boulton, 1993). Conversely, downward movement of water at downwelling zones may hinder invertebrates from entering the water column and drifting downstream. Indeed, this downward movement may actually carry benthic invertebrates deeper into sediments (Cooling and Boulton, 1993). The net effect may be movement of invertebrates from upwelling to downwelling zones and possibly faster recolonization following disturbance at downwelling zones (Boulton, 1993).

V. EFFECTS ON FISH

Some of the earliest work on subsurface exchange relates to effects on fish. Subsurface zones and hydrologic exchange influence fish by providing a refuge from drought, by changing oxygen concentrations in sediments where fish spawn, and by altering food availability via enhanced algal growth at upwelling zones.

Some adult fish (e.g., dipteriform lungfish) occupy sediments periodically as a refuge from drought. African and South American Lungfishes burrow into sediments during estivation, when their rates of metabolism slow down (Lagler *et al.*, 1977). Vertical exchange during this period is mainly downwelling, culminating in dry sediments. During periods of maximal dryness, the fish use a special tube to exchange respiratory gases with the atmosphere until water returns via downward percolation into sediments (Lagler *et al.*, 1977).

Eggs and larval fishes of some species may occupy sediments for varying periods. In North America, salmonids excavate stream bottom gravels to create depressions for egg laying, known as redds. Eggs buried in sediments hatch after several months and the resulting larvae remain in sediments for a month or more (Breder and Rosen, 1966). During development, eggs require oxygen provided by vertical exchange. Many salmonids locate redd sites near up- or downwellings that provide current movement and enhanced aeration (Vaux, 1962; Keller *et al.*, 1990). Occlusion of pore spaces by sedimentation of fine particles precludes reproduction by salmonids largely through reduction in sediment permeability and resultant limitation of flow, leading to a decrease in oxygen supply to eggs. Selection of groundwater upwelling sites for redds may also be adaptive because ground water prevents freezing of eggs in winter (Benson, 1953) and maintains cool temperatures in summer in thermally marginal streams. After hatching, larval fishes confined to sediments rely on hydrologic exchange with the surface stream both to supply food and to remove wastes. The reverse (i.e., an influence of larval-derived materials on surface processes), has not been investigated to our knowledge, but is a lucrative area for future research. Although only indirectly related to the phenomenon of vertical exchange, it should be noted that the act of redd building by salmonids sorts sediments locally, maintains and enhances permeability of nest sites, and increases heterogeneity of substrates in the stream as a whole, as finer sediments released during nest building are entrained in backwaters and eddies.

Because salmonids are so widespread and well known among stream fishes, we often assume egg burying is a common form of parental care. In fact, the practice is rare outside this family. However, other taxa do exhibit similar forms of parental care. For instance, catostomids mate in such a way that some eggs become buried in sediments (Breder and Rosen, 1966), and certain stream cyprinids bury eggs in pits which are then covered with sediments by males (e.g., *Semotilus* spp; Johnston and Ramsey, 1990) or "inject" eggs into sediments beneath saucer-shaped depressions *(Agosia chrysogaster* Girard; Johnston and Page, 1992). In these cases, egg incubation occurs in the hyporheic zone of the stream, an environment sensitive to vertical hydrologic connections.

In Sycamore Creek, Arizona, nest building by *A. chysogaster* is restricted to a small proportion of the streambed. Nests are arranged in clusters of a few to over one hundred. Experiments with dye injection reveal that these nests are invariably located at downwelling sites (Fig. 4). A survey of nest locations indicates that nests are strongly associated with the downstream ends of sandy runs (Table 1) where downwelling zones predominate. Since de-

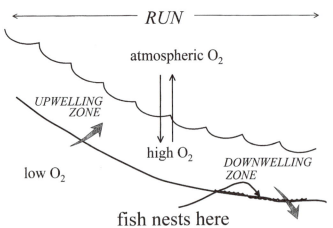

FIGURE 4   Location of nests of *Agosia chrysogaster* in relation to upwelling and downwelling zones. Upwelling water tends to be depleted in oxygen, which is replenished by diffusion from the atmosphere and transported into the subsurface at downwelling zones.

TABLE I   Location of Nests of Longfin Dace, *Agosia chrysogaster,* in Sand-Bottomed Runs of Sycamore Creek, Arizona[a]

Location	Total nest clusters present	Total nests present	Average number of nests (n = 5)
Upstream	1	7	1. 4
Middle	2	28	5. 6
Downstream	11	390	78. 0

[a]Nests and nest clusters were found in 5 of 10 runs examined and were enumerated for upstream, middle, and downstream thirds of each of these runs. Nests were significantly more abundant at downstream locations, where downwelling was prevalent ($p < .05$).

pressed oxygen is common in sediments beneath upwelling zones and oxygen concentrations are generally much higher at downwelling zones (Jones *et al.,* 1995b), we suspect fish are choosing sites of high oxygen availability. Even though selection of downwelling zones for nest construction is a very plausible adaptation in this species, how the fish detect vertical hydraulic gradient remains a mystery.

As adults, fish are usually restricted to the surface stream and burrowing in sediments is uncommon. Insofar as fish exist in trophic structures based on algae enhanced by nutrient subsidies at upwelling zones (Valett *et al.,* 1994), vertical exchange enhances fish populations. This would be especially true of algivorous fishes such as *Agosia chrysogaster.* Trophic linkages between hyporheic meiofauna and surface stream fishes are undocumented but may also play an important role, depending on rates of subsurface secondary production. A comparative study of foraging effort and gut content of fishes over upwelling and downwelling zones would begin to answer this interesting question.

VI. HETEROGENEITY AND SCALE

A. Temporal Heterogeneity

The effects of subsurface processes on surface biology are not constant; they vary over time both quantitatively and qualitatively. Much of the variation is due to changes in stream discharge, which alter the strength and in some cases the direction of hydrologic exchange between the surface and subsurface. Several studies have found that surface discharge is negatively correlated with storage zone volume and residence time, where the storage zone includes both subsurface sediments and regions where surface water is temporarily slowed, such as backwater pools (D'Angelo *et al.*, 1993; Wondzell and Swanson, 1996; Morrice *et al.*, 1997). Although the extent to which these storage zone studies capture the dynamics of subsurface sediments is unclear (Harvey *et al.*, 1996), direct hydrometric studies have also found that hyporheic exchange is reduced during high flows (Harvey *et al.*, 1996; Wroblicky *et al.*, 1998). Thus during high discharge periods, the effects of subsurface processes on surface biology should be reduced. During periods of extended drought, however, upwelling zones may become downwelling zones, eliminating the effect of subsurface processes completely (Stanley and Valett, 1992). Thus streams may alternate over time between being more "pipe-like" and being more connected with vertical and lateral subsystems (Valett *et al.*, 1996).

Qualitative changes in subsurface processes also occur. For example, in a mountain stream, gravel bars were a significant source of nitrate for the surface stream during fall, when seasonal increases in discharge caused water tables to rise into the rooting zone of alders growing on the gravel bars (Wondzell and Swanson, 1996). The source of nitrate was suggested to be mineralization and nitrification of organic nitrogen from root turnover or leaf fall. During summer when the water table was below the rooting zone, gravel bars were no longer a source of nitrate. Seasonal changes may also occur in concentrations of oxygen and other redox-sensitive solutes, leading to differences in subsurface processes (Dahm *et al.*, 1998).

B. Spatial Heterogeneity

Thus far our discussion has focused on how individual upwelling or outwelling zones may influence organisms and processes in stream surface waters. Over a longer stream segment, localized areas of upwelling and outwelling interspersed with downwelling or nonwelling produce a patchwork of local habitats that differ in physical and chemical characteristics such as nutrient and oxygen concentrations, flow, and temperature. As we have seen in this chapter, these physical and chemical properties affect many biological processes. Thus surface–subsurface exchange patterns help to create an environment in the surface stream that is spatially heterogeneous in terms of physical, chemical and biotic characteristics. How might this affect surface biology?

Fisher *et al.* (1998a) suggest that segregation of upwelling and downwelling zones in nitrogen-limited streams might increase overall nutrient loads in the system by providing low nitrogen habitat where nitrogen-fixing cyanobacteria thrive. High algal productivity at upwelling zones reduces the concentration of nitrogen flowing downstream, resulting in low nitrogen availability at downwelling zones. These zones are then colonized by producers with low instream nitrogen requirements such as cyanobacteria that add nitrogen to the stream system through fixation of atmospheric nitrogen (N_2; Grimm and Petrone, 1997). Empirical data are required to test these hypotheses.

Although no information is yet available on how spatial heterogeneity caused by surface–subsurface interactions affects surface biology, some recent studies address the impact of spatial heterogeneity on stream organisms. Pringle (1990) examined the response of a periphyton community to enriched and ambient nutrient levels in substrata and in the water column. Community richness and diversity were highest when periphyton were grown on nutrient-diffusing substrates in nutrient-poor water, suggesting that nutrient-poor systems with spatially variable nutrient levels can maintain high diversity. These results are consistent with more general models showing that resource heterogeneity is positively related to community diversity (Tilman, 1982).

The size and arrangement of algal patches has substantial effects on the foraging behavior of some species of grazing invertebrates (Hart, 1981; Kohler, 1984); grazers tend to move less and change direction more often in patches with high algal abundance. These effects are mediated by heterogeneity in physical factors such as current (Poff and Ward, 1992) and by differences in species' mobility (Cooper *et al.*, 1997). The growth rates of two mayfly nymphs were lower in environments with patchily distributed resources than in homogeneous environments, although the effect was mediated by current velocity (Palmer, 1995). Grazer activity also can feed back to algal heterogeneity. Based on a simulation model, the effect of physical habitat heterogeneity on heterogeneity in algal biomass depends on the number of grazers present (Poff and Nelson-Baker, 1997). In the simulations, physical habitat heterogeneity referred to the number of boulders in the stream, a factor not likely to be affected by surface–subsurface exchange. Still, this study illustrates the potential complexity of determining the effects of heterogeneity, however it is produced, on surface biology.

Spatial variability is an important and understudied aspect of stream ecology (Palmer *et al.*, 1997). Surface–subsurface exchange can produce spatial heterogeneity in physical, chemical and biological features (e.g., Dent and Grimm, 1999), and thus may play a central role in the biology of surface stream communities—one that is not yet well understood.

C. Scale

The spatial organization of stream ecosystems can be viewed at multiple scales (Frissell *et al.*, 1986; Gregory *et al.*, 1991; Grimm and Fisher, 1992; Fisher *et al.*, 1998a). Although we have concentrated on reach-level connections between surface and subsurface, exchange also occurs at scales smaller than lateral inwelling and outwelling, and larger than vertical upwelling and downwelling (which tends to be located at nick points between runs and riffles; Harvey and Bencala, 1993). The scale to which an organism responds depends on how it perceives the environment. In some cases, responses may occur at multiple scales. In a multiscale study of the relationship between bull trout spawning sites and vertical hydraulic gradient, spawning sites were positively correlated with gaining reaches (reach-scale upwellings), but within these reaches, localized downwelling sites were most often selected (Baxter, 1997). Reach-scale upwelling may provide thermal benefits, whereas small-scale downwellings may provide superior incubation conditions such as ample dissolved oxygen levels or higher intragravel flow rates (Baxter, 1997).

A complete understanding of the effect of surface–subsurface exchange on surface biology will require a hierarchical perspective in which processes of interest are *constrained* by processes occurring at the next hierarchical level, and *mechanistic explanations* for those processes are found at lower hierarchical levels (O'Neill *et al.*, 1986; Wu and Loucks, 1995; Fisher *et al.*, 1998a, Dahm *et al.*, 1998). Stevenson (1997) introduced a hierarchical framework for explaining the causes and consequences of benthic algal heterogeneity and showed how large-scale factors can constrain the impact of smaller-scale factors. For example, nutrient availability affects average annual algal biomass in streams with relatively stable discharge, but in very flashy streams, nutrient effects on annual biomass cannot be detected (Fisher and Grimm, 1988; Biggs, 1995). A preliminary step toward establishing this type of framework for surface–subsurface exchange is a study by Valett *et al.* (1997) showing that catchment geology may represent a large-scale constraint on the magnitude of exchange.

VII. HUMAN IMPACTS

Human activity may disrupt interactions between surface and subsurface waters through land-use practices, flow regulation, or extraction of either ground or surface water (Table II). These activities act on a variety of space and timescales and may have large effects on ecological processes in both surface and subsurface waters.

Many streams fluctuate between periods of deposition, when sediments become clogged with fine particulate organic matter (FPOM) from high algal productivity or leaf decomposition, and periods of scour (i.e., floods), when FPOM is removed and exchange is reestablished. The balance between these processes can be upset by human activities that increase seston, sediment, nutrient, or organic matter loading. Input of sewage can lead to clogging of sediments via growth of algae or microbial layers, which may reduce exchange, or via direct sedimentation of organic matter covering the stream bottom and filling in sediment interstices (Schälchi, 1992; Brunke and Gonser, 1997). Land-use practices that increase erosion, such as agriculture

TABLE II Physical and Biological Effects of Human Activities

Human activities	Potential physical effects	Potential biological effects
Land-use practices (e.g., agriculture, forest cutting, and residential uses such as sewage inputs)	Increased sediment loading Clogging of stream sediments Reduction in exchange	Decreased nutrient and energy input Reduction in refugia Decreased fish reproductive success
Flow Regulation (e.g., flood control, sediment retention)	Altered mixing of surface and subsurface water Increased riverbed incision Fewer scouring floods which renew exchange	Decreased nutrient and energy input Reduction in refugia Influence on fish reproduction Altered invertebrate production and species composition
Water removal (e.g., municipal and agricultural use of ground and surface water)	Reduction in magnitude and extent of upwelling Increase in magnitude and extent of downwelling	Decreased nutrient and energy input Reduction in invertebrate drift Increased fish spawning sites

Note. See text for more details.

or forest cutting, also often lead to clogging of stream sediments (Karr and Schlosser, 1978). Based on the work described in this chapter, reductions in surface–subsurface exchange may have negative impacts on surface biology for several reasons. Nutrient and energy inputs may be reduced, affecting primary and secondary production and decomposition. Subsurface refugia may become unavailable, changing the responses of surface biota to disturbance, and the reproductive success of fish species dependent on hydrological exchange may be reduced. Finally, temporal and spatial heterogeneity may be diminished.

Most river systems experience some form of flow regulation somewhere along their lengths. Dam construction causes large-scale alterations of the flow regime, leading to significant changes in patterns of surface–subsurface exchange (Galay, 1983; Curry *et al.*, 1994; Golz, 1994). Dams retain sediments, which can lead to increased riverbed incision, reducing the size of subsurface zones. Sediment retention by dams may also prevent clogging of sediments downstream, increasing surface–subsurface exchange, while at the same time reducing the frequency of scouring floods that can restore exchange rates that have declined due to natural sedimentation (Collier *et al.*, 1996). The extent to which dams alter surface–subsurface exchange and thereby influence structure and function of stream ecosystems is not well known and would be a major contribution to our understanding of the environmental effects of flow regulation.

Another human activity influencing hydrological exchange is the removal of water from stream channels for residential or agricultural use. This is especially important in arid climates and often takes the form of groundwater pumping. The effects of groundwater pumping on water table depth is well known. In general, the removal of ground water forms a cone of depression around the well, lowering the surrounding water table significantly (Stromberg *et al.*, 1996). The resulting decrease in available water can severely affect riparian vegetation (Perkins *et al.*, 1984; Gremmen *et al.*, 1990). The effect of groundwater removal on hydrological exchange within the stream channel is not well known but can be inferred from the effects of transpiration on subsurface upwellings. In Sycamore Creek, a gradual decrease in the strength of upwelling over successional time after a flood was attributed to transpiration by riparian vegetation (Valett *et al.*, 1994). Groundwater pumping also removes water, causing it to flow away from the stream channel, and therefore should reduce or eliminate upwelling and increase downwelling. These changes could reduce nutrient and energy inputs to the surface stream, reduce invertebrate drift, and increase the spawning success of fish.

VIII. SUMMARY

We have reviewed studies relating to the influence of surface–subsurface interactions on biological pattern and process in streams. Many questions remain to be addressed by future research. For example, subsurface sediments can be either sources or sinks for the surface stream in terms of both nutrients and organic matter. What determines when and where subsurface sediments are sources or sinks, and how do these different roles affect primary production and microbial activity in the surface? Do the effects of subsurface nutrient or carbon subsidies transfer up to higher trophic levels such as invertebrates and fish? Under what circumstances are the effects of vertical upwellings different from lateral outwellings? How does hydrologic exchange affect fungi and protists? How do dams and other human alterations of river structure affect surface–subsurface interactions? The answers to these and other similar questions would significantly enhance our knowledge of how subsurface processes affect surface stream biology.

We argue that a perspective incorporating multiple scales and recognizing the heterogeneity caused by surface–subsurface exchange is necessary for a complete understanding of this topic (see also Dahm *et al.*, 1998). The distribution and abundance of organisms in streams, from microbial populations to fish, has been shown to vary spatially because of surface–subsurface exchange. Functional responses to heterogeneity caused by hydrologic exchange have been studied less often, yet there is evidence that processes ranging from decomposition and microbial nutrient transformation to secondary

Griffith, M. B., and S. A. Perry. 1993. The distribution of macroinvertebrates in the hyporheic zone of two small Appalachian headwater streams. *Archiv für Hydrobiologie* **126**:373–384.

Grimm, N. B., and S. G. Fisher. 1992. Response of arid land streams to changing climate. *In* "Global Climate Change and Freshwater Ecosystems" (P. Firth and S. G. Fisher, eds.), pp. 211–233. Springer-Verlag, New York.

Grimm, N. B., and K. C. Petrone. 1997. Nitrogen fixation in a desert stream ecosystem. *Biogeochemistry* **37**:33–61.

Hart, D. D. 1981. Foraging and resource patchiness: Field experiments with a grazing stream insect. *Oikos* **37**:46–52.

Harvey, J. W., and K. E. Bencala. 1993. The effect of streambed topography on surface-subsurface water exchange in mountain catchments. *Water Resources Research* **29**:89–98.

Harvey, J. W., B. J. Wagner, and K. E. Bencala. 1996. Evaluating the reliability of the stream tracer approach to characterize stream-subsurface water exchange. *Water Resources Research* **32**:2441–2451.

Hawkins, C. P., M. L. Murphy, and N. H. Anderson. 1982. Effects of canopy, substrate composition, and gradient on the structure of macroinvertebrate communities in Cascade Range streams of Oregon. *Ecology* **63**:1840–1856.

Hendricks, S. P., and D. S. White. 1988. Hummocking of lotic Chara: Observations on alterations of hyporheic temperature patterns. *Aquatic Botany* **31**:13–22.

Holmes, R. M., S. G. Fisher, and N. B. Grimm. 1994a. Nitrogen dynamics along parafluvial flowpaths: Importance to the stream ecosystem. *In* "Proceedings of the Second International Conference on Groundwater Ecology" (J. A. Stanford and H. M. Valett, eds.), pp. 47–56. American Water Resources Association, Herndon, VA.

Holmes, R. M., S. G. Fisher, and N. B. Grimm. 1994b. Parafluvial nitrogen dynamics in a desert stream ecosystem. *Journal of the North American Benthological Society* **13**:468–478.

Holmes, R. M., S. G. Fisher, N. B. Grimm, and B. J. Harper. 1998. The impact of flash floods on microbial distribution and biogeochemistry in the parafluvial zone of a desert stream. *Freshwater Biology* **40**:641–655.

Howarth, R. W., and S. G. Fisher. 1976. Carbon, nitrogen, and phosphorus dynamics during leaf decay in nutrient-enriched stream ecosystems. *Freshwater Biology* **6**:221–228.

Hynes, H. B. N. 1970. "The Ecology of Running Waters." University of Toronto Press, Toronto.

Jackson, J. K., and S. G. Fisher. 1986. Secondary production, emergence, and export of aquatic insects of a Sonoran Desert stream. *Ecology* **67**:629–638.

Johnston, C. E., and L. M. Page. 1992. The evolution of complex reproductive strategies in North American minnows (*Cyprinidae*). *In* "Systematics, Historical Ecology, and North American Freshwater Fishes" (R. L. Mayden, ed.), pp. 600–621. Stanford University Press, Stanford, CA.

Johnston, C. E., and J. S. Ramsey. 1990. Redescription of *Semotilus thoreauianus* Jordan, 1877; a cyprinid fish of the southeastern United States. *Copeia*, pp. 119–130.

Jones, J. B., R. M. Holmes, S. G. Fisher, and N. B. Grimm. 1994. Chemoautotrophic production and respiration in the hyporheic zone of a Sonoran Desert stream. *In* "Proceedings of the Second International Conference on Groundwater Ecology" (J. A. Stanford and H. M. Valett, eds.), pp. 329–338. American Water Resources Association, Herndon, VA.

Jones, J. B., S. G. Fisher, and N. B. Grimm. 1995a. Nitrification in the hyporheic zone of a desert stream ecosystem. *Journal of the North American Benthological Society* **14**:249–258.

Jones, J. B., S. G. Fisher, and N. B. Grimm. 1995b. Vertical hydrologic exchange and ecosystem metabolism in a Sonoran Desert stream. *Ecology* **76**:942–952.

Karr, J. R., and I. J. Schlosser. 1978. Water resources and the land-water interface. *Science* **201**:229–234.

Katznelson, R. 1989. Clogging of groundwater recharge basins by cyanobacterial mats. *FEMS Microbiology Ecology* **62**:231–242.

Keller, E. A., G. M. Kondolf, and D. J. Hagerty. 1990. Groundwater and fluvial processes; selected observations. *Special Paper—Geological Society of America* 252:319–340.

Kohler, S. L. 1984. Search mechanism of a stream grazer in patchy environments: The role of food abundance. *Oecologia* 62:209–218.

Lagler, K. F., J. E. Bardach, R. R. Miller, and D. R. M. Passino. 1977. "Ichthyology." Wiley, New York.

Lawson, D. L., M. J. Klug, and R. W. Merritt. 1984. The influence of the physical, chemical, and microbiological characteristics of decomposing leaves on the growth of the detritivore *Tipula abdominalis* (Diptera: Tipulidae). *Canadian Journal of Zoology* 62:2339–2343.

Marti, E., N. B. Grimm, and S. G. Fisher. 1997. Pre- and post-flood retention efficiency of nitrogen in a Sonoran Desert stream. *Journal of the North American Benthological Society* 16:805–819.

McArthur, J. V., and G. R. Marzolf. 1986. Interactions of the bacterial assemblages of a prairie stream with dissolved organic carbon from riparian vegetation. *Hydrobiologia* 134:193–199.

McDowell, W. H. 1984. Temporal changes in numbers of suspended bacteria in a small woodland stream. *Verhandlungen Internationale Vereinigung für Theoretische ünd Angewandte Limnologie* 22:1920–1925.

Meyer, J. L., and C. Johnson. 1983. The influence of elevated nitrate concentration on rate of leaf decomposition in a stream. *Freshwater Biology* 13:177–183.

Morrice, J. A., H. M. Valett, C. N. Dahm, and M. E. Campana. 1997. Alluvial characteristics, groundwater-surface water exchange and hydrologic retention in headwater streams. *Hydrological Processes* 11:1–15.

Mulholland, P. J., A. D. Steinman, E. R. Marzolf, D. R. Hart, and D. L. DeAngelis. 1994. Effect of periphyton biomass on hydraulic characteristics and nutrient cycling in streams. *Oecologia* 98:40–47.

O'Neill, R. V., D. L. DeAngelis, J. B. Waide, and T. F. H. Allen. 1986. "A Hierarchical Concept of Ecosystems." Princeton University Press, Princeton, NJ.

Palmer, M. A., C. C. Hakenkamp, and K. Nelson-Baker. 1997. Ecological heterogeneity in streams: Why variance matters. *Journal of the North American Benthological Society* 16:189–202.

Palmer, T. M. 1995. The influence of spatial heterogeneity on the behavior and growth of two herbivorous stream insects. *Oecologia* 104:476–486.

Perkins, D. J., B. N. Carlson, M. Fredstone, R. H. Miller, C. M. Rofer, G. T. Ruggerone, and C. S. Zimmerman. 1984. The effects of groundwater pumping on natural spring communities in Owens Valley. *In* "California Riparian Systems: Ecology, Conservation, and Productive Management" (R. E. Warner and K. M. Hendrix, eds.), pp. 515–527. University of California Press, Berkeley.

Pinay, G., N. E. Haycock, C. Ruffinoni, and R. M. Holmes. 1994. The role of denitrification in nitrogen removal in river corridors. *In* "Global Wetlands: Old World and New" (W. J. Mitsch, ed.), pp. 107–116. Elsevier, Amsterdam.

Poff, N. L., and K. Nelson-Baker. 1997. Habitat heterogeneity and algal-grazer interactions in streams: Explorations with a spatially explicit model. *Journal of the North American Benthological Society* 16:263–276.

Poff, N. L., and J. V. Ward. 1992. Heterogeneous currents and algal resources mediate in situ foraging activity of a mobile stream grazer. *Oikos* 71:179–188.

Poff, N. L., J. D. Allan, M. B. Bain, J. R. Karr, K. L. Prestegaard, B. D. Richter, R. E. Sparks, and J. C. Stromberg. 1997. The natural flow regime. *BioScience* 47:769–784.

Pringle, C. M. 1990. Nutrient spatial heterogeneity: Effects on community structure, physiognomy, and diversity of stream algae. *Ecology* 71:905–920.

Schälchi, U. 1992. The clogging of coarse gravel river beds by fine sediment. *Hydrobiologia* 235/236:189–197.

Sedell, J. R., G. H. Reeves, F. R. Hauer, J. A. Stanford, and C. P. Hawkins. 1990. Role of refugia in recovery from disturbances of modern fragmented and disconnected river systems. *Environmental Management* **14**:711–724.

Stanley, E. H., and A. J. Boulton. 1995. Hyporheic processes during flooding and drying in a Sonoran Desert stream. I. Hydrologic and chemical dynamics. *Archiv für Hydrobiologie* **134**:1–26.

Stanley, E. H., and H. M. Valett. 1992. Interaction between drying and the hyporheic zone of a desert stream ecosystem. *In* "Climate Change and Freshwater Ecosystems" (P. Firth and S. G. Fisher, eds.), pp. 234–249. Springer-Verlag, New York.

Stanley, E. H., S. G. Fisher, and N. B. Grimm. 1997. Ecosystem expansion and contraction in streams. *BioScience* **47**:427–435.

Stevenson, R. J. 1997. Scale-dependent determinants and consequences of benthic algal heterogeneity. *Journal of the North American Benthological Society* **16**:248–262.

Stromberg, J. C., R. Tiller, and B. Richter. 1996. Effects of groundwater decline on riparian vegetation of semiarid regions: The San Pedro, Arizona. *Ecological Applications* **6**:113–131.

Suberkropp, K. F., and T. L. Arsuffi. 1984. Degradation, growth, and changes in palatability of leaves colonized by six aquatic hyphomycete species. *Mycologia* **76**:398–407.

Tank, J. L., and J. R. Webster. 1998. Interaction of substrate and nutrient availability on wood biofilm processes in streams. *Ecology* **79**:2168–2179.

Tilman, D. 1982. "Resource Competition and Community Structure," Monographs in Population Biology No. 17. Princeton University Press, Princeton, NJ.

Valett, H. M., S. G. Fisher, N. B. Grimm, and P. Camill. 1994. Vertical hydrologic exchange and ecological stability of a desert stream ecosystem. *Ecology* **75**:548–560.

Valett, H. M., J. A. Morrice, and C. N. Dahm. 1996. Parent lithology, surface-groundwater exchange, and nitrate retention in headwater streams. *Limnology and Oceanography* **41**:333–345.

Valett, H. M., C. N. Dahm, M. E. Campana, J. A. Morrice, M. A. Baker, and C. S. Fellows. 1997. Hydrologic influences on groundwater-surface water ecotones: Heterogeneity in nutrient composition and retention. *Journal of the North American Benthological Society* **16**:239–247.

Vaux, W. G. 1962. Interchange of stream and intragravel water in a salmon spawning riffle. *U. S., Fish and Wildlife Service, Special Scientific Report—Fisheries* **405**.

Wainwright, S. A., C. A. Couch, and J. L. Meyer. 1992. Fluxes of bacteria and organic matter into a blackwater river from river sediments and floodplain soils. *Freshwater Biology* **28**:37–48.

Wallace, J. B., and M. E. Gurtz. 1986. Response of Baetis mayflies (Ephemeroptera) to catchment logging. *American Midland Naturalist* **115**:25–41.

Ward, J. V. 1989. The four-dimensional nature of lotic ecosystems. *Journal of the North American Benthological Society* **8**:2–8.

Webster, J. R., and E. F. Benfield. 1986. Vascular plant breakdown in freshwater ecosystems. *Annual Review of Ecology and Systematics* **17**:567–594.

White, D. S. 1990. Biological relationships to convective flow patterns within stream beds. *Hydrobiologia* **196**:149–158.

White, D. S., S. P. Hendricks, and S. L. Fortner. 1992. Groundwater-surface water interactions and the distributions of aquatic macrophytes. *In* "Proceedings of the First International Conference on Ground Water Ecology" (J. A. Stanford and J. J. Simons, eds.), pp. 247–255. American Water Resources Association, Bethesda, MD.

Williams, D. D. 1977. Movements of benthos during the recolonization of temporary streams. *Oikos* **29**:306–312.

Wondzell, S. M., and F. J. Swanson. 1996. Seasonal and storm dynamics of the hyporheic zone of a 4th-order mountain stream. II: Nutrient cycling. *Journal of the North American Benthological Society* **15**:20–34.

Wroblicky, G. J., M. E. Campana, H. M. Valett, and C. N. Dahm. 1998. Seasonal variation in surface-subsurface water exchange and lateral hyporheic area of two stream-aquifer systems. *Water Resources Research* **34**:317–328.

Wu, J., and O. L. Loucks. 1995. From balance of nature to hierarchical patch dynamics: A paradigm shift in ecology. *Quarterly Review of Biology* **70**:439–466.

SECTION *FOUR*

*SUMMARY
AND SYNTHESIS*

17

Surface–Subsurface Interactions: Past, Present, and Future

Emily H. Stanley* and Jeremy B. Jones[†]

*University of Wisconsin
Center for Limnology
Madison, Wisconsin

[†]Department of Biological Sciences
University of Nevada Las Vegas
Las Vegas, Nevada

I. Introduction   405
II. Growth and Development Trajectories   406
III. Flowpaths and Interfaces   408
IV. The Spatial Context   409
V. Synthetic Models and Future Directions   412
References 414

I. INTRODUCTION

This book is a testament to the importance of surface–subsurface exchange processes in lotic ecosystems. It deals with one of the fastest growing research directions in stream ecology, but as is pointed out in several chapters, substantial gaps exist in our fundamental understanding of subsurface zones and exchange processes. In this final chapter, we would like to reflect on our position—that is, how have we progressed, and more importantly, where to from here? We will emphasize a series of themes that have emerged either explicitly or implicitly from the chapters in this book. These include growth and development trajectories of different disciplines captured within this book, the importance of flowpaths and interfaces, spatial context of study, and the challenging goal of development of synthetic models of sur-face–subsurface exchange in lotic ecosystems.

II. GROWTH AND DEVELOPMENT TRAJECTORIES

The growth in research related to surface–subsurface exchange pro-cesses (which captures investigations focusing on interactions between any two of the following subsystems: surface, hyporheic, riparian, and ground-water zones, and, as used in this book, processes occurring within hyporhe-ic zones) has mushroomed during the decade of the 1990s, particularly for papers dealing with physical (hydrological) and biogeochemical processes. The onset of this trend was documented by Valett *et al.* (1993), who used the number of abstracts related to hyporheic–interstitial dynamics at the an-nual meeting of the North American Benthological Society as a metric of growth. In the ensuing 7 years since the appearance of the Valett *et al.* (1993) article, abstract counts have fluctuated but have steadied to an average of 30–40 papers per meeting, and typically 1–2 special and/or contributed ses-sions dedicated to issues related to subsurface zones and/or interactions be-tween surface and subsurface environments. More recently, multiple review papers (a reasonable sign of field maturation) have appeared (Jones and Holmes, 1996; Brunke and Gonser, 1997; Boulton *et al.*, 1998). But it is equally clear that ecosystem–biogeochemical ("process–function"; *sensu* O'Neill *et al.*, 1986) and population–community level research have not pro-gressed at the same tempo. The former research area can be considered rel-atively new, emerging during the 1980s, with landmark papers by Grimm and Fisher (1984), Bencala *et al.* (1984), and Triska *et al.* (1989), and con-tinues to grow steadily. Rapid development of hydrologic expertise, and col-laborations between hydrologists and ecologists are clearly key driving forces behind the rapid growth in process–function aspects of subsurface–surface dynamics. The latter area of population–community research has a long tradition traced back to European studies of groundwater organisms and is the source of much of the now-familiar terminology (most notably the term "hyporheic"; *sensu* Orghidan, 1959). Yet, much of the work in this area continues on in the same way as it has for the past several years with de-scriptive studies of organismal distributions coupled with inferences regard-ing causes of these distributions (Boulton, Chapter 14 in this volume; Stray-er, 1994). Even research on microbes has languished, despite the central importance of prokaryotes in driving the biogeochemical processes that are currently drawing the lion's share of attention (Findlay and Sobczak, Chap-ter 12 in this volume).

There are undoubtedly several causes for this trend of slow progress in the study of interstitial population and community dynamics. One explana-tion is that research on meio- and macrofauna remains constrained by persistent gaps in the taxonomy of hyporheic fauna (Hakenkamp and Palmer, Chapter 13 in this volume; Boulton, Chapter 14 in this volume). Despite publications such as *Stygofauna Mundi* (Botosaneanu, 1986), hyporheic and groundwater fauna remain poorly characterized (Reid, 1992). Unfortunately, this problem is not likely to improve; the limited and declining number of individuals with taxonomic expertise is a pervasive problem throughout virtually all fields of ecology—an ironic trend in the face of the ever-increasing preoccupation with declining worldwide biodiversity. Hyporheic zones composed of coarse (well-oxygenated) sediments typically support diverse invertebrate communities (e.g., Stanford and Gaufin, 1974; Boulton *et al.*, 1992), but the limited number of studies and taxonomic resolution of existing research preclude us from assessing the contribution of interstitial environments to whole-system biodiversity. Because diversity is a broadly familiar concept outside of the discipline, it provides a means for ecologists to relate field-specific research to nonscientists. Information about interstitial fauna (including biodiversity, and the curious physical attributes of many obligate forms) represents an underused tool that could be employed in conservation efforts, education, and even fund acquisition. As an example of the use of organisms to increase public awareness, the occurrence of unusual organisms such as blind catfish and crayfish endemic to the Edward's Aquifer (Texas) have been used as centerpieces to increase public awareness of this unusual ecosystem (Bowles and Arsuffi, 1993). The diminished number of researchers drawn to taxonomy and population–community dynamics is a cause for concern both with respect to our general ecological understanding and in the context of a missed opportunity to popularize the discipline.

Beyond the fundamental determinants of distribution and abundance, significantly more remains to be studied about hyporheic organisms. Hakenkamp and Palmer (Chapter 13 in this volume) and Boulton (Chapter 14 in this volume) highlight several research avenues, including organic matter processesing, secondary production in both oxic and anoxic sediments, bacterivorey, trophic dynamics, and subsurface invertebrates as potential prey for benthic consumers. Interestingly, Hakenkamp *et al.* (1993) listed the elaboration of hyporheic energetics and trophic relations as completely unexplored and important avenues for research. To the best of our knowledge, only a handful of papers have even measured rates of bacterial production (see Findlay and Sobczak, Chapter 12 in this volume), fewer yet have considered energetics of microbes other than bacteria (e.g., Ellis *et al.*, 1998), and there are no quantitative studies of hyporheic food webs.

In stark contrast to the slow rate of change in population–community studies, hydrologic and biogeochemical research as related to exchange processes has shown stunning development over an extremely short period of time. The advent of solute injection techniques and the merging of tracer studies with nutrient uptake models have fueled much of this growth (see Harvey and Wagner, Chapter 1 in this volume; Packman and Bencala, Chapter 2 in this volume; Mulholland and DeAngelis, Chapter 6 in this volume). These valuable and inexpensive tools are now widely available to researchers, and a large number of estimates of transient storage parameters and nutrient uptake lengths exist in the literature. The wealth of information derived from standardized techniques and models now permits cross-system comparisons (e.g., Findlay, 1995) that have and will lead to broad, synthetic insights into subsurface and exchange processes. An excellent example of such an analysis is presented by Harvey and Wagner (Chapter 1 in this volume), in which estimates of storage zone residence is related to residence of the entire reach (Fig. 6 in Chapter 1) and relative size of the transient storage zone (the A_s/A ratio) is related to friction factor (Fig. 8 in Chapter 1) to develop a regional perspective of hydrologic retention in streams.

III. FLOWPATHS AND INTERFACES

One of the obvious themes that runs through almost every chapter in this book is the fundamental importance of understanding system hydrology. Clear changes in nitrogen (Duff and Triska, Chapter 8 in this volume), phosphorus (Hendricks and White, Chapter 9 in this volume), carbon (Kaplan and Newbold, Chapter 10 in this volume), terminal electron acceptors (Baker *et al.*, Chapter 11 in this volume), microbial activity (Findlay and Sobczak, Chapter 12 in this volume), meiofauna (Hakenkcamp and Palmer, Chapter 13 in this volume), and macroinvertebrates (Boulton, Chapter 14 in this volume) have been reported along subsurface flowpaths. Similarly, hydrologic exchange between surface and subsurface environments are resolvable in both the hyporheic and benthic zones with respect to physical (e.g., temperature), chemical (oxygen, carbon, nitrogen, phosphorus), and biological (microbial activity, meiofaunal, macroinvertebrate, and macrophyte densities and composition) patterns. At the stream surface, the ability to identify points and direction of hydrologic exchange is a powerful key for understanding such disparate ecological phenomena as primary production and fish behavior and, in a broader sense, for examining a fundamental cause of spatial and temporal heterogeneity in stream ecosystems (Dent *et al.*, Chapter 16 in this volume).

Quantifying the direction and magnitude of exchange has opened doors for studying linkages between as well as heterogeneity within stream subsystems. Unraveling the causes of spatial heterogeneity within subsurface environments is driven by elucidation of flowpaths and identification of changes that occur as water moves through the interstitial environment. The convergence of the changes along the length of the flowpath and the linkages between disparate subsystems has led to a second theme common to many chapters: the importance of interfaces between disparate subsystems. Much of the initial process–function interest in both riparian and hyporheic zones is related to the recognition of their ecotonal nature (Naiman and Décamps, 1989); by definition, riparian and hyporheic zones are themselves interfaces. Yet the scale at which the most dramatic changes in physical, chemical, and biological attributes occur appears to be relatively small. Pronounced physicochemical shifts over distances of centimeters have been documented for forms and concentrations of nitrogen (Duff and Triska, Chapter 8 in this volume), phosphorus (Hendricks and White, Chapter 9 in this volume), and organic carbon (Kaplan and Newbold, Chapter 10 in this volume), as well as in distribution of active bacteria (Holmes *et al.*, 1998) and microbial metabolism (Baker *et al.*, Chapter 11 in this volume).

Rates of change in transformations occurring at interfaces are reasonably described by first-order reaction kinetics (Kaplan and Newbold, Chapter 10 in this volume), meaning that relatively simple equations can be used to build models of ecological phenomena driven by hydrologic exchange, such as demonstrated by Mulholland and DeAngelis (Chapter 6 in this volume). However, substantial work remains in the examination interface dynamics. For example, we are far from understanding the fundamental hydrologic processes at surface–subsurface interfaces under field conditions (Packman and Bencala, Chapter 2 in this volume). Further, the interface between benthic and hyporheic zones has attracted the greatest attention to date. Less is known about layers where ground water enters the hyporheic zone (Holmes, Chapter 5 in this volume), and the riparian–hyporheic interface is virtually unstudied (Hill, Chapter 3 in this volume).

IV. THE SPATIAL CONTEXT

This book covers a broad range of environments and processes that have often been treated separately, despite the common theme of hydrologic exchange. For example, studies of the riparian zone are rarely combined with those dealing with the hyporheic zone (but see Naiman and Décamps, 1990). It is worth revisiting the point emphasized in Chapter 2: the target of study (surface stream, subsurface zone, or the interface) needs to be made explicit; different models are relevant to each of these entities, and conclusions derived from data will depend on the chosen target. Different chapters reflect the emphasis on different targets. For example, chapters by Boulton (Chapter 14), Hakenkamp and Palmer (Chapter 13), Findlay and Sobczak (Chapter 12), and Baker *et al.* (Chapter 11) focus on the subsurface, Dent *et al.* (Chapter 16) on surface environments, Martí *et al.* (Chapter 4) and Mulholland and DeAngelis (Chapter 6) on exchange between subsystems, and the remaining chapters consider some combination of all three perspectives. This point of clarification is particularly important when considering the challenge of extrapolating results from one scale to another, or generating broad conceptual models.

Most chapters, and indeed most research related to surface–subsurface exchange processes, deal with phenomena occurring at spatial scales of 10–500 m (with the notable exception of Pringle and Triska, Chapter 7). In this book strong arguments are made for smaller-scale work that hones in on specific mechanisms or processes (Chapters 1, 6–8, 12–15), and the need to understand exchange processes at larger spatial scales (Chapters 1, 5, 12, 16). The need for small-scale, mechanistically oriented studies is a direct result of the importance and activity of interfaces discussed earlier. We anticipate that these sorts of studies will become more common, particularly as tools and know-how from other fields are adapted to stream settings (Findlay and Sobczak, Chapter 12 in this volume; Hakenkamp and Palmer, Chapter 13 in this volume). The largest and most productive growth area for small-scale studies is likely to be in the realm of microbial dynamics, which can presently be considered a poorly understood area at best (Findlay and Sobczak, Chapter 12 in this volume).

Even though the future of studies focusing on small-scale phenomena may face their greatest challenge in the form of logistic constraints, consideration of surface–subsurface exchange processes at larger spatial scales faces both conceptual and logistic adversities. In general, the importance of subsurface processes in streams is a function of the extent of these subsurface zones and the nature of the hydrologic linkages between surface and subsurface environments. As a result, investigators who strive to expand the scale of study may do so by considering the spatial distribution of alluvial sediments in larger units of a drainage basin, by considering larger or longer hydrologic fluxes, or by doing both. In the first scenario, the spatial extent of consideration might be increased by starting with an examination of processes within gravel bars and then mapping the extent of gravel bars throughout a large channel section. Even though the physical template is expanded, no such assumption is made about hydrologic processes that link surface and gravel-bar environments. However, it is worth emphasizing that hydrologic flowpaths within a given channel section vary in length from centimeters to meters, and even kilometer-long flowpaths may originate or terminate within a short length of stream (Harvey and Wagner, Chapter 1 in this volume; Dent *et al.*, Chapter 16 in this volume). Thus, even if the intent is not to consider longer flowpaths, the possibility that longer water movements are incorporated into the study system increases with spatial scale.

The spatial scale of interest may alternatively be increased by examining hydrologic flowpaths that are longer and slower than those occurring at the scale of 10–100 m. In this case, however, larger or longer hydrologic fluxes by definition occur over larger areas. Hence, studies that consider larger hydrologic fluxes also consider progressively larger geomorphic units. That is, scaling up to consider larger hydrologic fluxes also involves scaling up to consider larger geomorphic units, but the reverse does not necessarily hold. Two hierarchies—a geomorphic hierarchy of increasing stream size and a hydrologic hierarchy of increasing flowpath length—are in place, and they are inextricably linked.

In contrast to the wealth of geomorphological classifications, few descriptions of hierarchies of subsurface flow or hydrologic exchange exist. One of the earliest considerations of hydrologic hierarchies was presented by Tóth (1963), who used a theoretical approach to demonstrate the existence of three relatively discrete scales of subsurface fluxes ("local, intermediate, and regional") in small watersheds. A similar scheme was also described by Fisher *et al.* (1998) and Dahm *et al.* (1998) based on empirical evidence. Several investigators have been able to distinguish between the effects of reach-level exchange patterns and larger-scale hydrologic fluxes associated with alluvial aquifers. For example, distribution of surface water (Stanley *et al.*, 1997), cyanobacteria (Grimm and Petrone, 1997), and nitrate (Dent and Grimm, 1999) that are not adequately explained by smaller (meter)-scale exchanges in desert streams can be attributed to kilometer-scale hydrologic fluxes. Similarly, base cation composition and invertebrate community structure in highly regulated riverine canals are driven by both meter- and kilometer-length flow patterns (Creuzé des Châtelliers and Regrobellet, 1990; Creuzé des Châtelliers and Marmonier, 1990). However, beyond individual channel sections or even drainage basins, studies involving regional groundwater/aquifer dynamics have traditionally been the purview of hydrologists and geochemists but not ecologists (but see Gibert *et al.*, 1994). Provocative examples of ecological relevance of these slower, large-scale fluxes are presented by Pringle and Triska (Chapter 7 in this volume) and Holmes (Chapter 5 in this volume) and include such phenomena as upwelling of solute-rich waters in volcanic landscapes and nutrient- and toxicant-loading to streams and rivers. Future collaboration between hydrogeologists and ecologists is likely to be a productive avenue for understanding exchange processes at a landscape scale.

Despite Tóth's (1963) description of classes of flowpaths of varying length, the develoment of a hydrologic hierarchy and scaling up from reaches to watersheds remains an extremely challenging prospect. Even though an individual flowpath can be treated as a discrete, unidirectional vector, all water movements in a basin are multidirectional (Bencala, 1993) and occur in a hydroscape that itself is not linear (Fisher, 1997). The collective diversity of directions and lengths of water movements and the shift from linear systems (such as a single flowpath or a single stream reach) to branched systems (several flowpaths or a drainage basin) present difficult conceptual hurdles to overcome in the pursuit of translating small-scale results to whole watersheds. However, this hesitancy may reflect our limited understanding of exchange in channels of different stream orders. Conceptual models of longitudinal exchange along a drainage continuum have been proposed by Stanford and Ward (1993) and White (1993) but are yet to be rigorously tested. In particular, information on large river systems is scarce, as most studies have focused on small headwater streams (e.g., Hill, Chapter 3 in this volume). Consequently, our understanding of exchange processes in larger rivers is strongly driven by a limited number of well-studied systems (Boulton *et al.*, 1998). The explanation for this dearth is obvious; the logistical challenges presented by increased river size are formidable. The use of solute injection techniques that have been so productive in small-order systems are intractable in larger channels. To date, we are aware of only one such investigation. The recent tracer study along the length of the Willamette River, Oregon (Wentz *et al.*, 1998) will provide an extremely useful case study for future investigations of surface–subsurface exchanges in high-order systems. An early lesson from this study is that multiple techniques for quantifying exchange need to be developed and implemented (Laenen and Bencala, 1997). One such promising approach may be the examination of rivers during floods. Groundwater-derived water on floodplains can be distinguished from channel-derived water on the basis of differences in suspended sediment loads (Mertes, 1997), thus upwelling areas in floodplains can be identified as those areas with relatively clear water (in contrast to sediment-laden water derived from the river channel). As differences in suspended sediment loads of water are distinguishable in remotely sensed images (Mertes *et al.*, 1993), this approach may represent one alternative means of examining exchange processes across large areas and in large rivers.

V. SYNTHETIC MODELS AND FUTURE DIRECTIONS

Findlay has repeatedly challenged his colleagues with the development of synthetic models describing hyporheic and exchange processes in streams and their ecological implications (e.g., Findlay, 1995; Findlay and Sobczak, Chapter 12 in this volume). What conceptual underpinnings need to be incorporated to develop broad generalizations and how do we do this? The first steps are being taken through cross-site comparisons (e.g., Harvey and Wagner, Chapter 1 in this volume) and conceptualizations (Fig. 9, Hill, Chapter 3 in this volume). To reiterate the point made in the previous section, however, these cross-site comparisons are currently restricted to headwater streams. Thus, another obvious need in the pursuit of general models of exchange is to improve our understanding of the hydrology and ecology of subsurface and exchange processes in larger channels.

As demonstrated by the diversity of topics included in this book, both within and among chapters, our understanding of streams has benefited enormously from the cross-fertilization between hydrology and biology. We feel that the next leap forward in understanding exchange processes will come from a similar cross-fertilization between the fields of ecology and geomorphology. Familiarity with geomorphology is likely to broaden our understanding of where and when subsurface processes exert large influences on streams. Convergence of geomorphology and hydrology has already proven to be immensely productive in terms of understanding groundwater–surface-water interactions (Brown and Bradley, 1995); pursuing the next logical step of weaving ecological phenomena into these sorts of studies cannot be far behind. As channel form is a product of regional climate and geology, characteristic regional channel types, sediment composition, and longitudinal profiles can be, and have been, described with some degree of reliability (Brussock *et al.,* 1985). Although these authors focused on surface processes, it is easy to extrapolate the geographic patterns in channel structure and slope to consider surface–subsurface interactions. For example, surface–hyporheic interactions are likely to be limited in streams and rivers of the Atlantic coastal and midwestern regions because of reduced channel gradients and sediment porosity. In contrast, extensive in-channel interactions should occur in mountainous glaciated areas where channels are broad due to glacial scour and melt, have steep slopes, and contain coarse sediments. Knowledge of regional geomorphology has been used to determine occurrences of streams with well-developed hyporheic zones and surface–surface exchange and to make predictions regarding the spatial distribution of hyporheic invertebrates (Hunt and Stanley, 1999).

Although fluvial geomorphology is useful in the consideration of within-channel hydrology, knowledge of regional geomorphology should also provide insights into hillslope and regional aquifer hydrology. As a simple and well-known example, the occurrence of direct groundwater–surface-water exchange (i.e., springs) can be predicted from knowledge of parent geology (e.g., occurrences of karst formation, or strata of high hydraulic conductivity sandwiched between more impermeable layers). Further, differences in parent geology affect transient storage zone size, the extent of hydrologic exchange (Morrice *et al.,* 1997), nutrient uptake lengths (Valett *et al.,* 1996), and groundwater inputs of dissolved gases such as methane (Jones and Mulholland, 1998) in small streams. An excellent example of the use of geological and geomorphic information to understand exchange processes and their coupled ecological implications is provided by Lowrance *et al.* (1997). The efficacy of riparian buffer strips was shown to be dependent upon regional variation in geology and soil type, which constrained water movements between uplands and streams in the Chesapeake Bay drainage. The promise of studying pattern to understand process (*sensu* Turner, 1989) to develop synthetic, broad-scale models in this field is also illustrated by Hill (Chapter 3 in this volume). This author emphasizes the fundamental relationship between hillslope geomorphology and hydrology and thus is able to identify times and places in which the riparian zone can be expected to be an effective nutrient filter (see Fig. 9 in Chapter 3). This point is underscored by comparing Hill's (Chapter 3 in this volume) models of stream–riparian interactions to that presented by Martí *et al.* (Chapter 4 in this volume), in which climate and geomorphology interact to create fundamentally different linkages between riparian and stream subsystems.

Beyond attempts to develop broad, synthetic models of exchange processes in streams and rivers, several authors have emphasized the lack of information in their specific areas of study. Most notable are holes in microbial, meio- and macrofaunal dynamics, anaerobic processes, and food web and energetics of subsurface environments. It is also surprising to find only limited consideration of anthropogenic impacts on subsurface zones and exchange processes in this book. Pringle and Triska (Chapter 7 in this volume) list several landscape-scale processes that are expected to interfere with surface–subsurface relationships, such as sedimentation, nonpoint nutrient pollution, movements of agricultural pesticides between regional aquifers, and streams and channel alterations (e.g., dam and levee construction, wetland drainage). Other landscape-level impacts on drainages that will be modulated by exchange processes include groundwater withdrawal (Dent *et al.*, Chapter 16), processing of point-source inputs of sewage effluents (Rutherford *et al.*, 1991) and other contaminants, and saline intrusions into lotic systems in coastal regions. Even though the study of groundwater contaminant movement is well established, investigations of this same topic for near-stream environments and for the transmission of contaminants between groundwater and streams (in either direction) are rare. This is surprising, given that relatively early investigations of exchange processes were driven by the desire to understand the impacts of pollutants (e.g., Mestrov and Lattinger-Penko, 1977), and that degredation of substances such as pesticides is substantially greater in hyporheic zones than in ground water (Schwarzenbach *et al.*, 1983). By necessity, we expect that significant advances in our understanding of exchange processes will be driven by investigation of anthropogenic impacts on surface and subsurface habitats.

In summary, a key step to advancing our understanding of stream-water–groundwater interactions will be the integration of hydrology, geomorphology, and ecology. Central to this integration is a need to expand our study beyond individual sites. While we know that exchange of water between surface water and ground water is an important regulator of system biogeochemistry and ecology, we do not know how the parameters change across streams. This broader perspective is important not only for generalizations about subsurface processes but, more fundamentally, for advancing our conceptual models of streams.

REFERENCES

Bencala, K. E. 1993. A perspective on stream-catchment connections. *Journal of the North American Benthological Society* **12**:44–47.

Bencala, K. E., V. C. Kennedy, G. W. Zellweger, A. P. Jackman, and R. J. Avanzino. 1984. Interactions of solutes and streambed sediments. 1. An experimental analysis of cation and anion transport in a mountain stream. *Water Resources Research* **20**:1797–1803.

Botosaneaunu, L., ed. 1986. "Stygofauna mundi." Brill, Leiden, The Netherlands.

Boulton, A. J., S. Findlay, P. Marmonier, E. H. Stanley, and H. M. Valett. 1998. The functional significance of the hyporheic zone in streams and rivers. *Annual Review of Ecology and Systematics* **29**:59–81.

Boulton, A. J., H. M. Valett, and S. G. Fisher. 1992. Spatial distribution and taxonomic composition of the hyporheos of several Sonoran Desert streams. *Archiv für Hydrobiologie* **125**:37–61.

Bowles, D. E., and T. L. Arsuffi. 1993. Karst aquatic ecosystems of the Edwards Plateau region of central Texas, USA: A consideration of their importance, threats to their existence, and efforts for their conservation. *Aquatic Conservation: Marine and Freshwater Ecosystems* **3**:317–329.

Brown, A. G., and C. Bradley. 1995. Geomorphology and groundwater: convergence and diversification. *In* "Geomorphology and Groundwater" (A. G. Brown, ed.), pp. 1–20. Wiley, Chichester.

Brunke M., and T. Gonser 1997. The ecological significance of exchange processes between rivers and groundwaters. *Freshwater Biology* **37**:1–33.

Brussock P. P, A. V. Brown, and J. C. Dixon. 1985. Channel form and stream ecosystem models. *Water Resources Bulletin* **21**:859–866.

Creuzé des Châtelliers M., and P. Marmonier. 1990. Macrodistribution of Ostracoda and Cladocera in a by-passed channel: Exchange between surficial and interstitial layers. *Stygologia* **5**:17–24.

Creuzé des Chatelliers M., and J. L. Regrobellet. 1990. Interactions between geomorphological processes, benthic and hyporheic communities: First results on a by-passed canal of the French Upper Rhône River. *Regulated Rivers: Research and Management* **5**:139–158.

Dahm, C. N., N. B. Grimm, and P. Vervier. 1998. Nutrient dynamics at interface between surface waters and groundwaters. *Freshwater Biology* **40**:427–452.

Dent, C. L, and N. B. Grimm. 1999. Spatial heterogeneity of stream water nutrient concentrations over successional time. *Ecology* **80**:2283–2298.

Ellis, B. K., J. A. Stanford, and J. V. Ward. 1998. Microbial assemblages and production in alluvial aquifers of the Flathead River, Montana, USA. *Journal of the North American Benthological Society* **17**:382–402.

Findlay, S. 1995. Importance of surface-subsurface exchange in stream ecosystems: The hyporheic zone. *Limnology and Oceanography* **40**:159–164.

Fisher, S. G. 1997. Creativity, idea generation, and the functional morphology of streams. *Journal of the North American Benthological Society* **16**:305–318.

Fisher, S. G., N. B. Grimm, C. L. Dent, E. Martí, and R. Gomez. 1998. Hierarchy, spatial configuration, and nutrient cycling in a desert stream. *Australian Journal of Ecology* **23**:41–52.

Gibert, J., D. L. Danielopol, and J. A. Stanford, eds. 1994. "Groundwater Ecology." Academic Press, San Diego, CA.

Grimm, N. B., and S. G. Fisher. 1984. Exchange between surface and interstitial water: Implications for stream metabolism and nutrient cycling. *Hydrobiologia* **111**:219–228.

Grimm, N. B., and K. C. Petrone. 1997. Nitrogen fixation in a desert stream ecosystem. *Biogeochemistry* **37**:33–61.

Hakenkamp, C. C., H. M. Valett, and A. J. Boulton. 1993. Perspectives on the hyporheic zone: Integrating hydrology and biology: Concluding remarks. *Journal of the North American Benthological Society* **12**:94–99.

Holmes, R. M., S. G. Fisher, N. B. Grimm, and B. J. Harper. 1998. The impact of flash floods on microbial distribution and biogeochemistry in the parafluvial zone of a desert stream. *Freshwater Biology* **40**:641–654.

Hunt, G. W. and E. H. Stanley. 1999. Environmental factors influencing the composition and distribution of the hyporheic fauna in Oklahoma streams: Variation across ecoregions. *Archiv für Hydrobiologie* (in review).

Jones, J. B., and R. M. Holmes. 1996. Surface-subsurface interactions in stream ecosystems. *Trends in Ecology and Evolution* **11**:239–242.

Jones, J. B., and P. J. Mulholland. 1998. Methane input and evasion in a hardwood forest stream: Effects of subsurface flow from shallow and deep pathways. *Limnology and Oceanography* **43**:1243–1250.

Laenen, A., and K. E. Bencala. 1997. Transient storage assessments of dye-tracer injections in the Willamette River basin, Oregon. *American Society of Limnology and Oceanography Aquatic Sciences Meeting*, Santa Fe, NM.

Lowrance R., L. S. Altier, J. D. Newbold, R. R. Schnabel, P. M. Groffman, J. M. Denver, D. L. Correll, J. W. Gilliam, J. L. Robinson, R. B. Brinsfield, K. W. Staver, W. Lucas, and A. H. Todd. 1997. Water quality functions of riparian forest buffers in Chesapeake Bay watersheds. *Environmental Management* **21**:687–712.

Mertes, L. A. K. 1997. Documentation and significance of the perirheic zone on inundated floodplains. *Water Resources Research* **33**:1749–1762.

Mertes, L. A. K., M. O. Smith, and J. B. Adams. 1993. Estimating suspended sediment concentrations in surface waters of the Amazon River wetlands from Landsat images. *Remote Sensing of Environment* **43**:281–301.

Mestrov, M., and R. Latinger-Penko. 1977. Ecological investigations of the influence of a polluted river on surrounding interstitial underground waters. *Journal of Speleology* **9**:331–335.

Morrice, J. A., H. M. Valett, C. N. Dahm, and M. E. Campana. 1997. Alluvial characteristics, groundwater-surface water exchange and hydrologic retention in headwater streams. *Hydrological Processes* **11**:253–267.

Naiman R. J., and H. Décamps. 1989. The potential importance of boundaries to fluvial ecosystems. *Journal of the North American Benthological Society* **7**:289–306.

Naiman R. J., and H. Décamps, eds. 1990. "The Ecology and Management of Aquatic-Terrestrial Ecotones." UNESCO and Parthenon Publishing Group, Paris.

O'Neill, R. V., D. L. DeAngelis, J. B. Waide, and T. F. H. Allen. 1986. "A Hierarchical Concept of Ecosystems." Princeton University Press, Princeton, NJ.

Orghidan, T. 1959. Ein neuer Lebensraum des unterirdischen Wassers: Der hyporheeische Biotop. *Archiv für Hydrobiologie* **55**:39–414.

Reid, J. W. 1992. Taxonomic problems: A serious impediment to groundwater ecological research in North America. *In* "Proceedings of the First International Conference on Ground Water Ecology" (J. A. Stanford and J. J. Simons, eds.), pp. 133–142. American Water Resources Association, Bethesda, MD.

Rutherford, J. C., R. J. Wilcock, and C. W. Hickey. 1991. Deoxygenation in a mobile-bed river-I. Field studies. *Water Research* **25**:1487–1497.

Schwarzenbach, R. P., W. Giger, E. Hoehn, and J. K. Schneider. 1983. Behavior of organic compounds during infiltration of river water to groundwater. Field studies. *Environmental Science and Technology* **17**:472–479.

Stanford, J. A., and A. R. Gaufin. 1974. Hyporheic communities in two Montana rivers. *Science* **185**:700–702.

Stanford J. A., and J. V. Ward. 1993. An ecosystem perspective of alluvial rivers: Connectivity and the hyporheic zone. *Journal of the North American Benthological Society* **12**:48–60.

Stanley E. H., S. G. Fisher, and N. B. Grimm. 1997. Ecosystem expansion and contraction in streams. *BioScience* **47**:427–435.

Strayer, D. L. 1994. Limits to biological distributions in groundwater. *In* "Groundwater Ecology" (J. Gibert, D. L. Danielopol, and J. A. Stanford, eds.), pp. 287–310. Academic Press, San Diego, CA.

Tóth, J. 1963. A theoretical analysis of groundwater flow in small drainage basins. *Journal of Geophysical Research* **68**:4795–4812.

Triska, F. J., V. C. Kennedy, R. J. Avanzino, G. W. Zellweger, and K. E. Bencala. 1989. Retention and transport of nutrients in a third-order stream in northwestern California: Hyporheic processes. *Ecology* **70**:1893–1905.

Turner, M. G. 1989. Landscape ecology: The effect of pattern on process. *Annual Review of Ecology and Systematics* **20**:171–197.

Valett, H. M., C. C. Hakencamp, and A. J. Boulton. 1993. Perspectives on the hyporheic zone: Integrating hydrology and biology. Introduction. *Journal of the North American Benthological Society* **12**:40–43.

Valett, H. M., J. A. Morrice, and C. N. Dahm. 1996. Parent lithology, surface-groundwaer exchange, and nitrate retention in headwater streams. *Limnology and Oceanography* **41**:333–345.

Wentz, D. A., B. A. Bonn, K. D. Carpenter, S. R. Hinkle, M. L. Janet, F. A. Rinella, M. A. Ulrich, I. R. Waite, A. Laenen, and K. E. Bencala. 1998. Water quality in the Willamette Basin, Oregon. *Geological Survey Circular (U.S.)* **1161.**

White D. S. 1993. Perspectives on defining and delineating hyporheic zones. *Journal of the North American Benthological Society* **12**:61–69.

Index

Advection-dispersion, 17, 48, 51, 54, 57, 150, 184, 212, 215
Agrio River, 174
Algae,
  acidophilic, 172
  anoxia, 260
  biomass, 169, 387
  chemical indicators, 172
  colonization, 278
  distribution, 382–386, 388
  exudates, 209, 251, 272, 294
  flooding, 349
  production, 113, 174–175, 210
    mats, 150
    nutrient assimilation, 169, 215, 222, 278, 383, 393
      nitrogen, 277
      phosphorus, 227, 230

Alkalinity, 142, 374
Allequash Creek, 247–248
Alpbacher Ache, 318
Amazon River, 241, 366
Anthropogenic impacts,
  acid mine drainage, 278
  acid precipitation, 142, 185
  agriculture, 170, 178, 201
    fertilizers, 177, 183
  desertification, 182
  eutrophication, 140, 182, 222, 368, 372, 376
  landuse, 169, 170, 172, 175–189, 222, 338
  pesticides, 183
  pollution, 302, 319, 338, 345, 347, 349, 354–355, 411, 414
    organic, 349

Anthropogenic impacts (*continued*)
  pulp, wood, 178
  sewage effluent, 212–214, 394
Arkansas River, 355
Aspen Creek, 73–74, 272
Aufre River, 174
Azul River, 174

Bacteria, 292–293
  biomass, 232, 268, 324, 387
  carbon supply, 303
  chemical indicators, 172
  dispersal, 350
  flow effects, 293
  growth, 350
  nutrient assimilation, phosphorus, 300
  production, 144, 232, 293, 386, 407
  see also microbial
Barro Branco Watershed, 88
Bear Brook, 142, 204
Big Canyon Creek, 318
Biological Oxygen Demand, 69
Bioturbation, 324, 327–328
Black Creek, 275
Breitenbach, 249, 293, 294, 386
Bryophytes, 182
Buzzards Branch, 296, 315, 318

Calcium, 100, 143
Caliente River, 174
Carbon,
  inorganic,
    dissolved, 142–143, 264–265
    uptake, 368, 374
  organic, dissolved,
    anoxia, 242
    carbohydrates, 94
    decomposition, 16, 152, 292, 375
    excretion, 352
    fatty acids, 94
    flocculation, 249
    flux, 297, 386
    ground water, 151, 240–241, 299
    humic, 248
    hydrolysis, 244, 245, 251
    hyporheic, 211, 215, 226, 241–253
    mineralization, 130
    nutrient interactions, 252
    organic acids, 215
    riparian, 93–95, 102–103, 106, 107,
      118, 123–129, 208
    soils, 238, 240

    sorption, 238, 244, 245, 248, 249, 268,
      299
    sources, 237, 295
    storm flow, 105, 239–240, 242
    surface stream, 299
    terrestrial inputs, 215, 237, 239, 251,
      387
    wetlands, 238–239
  organic, particulate,
    burial, 246, 250
    leaching, 246
    decomposition, 297
  organic, *see also* organic matter
Chemoautotrophy, 199, 200, 203, 264,
  276–278, 279, 321, 387
Chloride, 89, 91
Climate change, 107, 184–185
Colorado River, 318
Competitive exclusion, 321
Congo River, 273–274
Copper, 68
Cyanobacteria, 392, 411

Danube, 315
Debris dams, 25, 152, 247, 260, 300, 323
Decomposition, 174, 175, 350
Deposition, sediment, 150
Detroit River, 366
Disturbance, 113, 131, 338, 345, 348
Donau, 318
Dryman Fork, 294
Duffin Creek, 202, 204, 212, 215

Edward's Aquifer, 407
Exoenzymes, 289

Falling Springs Creek, 143
Firehole River, 144
Fish,
  phreatic, 407
  predation, 322, 327, 354
  production, 144
  refugia, 170, 389
  spawning, 170, 184, 356, 389–391,
    393
Flathead River, 128, 326, 344, 352, 353,
  355, 388
Flood-Pulse Concept, 115, 129, 130, 349
Fungi, 290–291, 300, 303, 345, 352, 374,
  375, 387; *see also* microbial

Gallina River, 272
Garrone River, 242, 246
Gata River, 174
Geologic influences, 168, 172
  calcite, 143
  crystalline rock, 240
  dolomite, 141
  evaporite, 130
  geothite, 248
  gneiss, 267–268
  granite, 267–268
  igneous, 205
  limestone, 143, 240
  sandstone, 240, 267–268
  sedimentary, 205
  siltstone, 267–268
  volcanic tuff, 267–268
  weathering, 143
Geomorphology, 412–413
Glen Major Creek, 89, 91, 93, 100, 202,
    204, 212, 215
Goose Creek, 296, 309, 315, 327
Grazing, 290, 301, 324, 327, 351, 352, 393
Ground water,
  defined, 138–139, 240
  quality, 354

Habitat Index, 178
Harp Swamp, 94, 97, 98, 102–103
Hubbard Brook, 295
Hugh White Creek, 152–153
Hundsheimer Haufen, 318
Hydraulic conductivity,
  flow velocity, 11, 13, 24, 60, 62, 69, 75,
    153, 241
  ground water, 144
  hyporheic storage, 84, 202, 215
  infiltration, 300
  macroinvertebrates, effects on, 346–347,
    355
  macrophytes, effects on, 370
  meiofauna, effects on, 325–328
  redox potential, 267–268, 269, 274, 290
  riparian, 120
  soils, 87, 170
Hydrologic modeling and methods, 47
  assumptions, 21
  Damkohler number, 28, 29, 30, 33, 35, 36
  dilution gauging, 9
  donor controlled, 155
  experimental design, 32, 33, 52
  forward modeling, 19, 39

inverse modeling, 19–21
MODFLOW, 11, 73
MODPATH, 74
Monad kinetics, 155, 164
Navier-Stokes equation, 55
OTIS, 6, 21
parameter fitting, 20–21, 24, 28, 29, 34,
    35, 37, 48
particle deposition, 70, 71
pumping exchange 57, 58–62
reactive solutes, 18–19, 39, 40, 51, 66–
    72, 151, 152, 158–163
  ammonium, 207
  nitrate, 207
Reynolds numbers, 58
Reynolds stress, 56
slip velocity, 54, 55, 56
solute transport, 6, 48
sorption, 67
tracers, 5, 10, 14–17, 28, 48, 52, 62, 86,
    150, 152, 207, 225, 267, 407
  bromide, 16, 17, 152
  chloride, 16, 17, 152, 207, 211, 214
  dye, 62, 65, 75
  hyporheic, 208
  kaolinite, 70, 71
  lithium, 62, 64, 68, 71
  radon, 16
  silicate, 214
  temperature, 75
  trituim, 152
transient storage, 17–18, 22–26, 34, 37,
    40, 47–51, 150–162, 203
uncertainties, 29, 37, 40
velocity gauging, 9
velocity profile, 56
Hydrology,
  contaminant transport, 171, 182
  Darcy flow, 10, 11, 53, 54, 55, 57, 58, 60,
    85, 249
  diel fluctuations, 5, 144
  drought, 113, 185, 338, 347, 388, 392
  drying, 348
  evapotranspiration, 101, 105, 107, 112,
    118, 121, 123, 126, 144, 184
  flow regulation, 179, 347, 349, 396
    dams, 183, 188, 395
    irrigation, 182–183
    water distribution, 355
  flowpath scales, 6, 25, 26–32, 36, 38, 39,
    46, 49, 51, 85, 168
  flowpath types, 5, 46, 382
  geothermal, 96, 141, 144, 171–175

Hydrology (*continued*)
  hyporheic, 11, 12, 243–244
  macrophytes, effects of, 364, 367–371, 385
  overland flow, 86–87, 93, 96, 100–106, 114–115, 126, 139, 239
  riparian, 85, 112, 396
  seasonal fluctuations, 11, 117–118, 130
  snow melt, 273
  solute transport, 328
  storage, 202, 214, 225, 227, 266, 268, 269
  storm flow, 86–87, 93, 96, 100–106, 112–133, 139, 296, 338, 345–348, 353, 376, 388, 392
  see also hydraulic conductivity
Hyporheic Corridor Concept, 353
Hyporheic, defined, 5, 46, 288, 382
Hyporheos, defined, 338

Index of Biotic Integrity, 178
Intermediate Disturbance Hypothesis, 112, 132
Iron, 269, 277, 373
  reduction, 87, 94, 272, 273–274
  sampling, 265
Isotopes,
  radioactive, 152, 251, 266, 268, 276
  stable, 100, 141, 216

Karst, 143, 344, 354
Kesterson Reservoir, 183

Lake Kinneret, 273–274
Lake Tahoe Watershed, 89
Langmuir Adsorption Model, 248
Leaf litter, 388
Ledbetter Creek, 228
Lithium, 68
Little Lost Man Creek, 16, 49, 50, 152, 205–209, 212, 215
Little Rock Lake, 275

Macroinvertebrates,
  amphibites, 353–354
  bactivory, 345
  biomonitoring, 354–355
  competition, 345, 347
  defined, 338

dispersal, 351
distribution, 348, 350, 388
drift, 327, 347
drying, 348
epigean, 337, 347, 348, 352–354
growth, 344
hydrology, effects of, 345
life history, 344
migration, 353
oviposition, 388
phreatic, 348
production, 144, 338, 347, 354
predation, 322–323, 345, 347, 350, 351, 353
refugia, 338, 388
reproduction, 344, 347
sediment size, effects of, 326–328, 345
stygobites, 344, 347, 348, 354
stygophiles, 344
stygoxenes, 344
Macrophytes, 169, 177, 182, 211, 225, 384
  deposition, effects on, 369
  flooding, effects of, 367
Magnesium, 68, 100
Manganese, 16, 87, 94, 265, 272, 277, 373
Maple River, 228, 293, 294
Mass balance, 7–9, 48, 84, 210, 250
McRae Creek, 74, 76, 203, 212, 215
Meiofauna,
  abundance, 309–319
  anoxia, 325
  bactivory, 324
  defined, 307–308, 338
  dispersal, 323
  distribution, 308
  diversity, 308, 309
  drying, 348
  hydrology, effects of, 319–320
  morphology, 319, 321–322
  predation, 327
  refugia, 323
  research history, 308
  resources, 324
  sampling, 316–319
  sediment size, effects of, 309, 316, 320, 325, 326–328
Mercury, 177, 182
Metals, toxicity, 278, 355
Methane, 87, 247, 271, 273–274
  methanogenesis, 224, 260–262, 275–276
  oxidation, 276–277
  sampling, 265
Methylmercury, 97, 99, 100, 184

Microbial,
    activity, 244–250
    biofilms, 289, 352
    competition, 368
    diversity, 301–302, 303, 352
    ecology, defined, 288
    loop, 287, 301, 345, 352
    nutrient limitation, 324, 325
MINTEQ, 265
Mirabel Canal, 348
Mississippi River, 177
Morava River, 353

Nitrogen, 87–93, 139–141
    ammonium,
        hyporheic, 151, 212, 213, 372–373
        riparian, 88, 90, 93, 106, 118–131
        sorption, 204–205, 208, 211, 368
        storm flow, 93
        surface, 212, 213
        upland, 88
    assimilation, 198, 214, 368, 371, 374,
        387
    denitrification, 200, 203–204, 224, 260
        ground water, 140
        hyporheic, 152, 207, 208, 210, 214,
            373
        riparian, 87, 89, 90, 101, 128, 130, 170
    excretion, 352
    fixation, 89, 200
    ground water, 151, 177, 182, 210, 211
    mineralization, 151, 198, 200, 204, 208,
        209, 211
    nitrate,
        agricultural inputs, 186–188
        dissimilatory reduction, 200
        ground water, 140
        hyporheic, 113, 153, 212
        reduction, 204
        riparian, 88–93, 101–103, 105–106,
            118–131, 208
        soils, 88
        sources, 392
        surface, 152, 411
    nitrification, 198, 200, 204, 277
        hyporheic, 151–152, 207, 209, 211,
            212, 373, 387
        riparian, 87, 89, 128, 130
    nitrogen loading, 107
    organic, 88, 90, 93, 105, 118–131, 201,
        208
    terrestrial sources, 201, 215

    transformation, 375
    transport, 115, 126, 128, 198
    uptake, 16, 151–152, 170
    valence states, 199
    see also nutrient
Nutrient,
    assimilation, 392
    atmospheric inputs, 129, 175–185
    drying, 348
    plant uptake, 130, 139, 140
    retention, 113, 129, 131–133, 140, 141,
        150, 162
    spiraling, 150, 152, 202, 250–253
    transport, 326–328, 347
    uptake length, 152–163, 202, 251

Oberer Seebach, 309, 318
Ogallala Aquifer, 179
Ogeechee River, 294, 296
Organic matter,
    decomposition, 327
    dissolved, 141–142
    particulate, 394
    transport, 326–328, 386
    see also carbon, organic
Oxygen,
    ground water, 141
    hyporheic, 5, 68–70, 152–153, 198, 203,
        208, 210–211
    meiofauna, effects on, 320–321
    phosphatases, effects of, 224–225
    photosynthesis, effects of, 373
    probes, 264
    respiration, effects on, 16, 87, 244–250
    riparian, 89, 118–131
    seasonal patterns, 208, 272
    transport, 252, 301, 326–328, 347, 391,
        394

Pantano Stream, 96
Panther Creek, 228
Parafluvial, defined, 382
Patch dynamics, 324
Phosphorus,
    assimilation, 224, 368, 371, 374, 387
    decomposition, 229
    excretion, 352
    forms, 222
    geologic sources, 222
    ground water, 139–141, 174, 177
    hyporheic, 372–373, 375

Phosphorus (*continued*)
mineralization, 153, 244
phosphate, 182
precipitation, 222, 223, 230, 232
riparian, 93, 94, 96, 105, 106, 118–131
soils, 222
sorption, 94, 211, 222–223, 230, 278, 368
surface, 272
uptake, 152
see also nutrient
Photosynthesis,
carbon assimilation, 143
meiofuana, effects of, 325
nutrient limitation, 141
nutrient uptake, 162, 208
oxygen, 260, 268
Phreatic, defined, 240
Piezometeric surface, 10, 58–59, 85, 91, 96–98, 101–105, 116, 118–131
Pinal Creek, 24
Plastic Swamp, 96, 97, 98, 102–103
Platt River, 206, 212–214
Porewater sampling, 15
Potassium, 100, 101
Precipitation, 86, 115, 117–118, 129, 139, 184, 239
Pregolya River, 273–274
Primary production, 117, 133, 139, 144, 388, 392
Processing efficiency, 250

Queets River, 383

Raison River, 178
Redox potential, 203–204
anoxia, 69, 89, 94, 96, 141, 184
ground water, 177
hyporheic, 212, 214, 215, 222, 223, 227, 230, 232, 242, 247
meiofauna, 321
molecular speciation, 269
rhizosphere, 370, 371, 373, 392
riparian, 94, 101, 105–107, 129
secondary production, 407
terminal electron acceptors, 87, 290, 303
Reedy Creek, 93, 96, 97, 102–103
Respiration,
anaerobic, 129, 215, 239, 259–284
fermentation, 215, 247, 260–262
measuring, 266
sediment size, effects of, 268, 269

terminal electron acceptors, 261–262, 301
controls, 303
hyporheic, 203, 209, 215, 229, 244, 245, 293–295, 299
macroinvertebrates, 349–350
meiofauna, 325
microbial diversity, 302
organic carbon, 249, 250, 252, 294, 295, 297–299
riparian, 117, 128, 129
sediment size, effects of, 300
Retention efficiency, 85, 252–253
Rhine River, 177, 182
Rhode River, 102–103, 112, 126
Rhone River, 292, 347, 348, 384, 386
River Glatt, 272
Rio Calveras, 73–74, 202, 272, 273, 275
Riparian,
buffer, 413
defined, 83–84, 382
filter, 139
flooding, 349
nitrogen fixation, 89, 203, 392
vegetation, 90, 93, 101, 112–113, 170, 178, 179
River Continuum Concept, 186, 188

St. Kevin Gulch 49, 50
Salitral River, 174
Salto River, 96, 106
San Domingos River, 178
Seepage meters, 10, 13
Selenium, 183
Serial Discontinuity Concept, 188
Shear stress, 323, 326
Shingobee River, 206, 210–212, 215
Snake River 49, 50
Sodium, 100
Soils, influences, 169
Specific conductance, 16, 118, 119, 123–125, 127, 129–130
Speed River, 318
Steina River, 246, 292, 294, 295, 296
Stillaguamish River, 242
Stripping Hypothesis, 248–250
Subsidy-Stress Gradient Model, 133
Sulfur, 105
deposition, 185
oxidation, 185, 277, 373
sampling, 265
sulfate, 87, 96, 97, 98, 101, 102–103, 105, 107, 223, 232, 271

sulfate reduction, 87, 172, 260–262, 273, 275, 321
Svalberget Watershed, 99
Sycamore Creek, 112, 115, 117–131, 203, 206, 209–210, 212, 215, 229, 240, 241, 275, 276, 277, 293–294, 296, 327, 348, 383, 384, 385, 386, 387, 388, 391, 396
Symbiosis, 324, 374

Tarawera Stream, 69
Telford Swamp, 100, 101, 102–103
Temperature, 5, 143–144, 170, 179, 185, 290, 353, 366, 368, 393
Thermodynamics, 260
Tonnerjoheden Watershed, 90
Trophic interaction, 142, 291, 302, 322, 324, 325, 338, 345, 352, 353, 355, 391, 407

Upper Clark Fork River, 272
Upper Three Runs Creek, 367
Uvas Creek 49, 50

Vadose, defined, 240

Waiotapu Stream, 69
Walker Branch, 106, 141, 152–153, 204, 227–229, 240
Wappinger Creek, 242, 246, 295, 296, 299, 315
Wetlands, 179, 182
White Clay Creek, 242, 251–252, 291, 293, 294, 315
Willamette River, 412

Zinc, 68